Lecture Notes in Computational Science and Engineering

13

Editors
M. Griebel, Bonn
D. E. Keyes, Norfolk
R. M. Nieminen, Espoo
D. Roose, Leuven
T. Schlick, New York

Simulation and Visualization on the Grid

Parallelldatorcentrum
Seventh Annual Conference

Springer
Berlin
Heidelberg
New York
Barcelona
Hong Kong
London
Milan
Paris
Singapore
Tokyo

Björn Engquist Lennart Johnsson
Michael Hammill Faith Short (Eds.)

Simulation and Visualization on the Grid

Parallelldatorcentrum
Kungl Tekniska Högskolan
Seventh Annual Conference
Stockholm, Sweden
December 1999
Proceedings

With 118 Figures, 25 in Colour

Springer

Editors

Björn Engquist
Michael Hammill
Faith Short

Parallelldatorcentrum
Kungl Tekniska Högskolan
100 44 Stockholm, Sweden

e-mail: engquist@pdc.kth.se
mike@pdc.kth.se
faith@pdc.kth.se

Lennart Johnsson

Department of Computer Science
University of Houston
4800 Calhoun Road
Houston, TX 77204-3475, USA

e-mail: johnsson@cs.uh.edu

Cataloging-in-Publication Data applied for

Die Deutsche Bibliothek - CIP-Einheitsaufnahme
Simulation and visualization on the grid : Parallelldatorcentrum, seventh annual Conference, Stockholm, December 1999 ; proceedings / Björn Engquist ... (ed.). - Berlin ; Heidelberg ; New York ; Barcelona ; Hong Kong ; London ; Milan ; Paris ; Singapore ; Tokyo : Springer, 2000
(Lecture notes in computational science and engineering ; 13)
ISBN 3-540-67264-8

Front cover image courtesy Jason Leigh, Electronic Visualization Laboratory, University of Illinois at Chicago, USA.
Center "cube" image courtesy National Center for Supercomputing Applications Cosmology Group, USA.
"Simulation and Visualization on the Grid" logo by Frank, Etc., Stockholm, Sweden.

Mathematics Subject Classification (1991): Primary: 68U20,
Secondary: 65Y05, 65Y25, 68U05

ISSN 1439-7358
ISBN 3-540-67264-8 Springer-Verlag Berlin Heidelberg New York

Cover Design: Friedhelm Steinen-Broo, Estudio Calamar, Spain
Cover production: *design & production* GmbH, Heidelberg
Typeset by the authors using a Springer TEX macro package

Printed on acid-free paper SPIN 10699932 46/3143LK - 5 4 3 2 1 0

Preface

It is now 30 years since the network for digital communication, the ARPA-net, first came into operation. Since the first experiments with sending electronic mail and performing file transfers, the development of networks has been truly remarkable. Today's Internet continues to develop at an exponential rate that even surpasses that of computing and storage technologies. About five years after being commercialized, it has become as pervasive as the telephone had become 30 years after its initial deployment. In the United States, the size of the Internet industry already exceeds that of the auto industry, which has been in existence for about 100 years. The exponentially increasing capabilities of communication, computing, and storage systems is also reshaping the way science and engineering are pursued.

Large-scale simulation studies in chemistry, physics, engineering, and several other disciplines may now produce data sets of several terabytes or petabytes. Similarly, almost all measurements today produce data in digital form, whether from collections of sensors, three-dimensional digital images, or video. These data sets often represent complex phenomena that require rich visualization capabilities and efficient data-mining techniques to understand. Furthermore, the data may be produced and archived in several different locations, and the analysis carried out by teams with members at several locations—possibly distinct from those with significant storage, computation, or visualization facilities.

The emerging computational Grids enable the transparent use of remote instruments, computational and data resources. The Grids also facilitate interactive distributed collaboration among team members sharing significant computation, information, and visualization resources.

The seventh annual PDC Conference on high-performance computing, titled "Simulation and Visualization on the Grid," focused on applications and technologies for computational Grids. Fifteen contributed and nine invited presentations, along with a Panel session, addressed these issues. The vision of computational Grids expressed by the Panel ranged from distributed simulation and visualization for multinational corporate design teams to three-dimensional phone calls, and covered issues such as the open source movement and the central role the human plays in the loop.

How fast will we get there?

According to Panel member Carl Kesselman "five years from now, if we're extremely successful, we'll see some real, persistent, wide-scale infrastructure that will solve real problems," and "in ten years, it should be way cool."

The Grid-technology topics specifically addressed by the conference contributions include distributed file I/O, clustering, CORBA software infrastructures, tele-immersion, interaction environments, visualization steering and virtual reality. The contributions in this proceedings also include descriptions of applications in biology, chemistry, and physics which will benefit from computational Grids, and the techniques used in such applications to realize the power of Grids.

The venue for the conference, which had 85 participants from 10 countries, was the Parallelldatorcentrum (PDC). PDC is the main Swedish national center for high-performance computing. It has participated in GUSTO, the Globus Ubiquitous Supercomputing Testbed Organization, since the fall of 1997. The Globus testbed was first demonstrated at SC97 with over 3,600 processors on five continents. During the Alliance '98 meeting, PDC also participated in a transatlantic, collaborative, virtual-reality demonstration with four sites in the United States, as a member of the CAVE Research Network (CAVERN). High-speed networking, which was provided by SUNET, Nordunet, and the US National Science Foundation (the vBNS), and the VR facilities were critical Grid elements for this Grid-technology demonstration featuring an industrial application from Carterpillar. Thus, PDC was a fitting location to discuss simulation and visualization on the Grid.

A good starting point in this book is the chapter, "The Grid: What's Really Going On?" This represents the lively Panel discussion on the last day of the conference, featuring experts in networking and applications on the Grid. The participants addressed current successes and pitfalls of the high-performance computing Grid, as well as issues of the next millennium.

The PDC Conference has been held annually since 1993. Previous themes include *High-Performance Computing in Visualization and VR* (1998), *Future Directions in High Performance Computing* (1997), *Software for Parallel Computing* (1996), *Parallel Computing in Science and Industry* (1994), and *Parallel High-Performance Applications* (1993). One goal of the conference series is to gather high-performance computing experts for an exchange of ideas and information in leading-edge topics. More information about the PDC Conference in year 2000 can be found on the Web at `<http://www.pdc.kth.se/conference/2000/>`. General questions about the conference may be directed to `<conf-2000-info@pdc.kth.se>`.

So, how do we get to a Grid that will offer persistent, wide-scale infrastructure for solving real problems five years from now? Though the path is likely to have many twists and turns, we hope the reader will agree with Maxine Brown that "we're going to have fun getting there!"

January 2000

Björn Engquist
Lennart Johnsson
Michael Hammill
Faith Short

The PDC Annual Conference

The seventh annual PDC Conference, "Simulation and Visualization on the Grid," was organized by the Parallelldatorcentrum (PDC), located on the campus of Kungl Tekniska Högskolan (KTH), Stockholm, Sweden.

Steering Committee

Björn Engquist	PDC, KTH (Sweden)
Lennart Johnsson	University of Houston (USA)

Program Committee

Björn Engquist	PDC, KTH (Sweden)
Lennart Johnsson	University of Houston (USA)
Michael Hammill	PDC, KTH (Sweden)
Per Öster	PDC, KTH (Sweden)

Organizing Committee

Conference Chair:	Björn Engquist (PDC, KTH)
Conference Coordinator:	Michael Hammill (PDC, KTH)
Local Organizing Committee:	Gunilla Appelquist (Parallel and Scientific Computing Institute, KTH)
	Faith Short (PDC, KTH)
	Per Öster (PDC, KTH)

Sponsoring Institutions

Kungl Tekniska Högskolan (KTH)
Swedish Council for Planning and Coordination of Research (FRN)
Parallel and Scientific Computing Institute (PSCI)

Acknowledgments

We acknowledge the enthusiastic work of the local organizing committee. We also acknowledge the following people for the assistance and support in the organization of the PDC Conference: Harald Barth, Stefan Bindzau, Erik Bågfors, Simon Berg, Johan Danielsson, Per Ekman, Gunvor Engmarker from the KTH Conference Service, Daniel Linder, Jia Liu, Maria Mossberg, Jesper Oppelstrup, Carl Tengwall, Peter Thomas, and Tomas Wilhelmsson. The PDC Conference would not have been possible without the personal involvement of all these people. Finally, we would also like to thank the sponsoring institutions for their generous financial support, and the conference presenters for making this a stimulating and rewarding conference.

Table of Contents

Underlined names denote speakers. Bold names denote invited speakers.

Grid Technologies

Grid Visualization and Virtual Reality

Biology and Chemistry

Physics

Panel Discussion

Efficient Distributed File I/O for Visualization in Grid Environments

Werner Benger[1,2], Hans-Christian Hege[1], André Merzky[1], Thomas Radke[2], and Edward Seidel[2,3]

[1] Konrad-Zuse-Zentrum für Informationstechnik Berlin (ZIB); Germany
[2] Max-Planck-Institut für Gravitationsphysik Potsdam/Golm Albert-Einstein-Institut (AEI); Germany
[3] National Center for Supercomputing Applications (NCSA); Champaign, IL; USA

Abstract. Large-scale simulations running in metacomputing environments face the problem of efficient file I/O. For efficiency it is desirable to write data locally, distributed across the computing environment, and then to minimize data transfer, that is, reduce remote file access. Both aspects require I/O approaches that differ from existing paradigms.

For the data output of distributed simulations, one wants to use fast local parallel I/O for all participating nodes, producing a single distributed logical file, while keeping changes to the simulation code as small as possible. For reading the data file, as in postprocessing and file-based visualization, one wants to have efficient *partial* access to remote and distributed files, using a global naming scheme and efficient data caching, and again keeping the changes to the postprocessing code small. However, all available software solutions require all data to be staged locally (involving possible data recombination and conversion), or suffer from the performance problems of remote or distributed file systems.

In this paper we show how to interface the HDF5 I/O library via its flexible Virtual File Driver layer to the Globus Data Grid. We show that combining these two toolkits in a suitable way provides us with a new I/O framework, which allows efficient, secure, distributed and parallel file I/O in a metacomputing environment.

1 Introduction

The varied needs of large-scale simulations are driving the computational science community to a computing paradigm that uses whatever resources can be brought to bear on the problem at hand. Very often local computing resources are insufficient to solve the problem on the desired scale. Remotely accessible resources at individual sites may individually be too small as well. However, the combined distributed resources available to a research group, which may itself be distributed across the world, are clearly much larger than any single resource. In this metacomputing environment, one seeks to develop efficient techniques to harness a variety of computing resources across the Grid [15], and to use them as a logically single machine. There are many issues to be addressed to make such simulation effective and efficient, for example, the scheduling and acquisition of the different resources must be convenient, the simulation code must run efficiently on the different machines, the algorithms must be appropriate for networked computers, and issues of latency and bandwidth must be considered.

In this article, we focus on approaches to file I/O in a metacomputing environment. The kinds of large-scale simulations that require a Grid-based solution often generate large-scale data. The data need to be efficiently handled across the Grid, often by a group of researchers who are themselves distributed across the Grid. This brings up many new issues in I/O, including data layout, writing distributed files, and access to the distributed data for postprocessing.

A simulation running in a metacomputing environment has several ways to handle its file I/O. There are various points to be considered in choosing the most appropriate approach to I/O.

- One must balance the needs of simulation speed and ease of postprocessing, for example, at what stage is data conversion accomplished, are expensive postprocessing steps included in the simulation code?
- Are there remote file systems (NFS, AFS, etc.) available on all participating nodes, and are they compatible with each other?
- Do data have to be located on one file system finally, or can they be distributed?
- Do we want to end up with a single data file, or can we split data over multiple files?
- Which I/O libraries are available at the different sites?
- Which file format is useful and accepted by postprocessing routines?

Often it turns out that the speed of the simulation is the most important factor (and usually the most expensive), and thus everything, including file I/O, is optimized in regard to this. Even in a standard computing environment, I/O can be a major bottleneck in the simulation. In a metacomputing environment, it can become crippling if the standard local I/O paradigms are applied. Creating a single output file for all nodes necessarily involves a network transfer of major portions of the data, either explicitly by collecting data on one node and writing a single local file, or implicitly by using remote file systems. Remote file systems like NFS and AFS often introduce additional problems: their performance is usually lower than that of comparable local file systems (due to some administrative overhead), and they presuppose a tight organizational structure between all hosting institutions to allow access over the whole Grid. For systems that need quite a bit of administration, such as DCE [19], this often turns out to be a real problem, and it contradicts the flexibility of an ideal Grid environment.

If we put aside the desire to have a single compact output file, all compute nodes, regardless of their location, can use optimized local I/O libraries and spread the data over many files, distributed over the Grid. This speeds up the simulation I/O routines, and subsequently the simulation. But then in postprocessing and visualization, all data from all machines must be moved to a single machine where they are recombined into some usual file layout. This would allow preexisting non-metacomputing software to process the generated data in the usual way, but at a huge cost. For datasets on the of order 10^9–10^{12} bytes, which can easily be generated in such an environment, this cost may be prohibitive.

It would therefore be desirable and much more efficient to make postprocessing mechanisms aware of the distributed character of the data. It would be even more efficient to then access only data subsets, which really are required for postprocessing, in order to avoid unnecessary network communication. For example, some visualization procedure may only want to operate on a down-sampled dataset or sub-domain, demanding a much lower network traffic than a complete data staging.

There exist approaches to solve all these problems on the file system level, introducing the concept of distributed parallel file systems. In contrast to merely remote accessible file systems—such as AFS, NFS, and DCE and their client-server concept,—distributed file systems are physically distributed, and all participating nodes are equivalent. We want to mention ParFiSys, a parallel file system from Madrid [9], and DPSS, the distributed-parallel storage system from LBL [25]. Such approaches have the following features in common.

- To the application, they act as normal local Unix file systems, even if parts of the hosted files are located remotely.
- Work load is distributed over multiple I/O nodes, providing very fast I/O even compared to local disks.
- They provide a global name space to all clients, just like NFS, AFS, etc.
- Often they depend on special hardware infrastructure, and are limited to small local networks or Grids.

With the emergence of Grid infrastructures and corresponding toolkits, the management of distributed data is also assigned by Grid services. The Globus metacomputing toolkit [13,14], for instance, provides mechanisms [7] to access remote data transparently (by substituting `fopen` and `fclose` calls) on software level. Newer developments like the Data Grid API will provide efficient I/O interfaces to various storage systems, and will also allow partial access of remote and distributed files.

Our goal is to show the use of such Grid techniques on an application level. For our example of a distributed computational environment, which is described in the next section, we designed and implemented such a solution based on an additional software layer between our I/O library (HDF5) and the storage system (DPSS and Unix file system), which handles the data distribution over the Grid.

2 Software Components

The DFN Gigabit Testbeds in Germany [12] form a unique opportunity to develop and test Grid applications in a network environment, which is going to be the standard environment for German research groups during the next couple of years. One of the pilot projects initiated by this Test Grid is *TIKSL: TeleImmersion—Kollision schwarzer Löcher* (or *Tele-Immersion: Collision of Black Holes*) [24].

In the course of the TIKSL project, a distributed environment for visualization, the results delivered by the Cactus Computational Toolkit [8,3] [2,6,21] simulations will be developed. Cactus is an environment for collaboratively developing high-performance multidimensional numerical simulations; it was developed at AEI Potsdam, University of the Balearic Islands, NCSA, Washington University, and other cooperation partners. With the Cactus scientific team itself distributed across the world, it provides an excellent project for partnering in the development of techniques to make use of distributed computing. From its history, it focuses on Numerical Relativity; the scientific part of the TIKSL project is especially focused on the simulation of black hole collisions.

The coalescence of two black holes is considered to be one of the most promising sources of gravitational waves [5,22]. This problem is especially important now that LIGO [1] and other physical experiments designed to detect gravitational waves for the first time may be able to detect such events during the next few years. The numerical simulations performed by Cactus are aimed at predictions and calibrations for such gravitational wave detection experiments.

Simulating the complete set of Einstein Equations for black hole collisions numerically requires big computational resources even for "simple" problems: hundreds of equations and many dozens of grid functions have to be evolved on a sufficiently smooth and dense grid (sufficiently dense means resolutions of about 256 and more in each spatial dimension). Computing time for typical problems and boundary conditions are several hours to days. During such simulations, binary output data of tens to hundreds of GBytes are produced, at a peak rate of several 100 MBytes per second. These data are stored in local HDF5 files.

While Cactus is usually running on MPP systems like T3E and O2K, off-line visualization is usually performed on smaller visualization workstations. The HDF5 files are therefore staged to these remote locations; and for distributed Cactus runs, the data are recombined into a single data space and data file. They are then processed by Amira [4], an advanced visualization toolkit developed at ZIB Berlin, which directly reads HDF5 files via the HDF5 library and its Unix File Driver. Drawbacks of this solution are obviously the need for recombination of data originating from distributed runs, the (possibly multiple complete transfer of data),[1] and the locally required storage capacities.

The goal of the TIKSL project and the solution presented here therefore include the capability to

- read data files remotely, at the server(s) hosting the simulation,
- recombine the data on-the-fly,
- write/read directly via the HDF5 library (for minimal code changes on simulation and visualization side),
- read data selectively, or stage partial files (usually not all data are visualized, thus there is no need to transfer all data),
- read data without significant performance penalties (compared to local reads).

The next sections give a short overview about the crucial software components used in our environment, and describe our new approach for file I/O in detail.

2.1 The HDF5 Data Format and I/O Library

The basic file I/O in Cactus is focused on HDF5 [26,17] (Hierarchical Data Format, release 5). HDF5 is a file format and associated I/O library, designed for efficient, platform-independent file I/O. HDF5, like its predecessors, has been developed at the National Center for Supercomputing Applications.

HDF5 was chosen as I/O library for several reasons.

- The HDF5 storage model is very simple. It consists of only two fundamental object types: datasets, which contain N-dimensional arrays of arbitrary data; and groups, which may be used to arrange related datasets in a hierarchical structure (similar to a Unix file system).
- Datasets are self-describing, their size and data type are stored, along with grouping information and optional user-supplied data (via attributes attached to datasets or groups) as metadata in the HDF5 file. This makes it possible to interpret structure and content of a file without any external information.
- The library offers a variety of predefined standard data types as well as the capability of mapping arbitrary types or even structures into a portable file format representation.
- Its platform independence allows one to use HDF5 data files without the need of any explicit data conversion. Necessary type conversions are done on-the-fly when reading datasets—completely transparently to the application. This is especially useful on the Grid. HDF5 is ported to a wide variety of OS/machines and supports a large number of different hardware platforms.

[1] If multiple scientists want to visualize the data at different locations, the complete dataset has to be transferred to each of these locations.

- The HDF5 library implements sophisticated techniques for efficient and flexible data access/transfer, for example, data compression, parallel I/O using the MPI-2 file I/O extensions, and mapping of "hyperslabs" as arbitrary subspaces into the global data space. The availability of hyperslab selections for read and write operations allows one to minimize the amount of data actually transferred from or to the storage system, which is particularly important for remote file accesses.
- The library layout is very modular and clearly structured, with file I/O encapsulated in a Virtual File Driver (VFD) layer. This layer provides the mapping of basic I/O routines used by the HDF5 library onto the respective API of the underlying storage systems. Users can add their own VFD—for example, to implement transparent access to remote and distributed data, as shown in later sections.
- HDF5 already provides several approaches to process split (and potentially distributed and remote) files, like External File Lists (EFL), a split file driver, and a Unix file-system–like mounting mechanism.

All in all, HDF5 is a powerful file format and library. The library already provides a lot of functionality not available in other I/O libraries. It is quite likely that HDF5 will become a widely accepted standard in the scientific computing domain like its predecessors CDF and HDF4. By enhancing the HDF5 library with support for efficient remote file access, we will enable many application programs to transparently perform I/O with full HDF5 functionality also on remote (and possibly distributed) files.

2.2 Globus and Data Grid

There are many possible ways to develop and enroll an application in a metacomputing environment. And there are many tools and toolkits available [16,23,18,20] that are designed to support this task for programmers and users. The Globus metacomputing toolkit is certainly one of the most advanced toolkits available today. It provides a fundamental metacomputing infrastructure (like a global security layer, resource management tools, and communication libraries) as well as a flexible and secure administrative framework.[2]

The I/O scheme presented in this paper relies on one of the Globus components, the Data Grid [10,11]. The term "Data Grid" describes distributed systems that "support the integrated management and analysis of large (petabyte-scale) distributed data collections." The Data Grid Storage Client-API (SC-API) is designed to ease the development of Data Grid applications exploiting sophisticated techniques (e.g., network striping, parallel I/O, network protocol tuning) to achieve very high performance. The Globus

[2] For more information on the many different parts of the Globus Toolkit, see <http://www-fp.globus.org/details/>.

Data Grid toolkit is supposed to provide platform independence (that is, storage system independence) for such applications.

The Data Grid cannot be seen as a stand-alone toolkit. To build a Data Grid Application, the programmer will typically use functions from the Data Grid Storage Client-API as well as from other Data Grid or Globus components (e.g., mechanisms for access to metadata catalogs, security, and computing resources) to implement a particular application.

A file is usually regarded as an uninterpreted sequence of bytes that is located in a storage system. The term "logical file," as used by the Data Grid, refers to an entity that has a globally unique name and that might also have associated with it metadata and one or more physical file instances. Logical files may be organized in collections. In this context, a storage system is an entity that responds to requests to read and/or write named file instances. Note that the term "storage system" as used here denotes a logical construct and need not map directly to low-level storage. Additionally, the Data Grid provides the semi-automatic possibility of creating replicas and buffering logical files on faster or closer storage sites. Such operations cause updates of metadata catalogs, like the replica catalog.

The Data Grid concept for logical files not only provides a global name space for these files, but also allows physically distributed files as well. This is because a logical file access is no longer restricted to local physical files, but may also refer to one or more files somewhere in the Grid. Access to a certain part of this logical file results in access to the associated file instance, whose true location or storage layout remains completely hidden from the application.[3]

To enhance the HDF5 library with the ability to transparently access remote and distributed files, which are located somewhere in the Data Grid, we provide an HDF5 Virtual File Driver interfacing to the Globus Data Grid.

2.3 Amira Visualization System

Amira is a 3D visualization and volume-modeling system for interactive applications in natural sciences, engineering, and medicine [4]. The software system has been developed at the Konrad-Zuse-Zentrum für Informationstechnik Berlin (ZIB) and serves as a software platform for the development of working environments for specific applications domains (so-called virtual labs). A professional version of Amira is distributed commercially.

Amira is built in a modular, object-oriented fashion. It does not adhere to the data-flow concept. After loading data objects, they can be displayed or modified simply by selecting (from context menus) appropriate display or computational modules. Instantiated objects and their mutual dependencies

[3] The Data Grid toolkit is currently under development. Its functionality as stated in this paper is only partly implemented at the current state. This refers in particular to the replication and buffing mechanisms.

appear automatically as a graphical network. By modifying this representation and selecting suitable parameters for active modules, the visualization process can be controlled with minimal amount of user interaction.

The system's 3D graphics have been implemented using Open Inventor and OpenGL. Graphics hardware, if available, can be used to display large datasets at interactive speed. Amira displays its visual results in one or several graphics windows that allow users to interact directly, in three dimensions, with the depicted data objects. Different display techniques can be arbitrarily combined in a single window. All functions are scriptable via a TCL-command interface.

The system currently comprises about 25 different data-object types and more than 100 modules that can be dynamically loaded at runtime. It can be easily extended with other additional modules. Currently, Amira intrinsically supports various grid types like hexahedral, curvilinear, and unstructured tetrahedral grids. AMR data types are currently under development.

At present, Amira supports access to remote data by retrieving data carrying geometric information via a TCP socket connection. This method allows users to inspect data from a running simulation program. For example, the Cactus code is equipped with such an interface that allows users to send surfaces or line information like particle paths (see Fig. 1).[4] This was demonstrated during the SC99 conference in Portland, Oregon.

The current implementation uses a proprietary socket communication library and a data type specific API, which needs to be extended when the need for transferring new data types arises. A much more general concept is desirable for future online visualization of intermediate simulation results.

Several basic aspects of streaming huge datasets on a Grid are comparable to the file I/O issues addressed in this paper.

- Self-describing datasets based on simple primitive data types are favored to provide a future-safe coupling of simulation and online visualization programs.
- Data conversion has to be done transparently, fast, and on-the-fly, providing platform independence.
- Selection of partial data, such as hyperslabs, can reduce the actual network traffic.

All these issues are handled by the HDF5 library, but for file I/O only. It is therefore a reasonable consideration to use the HDF5 API as well for live data streaming as it is used for file access. In our work, we also plan to provide an HDF5-compatible network streaming interface, either implemented as an HDF5 replacement (which might still make use of some functionality of the original HDF5 library), or again as Virtual File Driver.

[4] Actually implemented are isosurfaces, which are computed on the supercomputer in parallel during the simulation run.

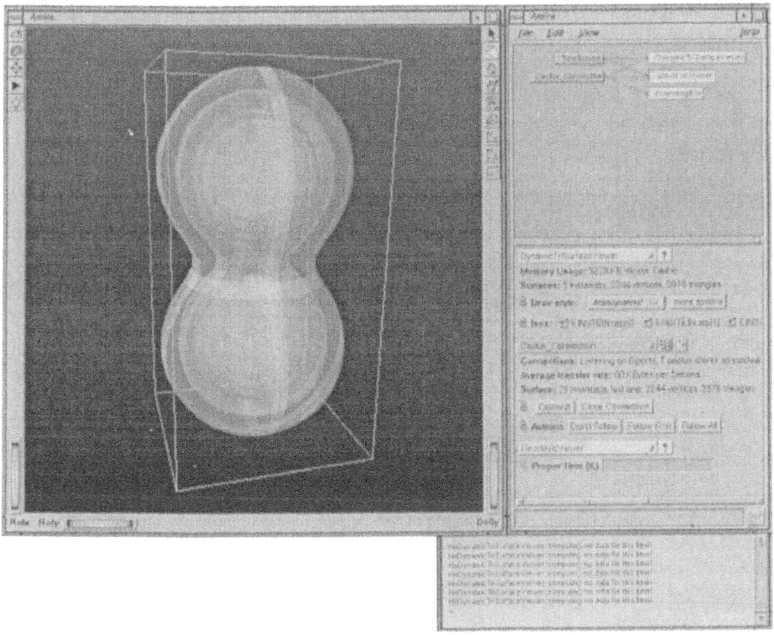

Fig. 1. Amira workspace visualizing isosurfaces, which are streamed from a remote Cactus simulation run.

Some of the various problems we are facing in developing such a generic remote streaming library include possible metadata inconsistencies inside the HDF5 library, the design of a sensible bi-directional protocol when employing a VFD-based solution, and the general fact that HDF5 was designed with file I/O in mind, not socket connections and asynchronous data requests. This work will be the topic of a future report.

For now, the presented enhancement of HDF5 at the level of file I/O makes Amira and every other HDF5-compatible postprocessing tool Grid-aware—in the sense that it allows simple and efficient handling of huge distributed datasets.

3 Adding Distributed File I/O to HDF5

As described before, the HDF5 library in its current distribution provides access to local files only. For use in a metacomputing environment, we want to extend its file I/O interface in order to handle remote data as well. Because of the clear functional structure of the HDF5 library, this means simply adding another Virtual File Driver that is capable of performing efficient remote file I/O. This driver would make use of the Globus Data Grid API and thus would provide the user with an interface that allows access to HDF5 files located anywhere in the Data Grid (see Fig. 2).

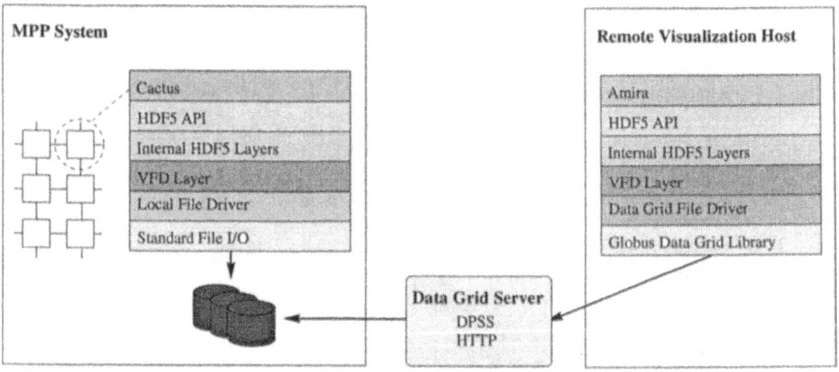

Fig. 2. Files written with a local File Driver of HDF5 are accessible off-line via the Globus/Data Grid File Driver from remote hosts, hiding their possibly split and distributed nature. The VFD Layer in HDF5 completely conceals the underlying I/O infrastructure to the application.

A second task is to support distributed file I/O. Under the premise that the simulation should be allowed to write datasets in the most efficient way (i.e., write chunked data into local files), the visualization must then be able to establish a global view on the complete data space as originally seen by the simulation program. There are two possible approaches to reconstruct this global data space. Both can be implemented using the features of the HDF5 library and the Globus Data Grid.

- The simulation handles the physical distribution of datasets within the global data space as a set of segments (offsets and sizes), which are written to one or more files local to that processor. These are then referred to in an HDF5 External File List (EFL). The EFL contains the filenames as URLs to qualify their distribution over the Data Grid. After the EFL is completely specified, it is automatically stored as metadata information in the HDF5 files, and it can be immediately used by the visualization program to read the distributed remote data via a Data Grid VFD—no additional preparations to address remote files are needed here.
  The disadvantage of this approach is the potential negative impact on the I/O performance when writing chunks of data in a heterogeneous environment: because external files do not contain any self-describing type information, the simulation must use a common data type representation throughout the whole application, which would make data conversions necessary on those nodes with a differing byte order. Right now, files

specified in EFLs do not also support hyperslab selections, and are re-opened for each access. This needs to be modified.

– Another possibility is to write each chunk of data independently as a single dataset into a separate HDF5 file. This gives the simulation the freedom to choose the easiest and most efficient method of outputting its chunked data (e.g., local optimizations can be applied by using parallel I/O for a shared file on all nodes that reside on the same machine).
  Of course, the simulation also has to provide the visualization with some description of the physical distribution of the files (analogous to the EFL). This can be done by storing the filenames as complete URLs in attributes attached to each of the HDF5 files or its datasets.
  On the visualization side, all chunked files are logically linked together to rebuild the original global data space. For linking, the mounting feature of HDF5 can be used, which allows one to mount external HDF5 files into the group hierarchy of the currently opened file (similar to mounting file systems in a Unix environment).

3.1 Current Implementation

As a first step towards the realization of this global picture, the HDF5 library was made Data Grid–aware by adding a new virtual File Driver to the VFD layer. This Data Grid VFD maps the basic I/O operations onto the Data Grid SC-API routines. Depending on the type of URL passed to the open call, the Data Grid Storage Client library automatically selects the appropriate low-level protocol driver for remote data transfers. The Data Grid supports several protocols, which can be addressed by the URLs passed to the Data Grid VFD.

– `<file://localhost/<path>>`
  for processing files located on the local machine's Unix file system.
– `<x-dpss://<dpss_host>/<dpss_path>>`
  for processing remote files that reside on a scalable, high-performance, distributed-parallel data storage system (DPSS).
– `<http://<www_host>/<www_path>>`
  for processing files that are accessible from any Web server.
  The Data Grid HTTP driver can currently read files only sequentially, but its next version will also provide random file access. For efficient partial file accesses (which are important for hyperslab selections), the driver will then make use of the extended features offered by HTTP 1.1.

The FTP protocol is supported as well by the Grid Storage API, but it cannot be used for our purposes because it does not allow random file access.

Before opening a remote HDF5 file, the Data Grid VFD has to be selected by modifying the default file access property list passed to the HDF5 file open call. Subsequently all I/O operations on this file—such as opening, closing, reading, and writing of both metadata and raw data—are directed to this VFD, which in turn calls the appropriate Storage Client API routines according to the file's URL and the low-level protocol specified therein.

As discussed above, we are not only interested in remote reading of distributed datasets. We want complete flexibility in choosing general subsections of files (downsampling, choosing arbitrary subregions, etc.), which are called hyperslabs. This might be important not only to reduce data transfer over a network, but also because for scientific or visualization reasons it may be necessary to examine arbitrary subparts of the dataset. When performing I/O on a hyperslab selection, it might happen that a single HDF5 API operation actually consists of many basic reads or writes, which are all passed to the VFD layer individually. For example, a hyperslab read of a 10x10 dataset with down-sampling applied to every second element in each dimension would cause the VFD to execute 50 individual reads of single elements (compared to only one block read when accessing the full dataset). Such hyperslab operations would probably show very poor performance if done in a conventional way, because the latencies of each individual read/write passed over the network would sum up to unacceptable delays. Therefore the hyperslab selection code inside the HDF5 library was extended to pass hints down to the VFD layer, which indicate if the current read/write call belongs to a certain hyperslab selection, and more associated reads/writes will follow. The Data Grid VFD can now make use of these hints and optimize remote I/O operations by accumulating all subsequent individual reads/writes into a single transaction, representing the complete hyperslab I/O transfer. Such a transaction is then passed to the Grid Storage client en bloc; which in turn sends it to the Grid Storage server, where it is executed at once; and the results are sent back to the client in a single data stream.

As soon as the Grid Storage API can handle such transaction requests in addition to individual reads/writes, the implemented code modifications in the HDF5 library can be enabled to improve I/O performance for hyperslab selections.

A prototype of the Data Grid VFD was demonstrated at the SC99 conference in Portland, Oregon, in October of 1999 (see Fig. 3). Amira used this driver to visualize datasets from a simulation of colliding neutron stars and collapsing gravitational waves. The datasets in gigabyte range had been previously generated by Cactus. The HDF5 output files were located on a Distributed Parallel Storage System (DPSS) at the Argonne National Laboratory, but they had also been transferred to the local visualization workstation (driving an I-Desk on the show floor at Portland).

It was demonstrated that during the read of individual time steps, one could switch between the remote file and its local copy. This showed complete transparency of remote and local I/O at application level, with network delay added only when fetching the data from remote. Due to the available bandwidth of the OC-12 network connection between Portland (Oregon) and Argonne (Illinois), and the overall latency minimization due to the optimiza-

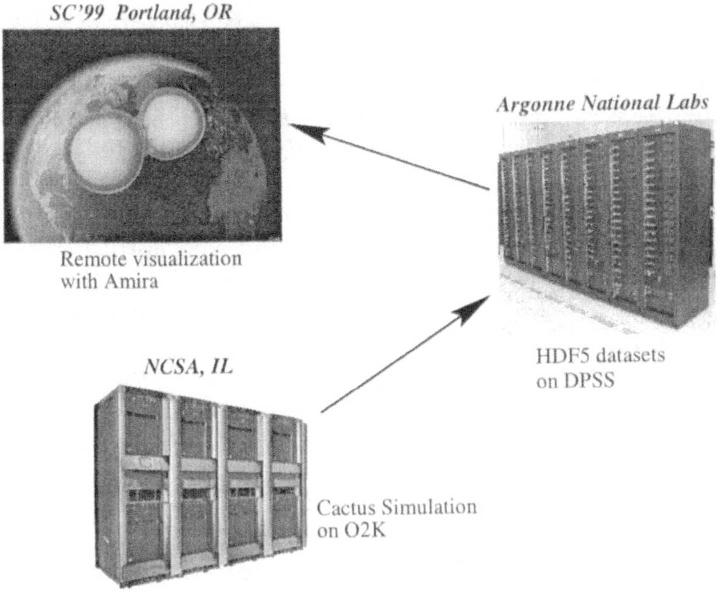

Fig. 3. The remote visualization demo as presented at SC99.

tion of HDF5's buffer usage for receiving and converting data, it was even possible to successfully present a smooth animation of the gravitational wave collapsing to a black hole.

3.2 Future Work

The current implementation is capable of accessing single remote files, but the datasets are still required to be written as unchunked data. To overcome this limitation, our future work will focus on an I/O scheme that allows the simulation to output its data as chunks into distributed files, using parallel I/O for node-local files. For the visualization, this means that it must be able to reconstruct the global data space from all chunks, without the need for staging any physical data. Read accesses are then performed on this global data space.

Because neither of the approaches described in Sect. 3 leading to such distributed I/O schemes has a clear advantage over the other, we decided to implement both of them. They require enhancements within the HDF5 library itself, as well as additional code functionality on top of it in order to establish a consistent view on the complete data space as seen by the simulation and the visualization.

Currently, an external file list may contain only local files, because all operations on an EFL file are mapped directly onto standard Unix I/O calls, bypassing the VFD layer of HDF5. This has to be changed in the future to allow EFL entries to refer to Data Grid URLs instead of local files.

Currently I/O operations on external files are implemented quite inefficiently: for each access the file is reopened, and then closed afterwards. Adding functionality for managing state information of already opened external files will help to improve performance dramatically.

The second approach of mounting independent chunked HDF5 files together requires a new abstraction layer for transferring distributed data, because the HDF5 data mapping can be applied to single datasets only. Because the global data space is now distributed over several datasets (chunks), the visualization would no longer be able to transparently select data from somewhere in this global data space; rather, the data distribution would have to taken into account. Thus an interface is necessary that is able to operate on such groups of chunked datasets, iterating over each chunk while applying the appropriate partial hyperslab selection.

Ideally such functionality would be integrated in an HDF5 wrapper on top of the HDF5 library. Most of the implementation work will probably comprise the re-implementation of the HDF5 hyperslab selection code. We hope that much of the existing code from the HDF5 library can be reused here.

4 Summary

We presented a solution to the problem of efficient file I/O in Grid Environments, where a large-scale scientific application running in a distributed computing environment produces huge amounts of data that are to be postprocessed.

Our approach is to allow the simulation program to output its data locally, using well-known efficient local I/O paradigms. This results in a data layout that is distributed over several physical files within the Data Grid.[5] For postprocessing, a global view on the original data is re-established, without the need of explicitly staging all data locally. Hyperslab selections can be applied on this global data space in order to fetch only those data which are really needed locally—for example, for visualization purposes.

We use the Globus Metacomputing Toolkit with its Data Grid component to manage the remote files, and HDF5 as the underlying I/O library and data format for accessing such distributed data. A new driver was added to the Virtual File Driver layer of the HDF5 library, which provides efficient and transparent access to remote files, based on the Data Grid library. A prototype version of this VFD was presented at the SC99 conference in Portland (Oregon), where we successfully demonstrated the visualization of datasets

[5] In terms of the Data Grid, the data are stored in a single logical file, which refers to multiple (distributed) file instances.

from a huge recombined remote HDF5 file located on a high-performance storage system at Argonne (Illinois).

Our future work will enable this version to handle datasets distributed over multiple files, using HDF5's features to specify a global data space either via an External File List, or by mounting separate HDF5 files together.

Acknowledgments

It is a pleasure to thank many colleagues and collaborators who contributed to this work, both directly and indirectly. Mike Folk and Quincey Koziol from the HDF5 development team have worked closely with us on this project, and hosted one of us (TR) for an extended visit to NCSA in advance of SC99. Ian Foster, Brian Toonen, Steve Tuecke, and others from the Globus team at ANL provided much help, particularly on the Data Grid work, and also hosted a visit from TR. Gabrielle Allen, Tom Goodale, Gerd Lanfermann, and Joan Massó of the Cactus development team at AEI and the University of the Balearic Islands have been essential in providing time and energy in Cactus support for the project. Detlef Stalling and Malte Zöckler of the Amira development team have been important collaborators in the visualization work. John Shalf at NCSA has been an excellent collaborator, providing many ideas and expertise in all areas, but especially related to this project, on general I/O issues. Jason Novotny at NLANR has also contributed, especially on issues involving Globus and user interfaces. Computing resources and general support, financial and otherwise, have been provided by AEI, ANL, NCSA, RZG, and ZIB. Finally, and most importantly, the project is directly supported by DFN/BMBF, both through financial support and through provision of the Gigabit Testbed.

References

1. A. A. Abromovici et al. Ligo: The laser interferometer gravitational wave observatory. *Science*, 256:325, 1992.
2. G. Allen, W. Benger, C. Hege, J. Massó, A. Merzky, T. Radke, E. Seidel, and J. Shalf. Solving Einstein's equations on supercomputers. In press in *IEEE Computer*, 1999.
3. G. Allen, T. Goodale, and E. Seidel. The cactus computational collaboratory: Enabling technologies for relativistic astrophysics, and a toolkit for solving PDEs by communities in science and engineering. In *IEEE 7th Symp. on the Frontiers of Massively Parallel Computation (Frontiers '99)*, 1999.
4. Amira: An advanced 3D visualization and volume modeling system, 1999. `<http://www.amiravis.com/>` or `<http://amira.zib.de/>`.
5. P. Anninos, J. Massó, E. Seidel, and W.-M. Suen. Numerical relativity and black holes. *Physics World*, 9(7):43–48, 1996.
6. W. Benger, I. Foster, J. Novotny, E. Seidel, J. Shalf, W. Smith, and P. Walker. Numerical relativity in a distributed environment. In *Ninth SIAM Conference on Parallel Processing for Scientific Computing Proceedings*, 1999.

7. J. Bester, I. Foster, C. Kesselman, J. Tedesco, and S. Tuecke. Gass: A data movement and access service for wide area computing systems. In *Sixth Workshop on I/O in Parallel and Distributed Systems*, 1999.
8. The cactus code server, 1999. `<http://www.cactuscode.org/>`.
9. J. Carretero, F. Pérez, P. de Miguel, F. García, and L. Alonso. ParFiSys: A parallel file system for MPP. *ACM SIGOPS*, 30(2):74–80, 1996.
10. A. Chervenak, I. Foster, C. Kesselman, C. Salisbury, and S. Tuecke. The data grid: Towards an architecture for the distributed management and analysis of large scientific datasets. Submitted to *NetStore '99*, 1999.
11. A. Chervenak, I. Foster, C. Kesselman, C. Salisbury, and S. Tuecke. Storage client API specification. Unpublished, 1999.
12. DFN gigabit testbeds, 1999. `<http://www.dfn.de/projekte/gigabit/home.html>`.
13. I. Foster and C. Kesselman. Globus: A metacomputing infrastructure toolkit. *Int. J. Supercomputer Applications*, 11(2):115–128, 1997.
14. I. Foster and C. Kesselman. The globus project: A status report. In *IPPS/SPDP '98 Heterogeneous Computing Workshop Proceedings*, pages 4–18, 1998.
15. I. Foster and C. Kesselman, editors. *The Grid: Blueprint for a New Computing Infrastructure.* Morgan Kaufmann, 1998.
16. A. S. Grimshaw, W. A. Wulf, J. C. French, A. C. Weaver, and P. F. Reynolds, Jr. A Synopsis of the Legion Project. Technical Report CS-94-20, University of Virgina Comp. Sci. Dept., 1994.
17. HDF5: API specification reference manual, 1999. `<http://hdf.ncsa.uiuc.edu/HDF5/doc/RM_H5Front.html>`.
18. M. Litzkow, M. Livny, and M. W. Mutka. Condor: A hunter of idle workstations. In *8th Int. Conf. of Distributed Computing Systems Proceedings*, pages 104–111, 1988.
19. The Open Software Foundation. *Introduction to OSF DCE.* Prentice Hall, Englewood Cliffs, NJ, 1988–1991. PTR.
20. M. Romberg. The unicore architecture: Seamless access to distributed resources. In *IEEE High Performance Distributed Computing Proceedings*, volume HPDC-8, 1999.
21. E. Seidel. Technologies for collaborative, large scale simulation in astrophysics and a general toolkit for solving PDE's in science and engineering. In T. Plesser and P. Wittenburg, editors, *Forschung und wissenschaftliches Rechnen.* Max-Planck-Gesselschaft, München, Germany, 1999.
22. E. Seidel. Black hole coalescence and mergers: Review, status, and "where are we heading?". In press in *Progress of Theoretical Physics*, 2000.
23. V. S. Sunderam. PVM: A framework for parallel distributed computing. *Concurrency: Practice and Experience*, 2(4):315–339, 1990.
24. Tele-immersion: Collision of black holes, 1999. `<http://www.zib.de/Visual/projects/TIKSL/>`.
25. B. Tierney, J. Lee, B. Crowley, M. Holding, J. Hylton, and F. Drake. A network-aware distributed storage cache for data intensive environments. In *IEEE High Performance Distributed Computing, Proceedings*, volume HPDC-8, 1999.
26. A user's guide for HDF5, 1999. `<http://hdf.ncsa.uiuc.edu/HDF5/doc/H5.user.html>`.

Performance Enhancements for HPVM in Multi-Network and Heterogeneous Hardware

Greg Bruno[1,2], Andrew A. Chien[1], Mason J. Katz[1], and Philip M. Papadopoulos[1]

[1] Department of Computer Science and Engineering; University of California, San Diego; USA
[2] INCEP Technologies, Inc.; San Diego; California; USA

Abstract. The Concurrent Systems Architecture Group (CSAG) has been investigating high-performance clustering of commodity PCs with gigabit interconnects and a commodity OS (Windows NT). This paper describes several of the major enhancements in the latest release of HPVM 1.9 (High Performance Virtual Machine). The first two of these are concurrent support of multiple gigabit fabrics and shared memory transport, and a self-tuning I/O scheme that manages performance disparities of various PCI implementations. In addition to these innovations, HPVM 1.9 also delivers both lower-latency (8.8 μs) and high-bandwidth (100+ MB/s) messaging. Another major contribution of HPVM 1.9 is the design and implementation a new unified polling strategy that allows simultaneous use of several networks (either VIA or Myrinet) and a shared memory transport without resorting to complex adaptive polling schemes. The implementation of our new high-performance shared memory transport is described in detail. Finally, several system and application benchmark results are given and interpreted to assess the value of these changes.

1 Introduction

High-performance PC clusters are rapidly becoming mainstream because of their favorable cost-to-performance ratios. Indeed, clusters of commodity PCs with gigabit networking (especially Myrinet, VIA, and to a lesser extent gigabit Ethernet) perform on par with significantly more expensive supercomputers. One focus of the CSAG group has been to explore methods for extracting high performance (100+ MB/s, 8.8 μs latency) out of the available hardware, and then translating that capability to applications through efficient layering of software. The hardware landscape has changed significantly during the past three years of this project. Low-latency gigabit hardware has achieved standardization through the Virtual Interface Architecture (VIA); SMPs (at least two-way) are increasingly attractive as nodes; and commodity motherboards with greater integration have significantly increased I/O performance variances across generations. Because of these rapid changes, greater flexibility and tunability was needed in the Fast Messages (FM) core messaging substrate of HPVM. This paper describes four enhancements to the latest version of HPVM, which improve its overall performance and increase performance stability over generations of motherboards.

Older versions of HPVM used the messaging hardware even for processes communicating on the same physical machine. Similar to older message passing machines like the Intel Paragon, this meant that node-to-node messaging performance was faster than on-node messaging performance. Furthermore, as the processor count per SMP increases, this design becomes increasingly disadvantageous. Instead of using the network hardware for every message, shared memory should be used for intra-machine transfers. While this optimization had been explored by other researchers [16], they employed complex adaptive polling schemes to manage the large cost differential between polling for intra- and inter-machine messages. The method presented here reduces the polling cost for inter-machine messages, eliminating the need for adaptive schemes without sacrificing performance on either transport. The resulting shared memory transport for FM is very efficient (delivering 3 μs latency, 200+ MB/s, see Sect. 5), with a poll cost of approximately 150 ns. A memory-based notification scheme for Myrinet network reduces polling cost to 100 ns, reducing FM point-to-point latency by nearly 15 percent. Parity in polling costs means that multiple interfaces can be polled in succession, eliminating the need for adaptive schemes and allowing additional transports to be easily incorporated. Section 4 describes the memory-based notification scheme and support for multiple network transports. Section 5 describes the design of FM's shared memory transport in detail.

Over the past few years, motherboard and specifically PCI chipset implementations have proliferated. In some cases, recent chipsets can deliver poorer programmed I/O performance than their predecessors. Because FM relies on efficient write combining and programmed I/O, these differences were directly reflected (negatively) in FM performance. To address this problem robustly, HPVM 1.9 includes self-tuning I/O code that empirically selects the optimum data burst size to obtain peak programmed I/O throughput. Section 6 describes this strategy and its associated performance.

In this paper, we summarize the relevant HPVM background and design in Sect. 2 and Sect. 3. We then describe the memory notification and shared memory implementation in Sect. 4 and Sect. 5. The observed performance differences across chipset generations and our adaptive I/O solution is described in Sect. 6, and Sect. 7 closes with a summary of our perspective on these results and conclusions.

2 Background

The plummeting price and increasing performance of both desktop computing systems and high-speed networks have made the use of cluster computing systems increasingly desirable. The Beowulf cluster approach [3], which relies on the pervasive Fast Ethernet technology (and essentially zero incremental cost) produces clusters that support coarse-grained, loosely coupled parallelism. The faster system-area network (SAN) technologies [1,2,4,10,13] provide hardware enablers for tightly coupled clusters. Significant advances from the research community in the past five years have made it possible to deliver hardware communication performance to the applications Fast Messages (FM) [17], Active Messages (AM) [5] U-Net [22], VMMC-2 [8], PM [20], BIP [18], MINI [12], and Osiris [7]. These efforts have forged a consensus on core requirements for network interfaces that was eventually crystallized in the Intel/Compaq/Microsoft standard for cluster interfaces—the Virtual Interface Architecture [21]. We briefly survey highlights from our efforts in this area during this period.

The HPVM research project has focused on crucial aspects of the design of high-performance messaging layers such as the trade-offs between protocol processing overhead and the functionalities exposed in the programming interface. The key design decisions taken for the original Fast Messages (FM) [17] for Myrinet implementation provided key communication services while retaining microsecond-level overhead. In April of 1995, Fast Messages 1.1 achieved short message latencies of 14 μs and a peak bandwidth of 17.6 MB/s Myricom's Myrinet. This performance represented 75 percent of the available 23 MB/s of I/O bandwidth on the Sparcstation 20 platform, as well as a performance improvement in both dimensions of approximately 5–10 times. The FM 1.1 implementation achieved $N_{\frac{1}{2}}$ of 54 bytes with a bandwidth of 17.5 MB/s available for messages as small as 128 bytes. FM 1.1 implemented the Berkeley Active Messages API [5].

Designed and implemented in autumn of 1996, Fast Messages 2.0 and 2.1 improved the API based on lessons learned from a Sockets-FM and MPI-FM implementation [14]. The new API provided streamed gather-scatter at both sender and receiver; per-message threaded handlers; and per-message flow control, which enabled the full performance of FM to be delivered to a wealth of application API's (MPI [14,15], Shmem Put/Get [11], Global Arrays, BSP). The performance achieved by FM 2.1 on 300 MHz Pentium II's and Myricom's Myrinet is 9 μs minimum latency, 92 MB/s peak bandwidth, with $N_{\frac{1}{2}} \approx 256$ bytes. These values represent high absolute performance, comparing to MPP interconnect performance and internal memory bandwidth.

Fast Messages 2.1 (as part of the HPVM 1.0 release) was used to build the 256-processor NT Supercluster, the world's largest Windows NT cluster. This system was demonstrated in April 1998, at the NCSA Alliance kickoff meeting, and has been in production use since autumn of 1998. The cluster has been upgraded several times, and current aggregate performance is approximately 200 gigaflops, 128 gigabytes of memory, and 2 terabytes of disk. Network bisection bandwidth delivered by Fast Messages running on Myrinet is $64 * 160$ MB/s = 10 GB/s.

The release of HPVM 1.2 in February 1999 added several important new functionalities to the system. First, we incorporated multiprocess and multiprocessor threading support, which allows FM to be used effectively in multiprocessor PC boxes. Separate processes using FM are safely isolated from each other. Second, peak bandwidth was increased from 92 MB/s to more than 100 MB/s. Third, improved dynamic configuration facilities (for messaging buffer pools) and an Installshield packaging made the entire system dramatically easier to install, manage, and use. Fast Messages and HPVM have also been used as a testbed for research on novel mechanisms for co-scheduling on a network of workstations [19], and for research on the introduction of QoS guarantees in a wormhole-routing interconnect [6]. HPVM is available for download on our Web site at <http://www-csag.ucsd.edu/>. The remainder of this paper summarizes the improvements to the HPVM system in HPVM 1.9.

3 HPVM Low-Level Design Review

Overall, the structure of FM is straightforward. Packets must traverse a pipeline across multiple I/O busses. Data integrity is ensured by guaranteeing that a receiving host has free memory for every incoming packet and by taking advantage of the hardware reliability of the network. Active-message–like semantics require that application programs are able to handle incoming messages while sending an outgoing message.

Reviewing the low-level architecture of the Myrinet LANai network interface and the basic structure of FM, the Myrinet LANai NIC houses a programmable 66 Mhz (Version 4.x) custom processor, and the code that runs on this processor is often called the LCP (LANai Control Program) or MCP (Myrinet Control Program). FM uses a custom LCP that only processes FM-formatted packets. The Myricom-supplied GM code also supports a TCP encapsulation. Working in conjunction with the LCP, a host library runs on the machine's CPU (x86 in this paper) to complete the software layout. Because the LANai is physically located on the PCI bus, memory transfers can occur in two ways: data can be DMAed *to/from* the LANai memory, or data can be PIOed (programmed I/O) *to* the NIC. The FM library always uses PIO for sending data to the LANai. DMA is always used to send data from the LANai to the host. One further feature/structure of the Myricom LANai is that packets must be DMAed to/from the physical network from NIC-local memory. This implies that a message packet must traverse a pipeline as shown in Fig. 1.

With proper software design, this entire process can be pipelined for high performance. All high-performance messaging implementations on Myrinet hardware must manage this pipeline to achieve their goals.

Data coming from the LANai to Host2 must use DMA to move data to host memory. DMA engines require that target memory be paged-in by the operating system. Usually, host memory is page-pinned so that the host operating system never pages-out this memory. While a variety of methods are possible, FM uses a static pinning policy. That is, the device driver pins one large region of memory when the device is loaded. The size of this region

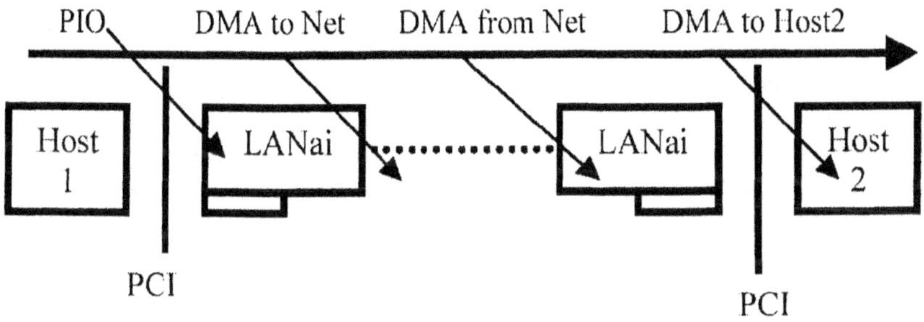

Fig. 1. Message packet traversing a pipeline.

is specified as a combination of maximum number of hosts in the cluster, the size of an FM packet (currently 2048 bytes of data for a total of 2080 bytes, including header), and the number of communication contexts. FM uses the term *communication context* to specify how many independent processes on a host must have simultaneous access to the high-speed network device. Special care (using the virtual memory protection mechanism of the host OS) keeps these contexts from interfering with each other. This is especially important because access to the LANai memory is memory-mapped into user space, allowing OS-bypass for common network interactions.

Because LANai memory is memory-mapped into one of multiple communication contexts, the LCP must distinguish incoming packets and place them in the correct location. This is accomplished by having the LCP inspect the FM header for a program ID. When a context first gains access to the device, its program ID (assigned from a global resource manager) and the instance number of that ID are placed into LANai memory.

The Myrinet network (and Giganet VIA networks) are considered to be hardware reliable, with very low bit-error rates (equivalent to memory bit-error rates). Taking advantage of this hardware reliability, FM is able to eliminate the software overhead of timeout/retransmission strategies. However, to ensure that data are received correctly, there must be available host memory at the receiver for every incoming packet. To ensure that memory is available, FM uses a credit-based flow control scheme. At startup, every node in a computation is assigned a certain amount of credit (measured in FM packets). All packets (full or partial) consume one credit. When a sender runs out of credit, it must wait for a credit refill. FM currently uses a static allocation policy, but the pinned receive region is actually a shared allocation. That is, the sum of all assigned credit across all nodes is shared from a single region. Changes to the FM credit allocation policy can be implemented without changing the MCP operation. In fact, the MCP knows nothing of credit, and dutifully acts as a packet forwarder. Incoming packets are placed in the next slot of the pinned memory region (with wrap at the region boundary). The FM credit policy ensures that packets will not be overwritten before being extracted from the network.

The FM library requires programs to extract packets from the network. FM is a handler-based system, so the packet type is associated with a "handler" code run on the destination host (active-message–like). The FM API call `FM_Extract()` allows the host library to examine packets that have arrived in the host's pinned memory region and to call the correct message handler. Because of this structure and in conjunction with the credit-based flow control scheme, when a sender runs out of credit, it is required to call `FM_Extract()`. This ensures message progression when two hosts are performing a head-to-head send. FM is completely thread-safe, so that multiple messages streams (say from different senders) can be handled safely.

4 Memory-Based Notification and Supporting Multiple NICs

4.1 Memory-Based Atomic Notification

In the previous versions of HPVM, applications detected the arrival of a new FM packet by calling an FM function that polled a memory location on the LANai interface. This is an expensive operation, because the request has to arbitrate for the CPU/memory bus, arbitrate for the PCI bus, then extract the value from the LANai's local memory and return it all the way back to a CPU register.

As described in Sect. 3, after the LANai control program receives a new packet from the network, it moves it into a main memory location, then the FM user-level library code extracts the packet from this location. The key is that the FM user-level library code and the LANai control program agree on the main memory location where the next packet is to be delivered. With this fact, it seemed clear that we could detect (atomic notification) by polling that location in memory rather than poll a control register across the PCI bus. This strategy has two key benefits. After the first poll, all subsequent polls can be satisfied in-cache; subsequent polls need not arbitrate or use the CPU/memory bus. The second benefit is that polling for FM packets doesn't require arbitrating for the PCI bus, so polls will not disturb any ongoing DMA transfers. The former ensures that polling scales on large SMPs, and the latter ensures fewer "turns" of the PCI busses, which is a critical factor when utilizing them at close to 100 percent (as FM does).

In essence, the hardware's cache coherence mechanism is used to signal the arrival of new packets. For example, if an application is actively polling for a new packet, on the first poll, the application will cache the poll value (which is a location in main memory where the LANai control program will deposit the next new packet). Subsequent polls will be satisfied by the cache—these polls will not generate any traffic outside the CPU the application is running on. When the LANai control program moves a new packet into the expected main memory location, the computer system's memory controller will determine that the application is caching the poll value and use the hardware-cache-invalidate mechanism to flush the poll value from the CPU cache. The next poll will cause the CPU to fetch the updated poll value from main memory. This value will be changed, thus indicating a newly arrived packet. This allows an application to actively poll for new packets in an efficient manner; regardless of the number of polls, at most two transfers from CPU to main memory will occur.

Finally, we achieve atomic notification with a single DMA transfer for variable-sized packets by using a novel custom packet layout. When a new packet arrives at the LANai, the LANai control program determines the packet's length and appends the length to the packet. The LANai control program then calculates the offset into the frame in main memory so that the newly appended packet length is aligned at the bottom of the frame. This allows applications to poll a common flag location at the end of the frame, regardless of the packet size. The use of the packet length allows the packet data to be contiguous with the flag, allowing all of it to be moved in a single DMA transfer. When the last value in the packet is written to main memory, the application knows that the entire DMA operation is complete. This simple optimization has a major impact on performance, improving latency for a zero-byte FM message over the previous HPVM implementation by 15 percent (10.4 μs vs. 8.8 μs).

4.2 Supporting Multiple NICs

HPVM 1.9 also allows several FM transports to be used concurrently: a shared memory transport (described in the next section) either a Myrinet transport or a Giganet transport. In our current implementation, switching between the Myrinet and Giganet transports only involves changing a registry variable and restarting the FM context manager service.

The shared memory transport enables efficient communication when FM tasks send messages to other FM tasks within the same SMP. The shared memory transport uses the same "memory notification" polling strategy as described above for the Myrinet interface. This has a subtle side effect, which is that polling for new packets costs the same for the shared memory and the Myrinet transports. The parity in polling costs allows multiple transports to be supported simply, and this architecture scales well for large SMPs, many network interfaces, and many different network types all within the same machine.

5 Shared Memory Transport

A deficiency of HPVM 1.2 was the implementation of the loopback message path, which traversed down to the Myrinet hardware layer. Although peak bandwidth between two Myrinet connected machines approached 100 MB/s, the bandwidth for two processes on the same machine was under 40 MB/s. The simple loopback design simplified the message layer; however, to better support larger SMP machines, we have designed and implemented a shared memory transport. In HPVM 1.9, each FM process is connected to Myrinet for inter-node communication, and to a shared memory region for intra-node communication. Both transport layers are supported concurrently, and because the FM design requires a central authority to map virtual node identifiers to physical nodes, each process can look up in the Global Resource Manager to decide which transport to use for peer communication. Further, due to the efficient receiver polling of the Myrinet transport, nothing more than adding a poll of shared memory is require to preserve low-latency communication for both transport layers.

The shared memory transport (hereafter referred to as SMT) uses the shared memory IPC mechanism provided by the host operating system. Although shared memory is straightforward, it does introduce the complication that pointer operations are unsafe, because there is no guarantee that separate processes will map the same memory region into the same address space. For this reason, all data structures used in SMT are array based. The basic design of SMT is for each process to control a message queue and permit other processes to atomically insert data into the queue. The implementation does not support any zero-copy IPC mechanism.

Mutual exclusion into shared message passing queues is low-latency and lock-free. The primary synchronization in SMT is a `fetch-and-increment` operation, which atomically fetches a value from memory and increments the value in place. By using a `fetch-and-increment` on the head and tail indices of a queue, lock-free mutual exclusion is guaranteed. The constraint that the queue length is a power of two removes any length overflow checks on indices when an appropriate bit mask is applied during index calculations. `Fetch-and-increment` is relatively inexpensive; on a 300 MHz Pentium II, our measurements show a single operation takes 85 nanoseconds (about 28 processor cycles).

The ordering property of `fetch-and-increment` is also used to assign SMT identifiers to processes. As processes attach to the shared memory region, they all `fetch-and-increment` a shared variable to determine their unique SMT identifiers. Just as a map of virtual node identifiers to physical node addresses is stored by the Global Resource Manager, a map of SMT identifiers to message queue locations is stored in the shared memory region. The map is used by senders to determine the message queue used by an SMT peer process.

Copying data into a frame in the message queue can be an expensive operation, and the host operating system may deschedule a process during this step. To allow for scheduler intervention without disrupting the sending and receiving in a message queue, the buffer filling and buffer pass-off steps are separated. Further, to allow for simple buffer reclamation, all buffers are controlled by the receiving process. Each process owns a free list and a receive queue. The free list is a ring buffer of 2K message frames, and the receive queue is a ring buffer of indices into the free list.

To send a message, a process must do the following.

1. Look up the receiving peer queue location using the SMT ID to queue location map.
2. `Fetch-and-increment` the head of the receiver's free list to consume a message frame.
3. `Test-and-set` an `inuse` flag in the frame. If the fetched value is non-zero, then go to step 2.
4. Copy data into the frame.
5. `Fetch-and-increment` on the tail of the receiver's receive queue.
6. Set the index at the fetched receive queue tail to the index of the consumed free list frame.

Note that under unfair scheduling, the sending process may traverse the free list in worst-case $\mathcal{O}(N)$ time, where N is the number of frames in the free list. The finite worst-case bound is a consequence of the fact that SMT is built on top of FM's credit-based flow control, which guarantees that a message frame will be present in the free list. However, assuming fair scheduling, or a single sender, the traversal time is $\mathcal{O}(1)$.

To receive a message a process must do the following.

1. Poll until the frame at the head of the receive queue is marked in-use.
2. `Fetch-and-increment` the head of the receive queue.
3. Index into the free list, and copy the data into the upper message layer if needed.
4. Unset the frame-in-use flag.

This scheme effectively reorders the receive queue according to the scheduling priority of the senders, but it requires an added `fetch-and-increment` and `test-and-set` operation, and adds some complexity to a single-queue scheme. This scheme will not protect from starvation when different process priorities are used. If a sender is scheduled at a higher priority, a lower priority sender may never acquire a message frame from the free list. The key advantage of this scheme is the reordering of senders to allow a sending process to effectively leapfrog over another, which keeps the receiver's message queue flowing. Without this property, a descheduled sender can block the receiver's message queue, thereby inducing bursts of high-latency communication directly correlated to the behavior of the host operating system scheduler.

To provide scalability to large SMP machines, the number of message queues a receiver owns is a parameter that can be adjusted to the amount of queue sharing desired. If the number of queues is set to one per receiver, then SMT has the same queuing structure as the Myrinet connectionless transport, where all senders transmit to the same device. The advantage of this scheme is that the receiver need only poll one receive queue, yielding low latency on the receiver side. The disadvantage is high contention due to numerous `fetch-and-increment` operations at the end points of the queue, which can result in high latency on the sending side. If the number of queues is set to 1 per receiver/sender pair, then SMT has the same queuing structure as the Giganet connection-oriented transport. The advantage of this scheme is that no synchronization primitives are needed in the queues, because this is a single writer/single reader scenario. The disadvantage is wasted memory, unless every permutation of process pairs communicates. More significantly, a receiver must poll a queue for every sender in order to extract messages, resulting in high latency on the receiver side.

If the number of queues can be set anywhere from one per receiver to N per receiver—where N is the number of SMT processes—then the amount of queue sharing, memory usage, and contention can be tuned for a given platform. If the number of receive queues is set at M—where $1 \leq M \leq N$—then the procedure for sending a message is the same as above, except that the receiver's free list and receive queue are now found as the receiver's $I \mod M$ free list and receive queue. Further, the receiver must now poll M receive queues.

The resulting performance of HPVM 1.9 with SMT improves on that of HPVM 1.2, both in delivered bandwidth (see Fig. 2) and in message latency (see Fig. 3). Incorporating the SMT into the FM system not only improves intranode communication performance, as discussed in Sect. 4, but it also does so with no penalty for internode communication.

Finally, to allow the SMT to run without a Myrinet network, we created a NULL transport to replace the Myrinet transport. This allows HPVM to run on a single workstation without any external transport hardware, resulting in a shared-memory-only implementation of HPVM. We envision this mode of operation becoming very common for HPVM users, because it allows them to develop HPVM software on their desktops.

6 PCI Chipset Pacing

HPVM (and FM in particular) was carefully tuned to use processor I/O and DMA asymmetrically to give highest performance. However, as PCI implementations have proliferated, we have found that PIO performance can be unpredictable. The following describes our experience and the adaptive tuning software we developed, which tests and selects good burst-size parameters to achieve good PIO performance.

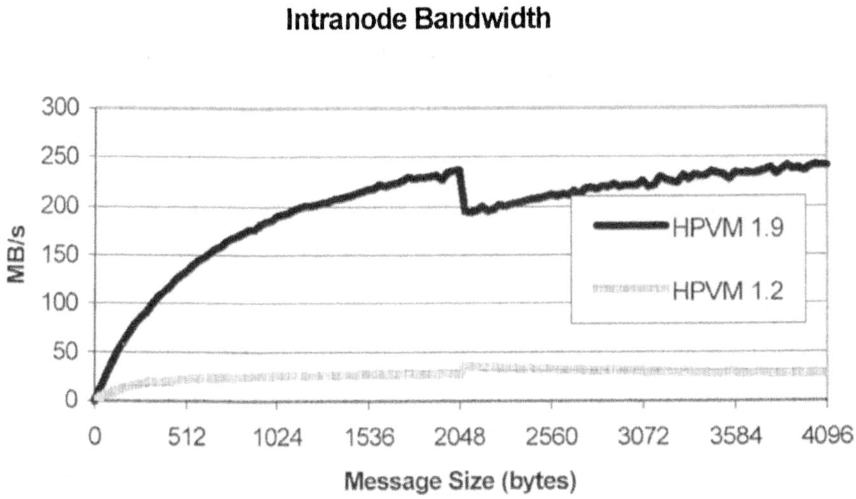

Fig. 2. Intranode FM messaging bandwidth, HPVM 1.2 vs. HPVM 1.9 (SMT).

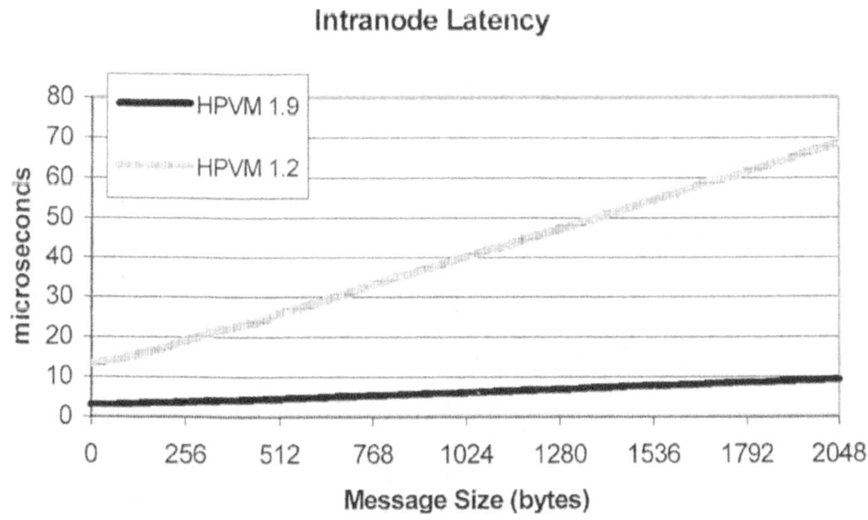

Fig. 3. Intranode FM messaging latency, HPVM 1.2 vs. HPVM 1.9 (SMT).

In December of 1998, we received a new 32-node cluster. Each node is an HP Netserver LPR that contains two 450 MHz Pentium IIs that are connected to a 100 MHz front-side bus and are coupled to main memory and I/O interfaces via the 440 BX chipset. Our previous cluster consisted of HP Kayak nodes. A Kayak node contains two 300 MHz Pentium IIs that are connected to a 66 MHz front-side bus and are coupled to main memory and I/O interfaces via the 440 LX chipset.

We were eager to compare FM performance for the clusters because the new cluster's processors and front-side buses are 50 percent faster (66 to 100 MHz). Preliminary FM micro-benchmarks (see Fig. 4) showed that the newer machines performance was better for smaller messages, but it also revealed that the newer machines performed worse for larger messages. The root of this behavior was inefficiencies of the PCI chipset at saturation, even though peak performance of the newer chipset is indeed higher than its older cousin. After adding sufficiently heavy instrumentation to the FM internals, it was discovered that the newer machines have worse PIO performance than the older machines for messages larger than 160 bytes. The slower machines' performance is robust, and performance continues to increase as message sizes increase; but the newer machines peak at 115 MB/s, and performance falls for messages of 160 bytes and greater. We verified that all configuration parameters (BIOS, chipset, and OS) between the two machines were identical.

Our working theory is that the queuing mechanism inside the 440 BX does not gracefully handle high-volume PIO scenarios (the queue in the 440 BX is 5 cache lines deep, that is, 160 bytes). With this theory, the following strategy was implemented: for messages more than 160 bytes, packets are "paced" so that the 440 BX chipset can perform optimally—that is, for messages larger than 160 bytes, data is PIOed across the bridge at 160 byte chunks. For example, if an application is sending a 400-byte message, the FM internal pacing code will PIO the message to the LANai in three transactions: two 160-byte chunks followed by an 80-byte chunk. This technique conditioned the newer machines' performance curves and expected performance as shown in Fig. 5.

To ensure that this performance is robustly attained in other systems, we inserted a PIO micro-benchmark in the distributed code (the same code used to uncover the 440 BX problem) that runs at boot time. This code analyzes the system's performance and chooses a pacing size for the system that attains the highest PIO performance.

7 Summary and Future Directions

The design of HPVM 1.9 builds on several generations of high-performance cluster software systems from the Concurrent Systems Architecture Group, originally at the University of Illinois and now at the University of California,

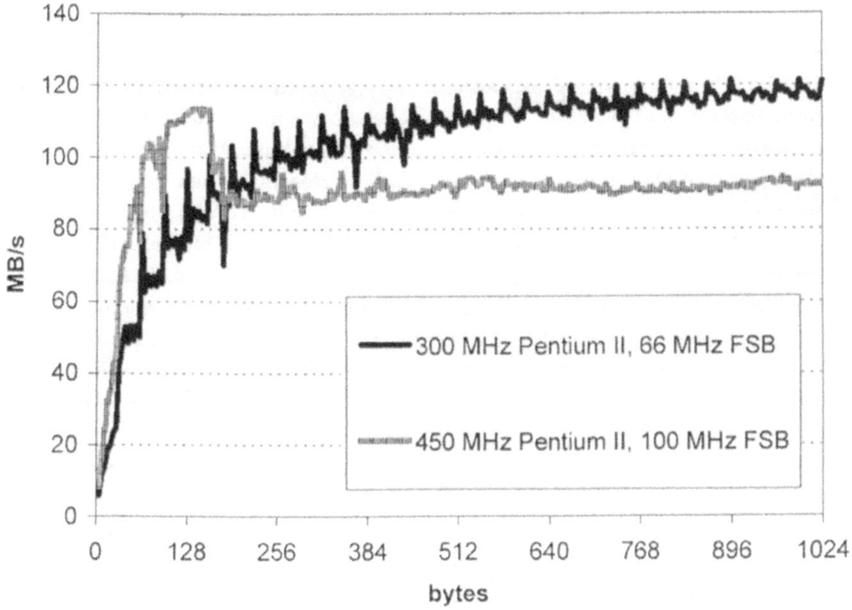

Fig. 4. FM micro-benchmarks without packet "pacing."

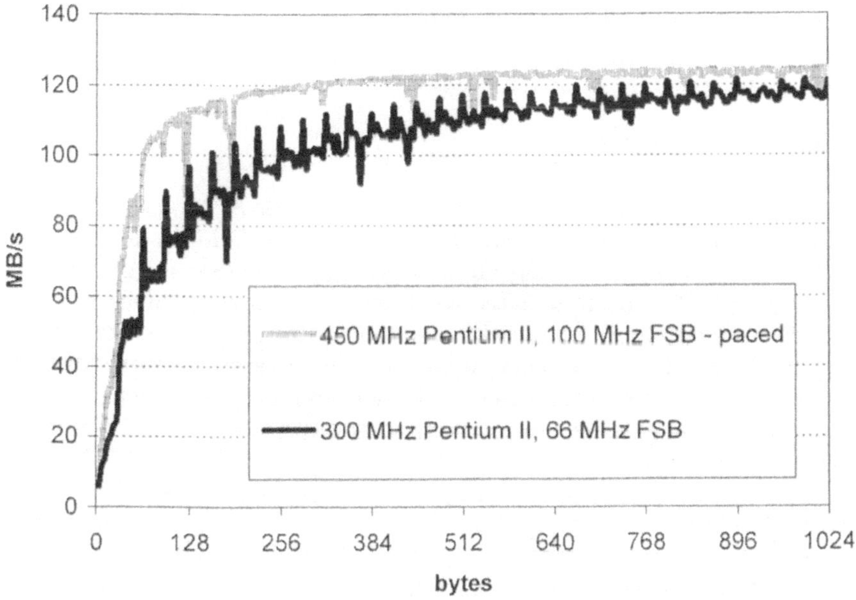

Fig. 5. FM micro-benchmarks results with packet "pacing."

San Diego. These previous systems have been used to build large production clusters and to deliver a wealth of application computational science. HPVM 1.9 adds a number of capabilities that further increase the flexibility, performance, and potential impact of the tools. Our efficient integration of multiple transports with high performance in a robust structure (a structure that is self-tuning and has low overhead) represents a new solution to an old problem [16,9]. Other specific new capabilities in HPVM 1.9 include support for multiple hardware network types (Giganet and shared memory), a workstation configuration for convenient software development, support for the Bulk-synchronous Programming API, and improved performance (8.8 μs latency, 100+ MB/s, see Fig. 2 and Fig. 3).

Currently, the NCSA Alliance NT Supercluster, which comprises more than five hundred processors, is using HPVM 1.9 (replacing HPVM 1.2 on that cluster). This cluster will continue to expand rapidly, because its delivered performance, reliability, and reasonable cost make it a compelling vehicle for computational science. We are exploring the use of HPVM 1.9 as a basis for a wide-area cluster federation (coupling over networks from OC-12 to OC-192, 10 gigabits/s and faster). We have successfully coupled the 128 processors of the UCSD cluster with the 256 processors of the NCSA NT Supercluster, using the OC-12 vBNS. We are exploring higher-speed network experiments, as well as the use of high-speed clusters to build a cost-effective network storage server. We have integrated a five-terabyte server and demonstrated a gigabyte per second of sustained external network bandwidth. We are working with other researchers to use this in computational Grid applications.

Acknowledgments

The research described is supported in part by DARPA orders E313 and E524, through the US Air Force Rome Laboratory Contracts F30602-99-1-0534, F30602-97-2-0121, and F30602-96-1-0286, and NSF Young Investigator award CCR-94-57809. It is also supported in part by funds from the NSF Partnerships for Advanced Computational Infrastructure—the Alliance (NCSA) and NPACI. Support from Microsoft, Hewlett-Packard, Myricom Corporation, Intel Corporation, Packet Engines, Tandem Computers, and Platform Computing is also gratefully acknowledged.

References

1. American National Standard for Information Systems. Fiber-distributed data interface (FDDI): Token ring media access control (MAC). Technical Report ANSI X3.139-1987, American National Standard for Information Systems, July 1987.

2. T. M. Anderson and R. S. Cornelius. High-performance switching with fibre channel. In *Digest of Papers Compcon 1992*, pages 261–268. IEEE Computer Society Press, 1992.
3. D. Becker, T. Sterling, D. Savarese, J. Dorband, U. Ranawak, and C. Packer. Beowulf: A parallel workstation for scientific computing. In *International Parallel Processing Symposium Proceedings*, 1995.
4. N. J. Boden, D. Cohen, R. E. Felderman, A. E. Kulawik, C. L. Seitz, J. N. Seizovic, and W. Su. Myrinet: A gigabit-per-second local-area network. *IEEE Micro*, 15(1):29–36, February 1995.
5. B. Chun, A. Mainwaring, and D. Culler. Virtual network transport protocols for Myrinet. In *Hot Interconnects V Proceedings*, August 1997.
6. K. Connelly and A. A. Chien. FM-QoS: Real-time communication using self-synchronizing schedules. In *SC97 Proceedings*, November 1997.
7. P. Druschel, L. L. Peterson, and B. S. Davie. Experiences with a high speed network adaptor: A software perspective. In *SIGCOMM '94 Symposium Proceedings*, pages 2–13, August 1994.
8. C. Dubnicki, A. Bilas, Y. Chen, S. Damianakis, and K. Li. VMMC-2: Efficient support for reliable, connection-oriented commnication. In *Hot Interconnects V Proceedings*, August 1997.
9. I. Foster, C. Kesselman, R. Olson, and S. Tuecke. Nexus: An interoperability layer for parallel and distributed computer systems. Technical Report Version 1.3, Argonne National Laboratory, December 1993.
10. D. Garcia and W. Watson. Servernet II. In *Parallel Computer Routing and Communications Workshop Proceedings*, LNCS. Springer-Verlag, 1997.
11. L. A. Giannini and A. A. Chien. A software architecture for global address space communication on clusters: Put/get on fast messages. In *High-Performance Distributed Computing Conference Proceedings*, 1998.
12. F. Hady and B. L. Menezes. The performance of crossbar-based binary hypercubes. *IEEE Transactions on Computers*, 44(10):1208–1215, October 1995.
13. IEEE. Standard for scalable coherent interface (SCI) specification. Std. 1596-1992, IEEE, August 1993.
14. M. Lauria and A. Chien. MPI-FM: High performance MPI on workstation clusters. *Journal of Parallel and Distributed Computing*, 40(1):4–18, January 1997.
15. M. Lauria, S. Pakin, and A. A. Chien. Efficient layering for high speed communication: Fast messages 2.x. In *High-Performance Distributed Computing Conference Proceedings*, 1998.
16. S. Lumetta, A. M. Mainwaring, and D. E. Culler. Multiprotocol active messages on a cluster of SMPs. In *SC97 Proceedings*, San Jose, CA, November 1997.
17. S. Pakin, M. Lauria, and A. Chien. High performance messaging on workstations: Illinois fast messages (FM) for Myrinet. In *SC95 Proceedings*, volume 2, pages 1528–1557, 1995.
18. L. Prylli and B. Tourancheau. Protocol design for high performance networking: A Myrinet experience. Technical Report 97-22, LIP, Ecole Normale Superieure de Lyon, July 1997.
19. P. Sobalvarro, S. Pakin, A. Chien, and W. Weihl. FM-DCS: An implementation of dynamic coscheduling on a network of workstations. In *ASPLOS-VII NOW/Cluster Workshop*, Cambridge, MA, October 1996.

20. H. Tezuka, A. Hori, and Y. Ishikawa. PM: A high-performance communication library for multi-user parallel environments. Technical Report TR-96-015, Tskuba Research Center, Real World Computing Partnership, November 1996.
21. *The Virtual Interface Architecture Version 1.0*, December 1997. Promoted by Intel, Compaq, and Microsoft and available from `<http://www.viarch.org/>`.
22. T. von Eicken, A. Basu, V. Buch, and W. Vogels. U-Net: A user-level network interface for parallel and distributed computing. In *15th ACM Symposium on Operating Systems Principles Proceedings*, pages 40–53, December 1995.

JACO3: A CORBA Software Infrastructure for Distributed Numerical Simulation

Stéphane Gouache and Thierry Priol

IRISA/INRIA—Projet PARIS; Campus de Beaulieu; F-35042 Rennes Cedex; France

Abstract. This paper gives a description of a software infrastructure that allows the execution of complex simulation applications on distributed systems. These systems involve several high-performance computing resources linked together by a wide area network. The targeted applications are made of several simulation codes that are in charge of simulating either different physics or several physical objects. The implementation of this software infrastructure is based on CORBA, an object-oriented middleware from the Object Management Group, that is widely accepted for the programming of distributed systems. The paper outlines different basic services that constitute the software infrastructure as well as several techniques to encapsulate both off-the-shelf commercial simulation codes and parallel simulation codes.

1 Introduction

The increasing performance of computers and networks will change the way numerical simulation is performed. It is now possible to foresee comprehensive simulations of the design of complex manufactured products using a set of computing resources linked together by high-performance networking technologies. The aggregating of computing resources allows the simulation of complex manufactured products being performed in a distributed way to meet either security or performance requirements as illustrated by Color Plate 1 on page 285.

As an example, distributed numerical simulation can be used to simulate the same physics on different objects, with each of them being simulated on a separate computing resource in order to avoid the exchange of confidential data (i.e., the model of the object to be simulated). Another application is the simulation of different physical behaviors that encompass multi-disciplinary aspects, such as structural mechanics, computational fluid dynamics, electromagnetism, noise analysis.

To make distributed simulation techniques available to end-users, a distributed software infrastructure has to be provided to hide as much as possible the distributed nature of the high-performance computing platform. Such an infrastructure has to offer a set of basic services for the execution of distributed numerical simulation applications. The aim of the EU-funded Jaco3 project is to address these needs, developing a high-performance distributed software environment based on the Common Object Request Broker Architecture (CORBA) [11] standard from the Object Management Group (OMG). This paper gives an overview of the Jaco3 distributed software environment.

The paper is organized as follows. Section 2 outlines the characteristics of the Jaco3 software environment. Section 3 and Sect. 4 describe the services that are included in the environment. Section 5 describes several techniques to encapsulate different kinds of simulation codes into CORBA objects. Section 6 summarizes related works on the topic. Finally, Sect. 7 draws some conclusions and perspectives on this work.

2 Overview of the Jaco3 Architecture

The Jaco3 software environment aims at programming a distributed system for coupled simulation applications. A typical distributed system is a set of servers, visualization systems, and supercomputers connected together using either an intranet or the Internet. The programming of such systems is difficult due to both the distribution of computing resources and the different administration domains with which computing resources are associated. The Jaco3 software environment is an attempt to hide all these complex problems to the users. It provides, in some sense, a *single system image* of the distributed system thanks to a set of services available at each computing node of the system. Communication between these different services has to be handled using a middleware that is able to manage the heterogeneity of a distributed system both in terms of computing and networking technologies. The Jaco3 project selected CORBA as the basic middleware for the development of the distributed software environment. CORBA is a modern object-oriented middleware that is widely used for distributed processing.

2.1 Common Object Request Broker Architecture

CORBA is a specification from the OMG (Object Management Group) [11] to support distributed object-oriented applications. Such applications can be seen as a collection of independent software components or CORBA objects. Remote method invocations are handled by an Object Request Broker (ORB), which provides a communication infrastructure independent of the underlying network. An object interface is specified using the Interface Definition Language (IDL). An IDL file contains a list of operations for a given object that can be remotely invoked. Figure 1 provides a simplified view of the CORBA architecture.

In this figure, an object located at the client side (Machine A) is bound to an implementation of an object located at the server side (Machine B). When a client invokes an operation of the object, communication between the client and the server is performed through the CORBA ORB thanks to the IDL stub (client side) and the IDL skeleton (server side). The stub and the skeleton are

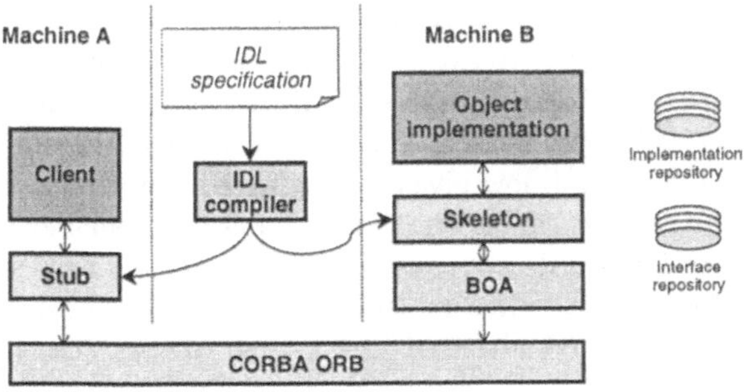

Fig. 1. CORBA system architecture.

generated by an IDL compiler taking as input the IDL specification of the object. Because CORBA is independent of the language used for the object implementation, an IDL compiler may generate stubs for different languages (e.g., Java, C++, Smalltalk). An object can thus be implemented in C++ and called by a client implemented in Java. The IDL specification of a given service can be stored in the interface repository so that any user is able to get this interface and to use the corresponding CORBA component. The implementation repository is used by the Basic Object Adapter (BOA) to execute the process that implements a particular object.

CORBA objects have an interface that is used to describe operations that can be remotely invoked. Object interface is specified using the Interface Definition Language (IDL). The following example shows a simple IDL interface:

```
interface myservice {
  void put(in double a);
  void get(out doubla a);
  double myop(inout long i);
};
```

In this example, there are three operations associated with the interface. Each operation has a single parameter. An operation parameter is assigned a type that is similar to a C++ type (e.g., a scalar, an array). A keyword added just before the type specifies if it is an input parameter, an output parameter, or both. IDL types are mapped to the language to be used at the server and the client side. IDL provides an interface inheritance mechanism so that services can be extended easily.

2.2 Jaco3 Software Architecture

The Jaco3 software environment is shown in Fig. 2. It is a set of tools and services to help in the programming of the distributed system. A typical use of the Jaco3 software environment starts with the description of the application using a code-coupling tool, called VisualORB [5]. This tool offers a graphical interface that allows users to express the coupling of their simulation codes. After VisualORB has been given this description, the tool is able to generate an application scheduler that schedules the execution of the simulation codes according to the control flow given by the user. The application scheduler can be seen as a client calling simulation codes encapsulated into CORBA objects using techniques described in Sect. 5. The execution of the application is supported by a set of services implemented as CORBA objects. We consider two kinds of services: the advanced services and the basic services.

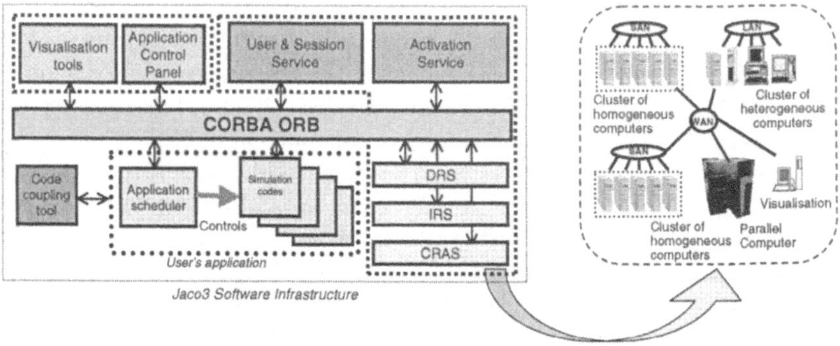

Fig. 2. The Jaco3 software environment for distributed numerical simulation.

The advanced services are the *User & Session Service (USS)* and the *Activation Service (AS)*. The USS aims at managing sessions associated with the execution of a simulation application, whereas the AS is used for the activation of simulation codes when the application is executed. The basic services are the *Data Repository Service (DRS)*, the *Information Repository Service (IRS)*, and the *Computing Resource Allocation Service (CRAS)*. The DRS is in charge of managing the flow of data between simulation codes that are executed on different computing resources. It migrates files between the computing resources on demand. It is also in charge of delivering data to the visualization tools running on the users' workstations. The IRS stores various information related to the configuration of the Jaco3 distributed system (availability of computing resources), the list of users able to connect to the Jaco3 environment, and the application configuration (a list of the simulation codes and other various information). These services are described in the following sections.

3 Jaco3 Advanced Services

3.1 User and Session Service

The USS aims at offering an entry point for users to execute their distributed simulation applications. It is accessed via a client application we call the Jaco3 shell. Users log on using the Jaco3 shell to ensure strong authentication of the actors and therefore strict limits of access to resources. The shell itself performs minimal processing; most commands are more or less direct calls to the USS, the IRS, and the DRS operations. From the user's point of view, after the login step, a session is created that is essentially a private filename space where a simulation takes place. Access to the contents of the session can be controlled by the owner of the session. The contents of the session are preserved until the owner deletes the session. The owner may leave the session and connect again later. Several users with read access to the session may watch the evolution of the session simultaneously. These users are called observers. It is possible for a single user to own multiple sessions at the same time.

Running an application using the Jaco3 shell requires the definition of this application using VisualORB, which is a graphical tool to specify the coupling of codes. This tool is fully integrated into the Jaco3 shell. The description of the application must contain the different simulation codes to be used, the coupling of these codes, the list of computing resources that are necessary to support the execution of the application, and the mapping of simulation codes to the computing resources. The mapping consists of associating each code with a particular computing resource.

3.2 Activation Service

The Activation Service is the main interface provided to the USS for the activation of the simulation codes that have to be executed on different computing resources. Activation of these codes depends on the computing resources. Some of them allow interactive execution of the codes, whereas others accept only the execution of codes through the use of scheduling systems (NQS, EASY, CODINE, etc.). The AS is connected to a local service that is available at each computing resource. This local service is in charge of managing the execution of the simulation codes on the computing resource. If a computing resource only accepts the execution of codes through a scheduling system, the local service is connected through an intermediate service, called CRAS, that is able to perform co-allocation of computing resources. Co-allocation is required when an application needs several computing resources at the same time that have to be allocated through scheduling systems [6]. The co-allocation mechanism is in charge of finding a suitable time slot for which all these computing resources are available. This co-allocation mechanism is implemented using a set of CORBA services, called LARS (Local Allocation

Resource Service), running on each supercomputer where a scheduling system is running.

4 Jaco3 Basic Services

4.1 Data Repository Service

The DRS manages the flow of data between the simulation codes that are executed. During the execution of a coupled simulation application, simulation codes have to exchange their data. Such transfer is not carried out by the CORBA ORB, because it requires making changes to the simulation codes and because it does not provide the level of performance that is needed to run a coupled simulation application efficiently. Moreover, the commercial simulation codes cannot be modified at all. Exchange of data within a distributed application has been identified as one of the major obstacles to achieving high performance from a distributed system. Several projects are investigating this problem and have proposed solutions either based on the use of a database, such as ADR [10], or specific software development, such as GASS [2]. We took an approach similar to the one that was chosen for the design of GASS in the Globus project [7] with some new functionalities. We designed a data repository service (DRS) that is able to transfer files between different computing resources.

The DRS behaves as follows. Before starting the execution of a simulation code, requests are sent to the DRS to get access to the files that are located somewhere in the distributed system. The DRS knows the location of each file in the distributed system. When a file is accessed by a simulation code, the whole file is moved to the machine where the code is running. Such policy ensures the best use of the available network bandwidth. The DRS provides a compression and a decompression mechanism when files are moved between computing resources. The DRS does not provide a remote read-from or a remote write-to a file to avoid the penalty of the usual high network latency. It provides a replication policy when files are read simultaneously by several codes. Such files will be replicated in each computing resource executing a simulation code that requests them. The DRS ensures the coherency when a replicated file is modified by one code. When a simulation code requests a file using either a write-only or a read-write access mode, copies of the file will be invalidated first before granting the access to the code.

Prior to the execution of a distributed application, input files have to be registered to the DRS using a set of commands provided by the Jaco3 shell. When these files are managed by the DRS, their movements in the distributed system are performed by the DRS. Shell commands are provided to get files from the DRS for post-processing.

Communication between simulation codes often involves a process aiming at converting a file produced by a code from a given format to another one suitable for the code that needs the data. This conversion process cannot be handled by the simulation code itself, because it does not know what code will be connected to it. In the Jaco3 project, we are committed to a component approach of the coupled simulation problem. It means that a component has to be designed in such a way that it can be assembled with another component very easily. Therefore, the conversion process has to be performed by the DRS instead of the simulation code itself. To tackle this problem, the DRS has to manage file types. A file type is associated with each file handled by the DRS. When a component requests a specific file, it has to provide its name, its access mode, and its access type. If the file exists in the DRS but with a different type, a conversion filter will be applied before the file movement process. A conversion filter is not only devoted to change the file data format; it may also extract only the useful information from the files that is needed by the code that requests the file access. Filtering of data is performed using Perl scripts. This allows the execution of the filtering process on any computing resource in the distributed system.

The DRS is also responsible for providing input data to visualization tools. It supports notification mechanisms to avoid polling by data consumers. One major advantage of this approach is to de-couple visualization tools from simulation codes.

4.2 Information Repository Service

The Information Repository Service (IRS) is a CORBA service that aims at gathering various persistent information that is required by other CORBA services such as the USS and the DRS. It stores the computing resource configuration that specifies which machines are involved in the Jaco3 environment, the list of available coupled simulation applications with their associated configuration (e.g., list of simulation codes, mapping of codes to computing resources), and a list of users who are allowed to use a particular application associated with a password.

4.3 Computing Resource Allocation Service

Initiating simulation codes on large shared resources (computer centre resources) requires that the Jaco3 environment be interfaced with batch environments. Further, for a globally distributed application, made of several simulation codes distributed over two or more resources of this type, a co-scheduling of resources has to take place. The Computing Resource Allocation Service (CRAS) provides this functionality. As a central service, CRAS is taking requests from the Activation Service to start concurrent batch simulation codes. CRAS is implemented using a set of local services (LARS) that are installed on each computing resource that runs a scheduling system. The negotiation of the job start time is carried out by the LARS. When jobs actually start successfully, a notification will be given to the "activation service." By then CRAS has fulfilled its job, and it is up to the "activation service" to tell the USS and the application scheduler that the simulation codes are ready.

5 Encapsulation of Simulation Codes into CORBA Objects

Within the Jaco3 software infrastructure, activation of the simulation codes is performed using the CORBA Object Request Broker (ORB). It is thus necessary to encapsulate all the simulation codes of a given application into CORBA objects so that they can be plugged to the ORB. However, most of the simulation codes were not designed for such use. For instance, commercial codes are only available as binaries and thus cannot be modified to act as CORBA objects. Moreover, some languages that are used for the development of in-house codes do not have an IDL mapping so that the codes cannot be easily transformed into CORBA objects. Another problem arises when dealing with parallel simulation codes. The encapsulation of parallel codes into CORBA objects has not yet been addressed. To address these encapsulation problems, we designed two techniques. The first one aims to encapsulate any binary code into a CORBA object, based on the use of a generic wrapper. Based on previous works [12,13], we proposed a technique to encapsulate MPI-based parallel codes into a new kind of CORBA object we called parallel CORBA object. These two approaches are described in more detail in the following sections.

5.1 Code Encapsulation Using a Generic Wrapper Object

Due to the unavailability of the source codes or suitable IDL mappings for the implementation languages, it is necessary to provide a technique to encapsulate simulation codes without any modification to them. We propose a generic C++ wrapper that aims at encapsulating a binary code into a CORBA object as shown in Fig. 3. The idea is to put a wrapper around the binary code that will be able to receive CORBA requests, and then interact with the binary code to perform the desired operations. When the encapsulation is completed, the whole functionality of the binary code will be available to clients via the CORBA ORB.

The generic aspect of the proposed technique comes from the use of a scripting language that describes the interaction between the code and the user. We selected the *Tcl* scripting language associated with the *Expect* library that provides additional commands for writing scripts interacting with programs. With such an approach, the wrapping of a given code only requires four steps. The first step consists of defining a set of operations that can be performed by the binary code. This will allow the definition of an IDL interface. The second step corresponds to the use of a specific pre-processor (`idl2wrapper`) that is able to generate a *Tcl* template from the IDL specification. In this template, a *Tcl* procedure is created for each IDL operation.

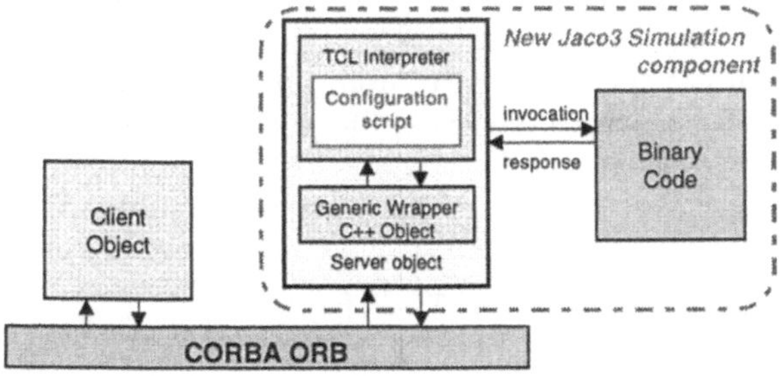

Fig. 3. Encapsulation of binary codes into a CORBA object.

This pre-processor is also able to generate the implementation of the CORBA object. Each IDL operation is associated with a corresponding C++ method in which several methods of the wrapper object are called. This particular object provides a set of methods to call a *Tcl* procedure with the corresponding parameters. The third step consists of compiling the generated codes with the wrapper object to act as a CORBA object. Finally, the last step is to use the *Tcl* template to write a configuration script providing the implementation of each procedure corresponding to an IDL operation.

When a client calls an operation on the server, it passes the request to the wrapper object. The wrapper object passes the request with the input arguments to the configuration script, which performs the corresponding operation by interacting with the binary code. The output from the binary code is passed back to the wrapper object, which sends back the output arguments to the server. Finally, the server answers the client's request.

5.2 Code Encapsulation Using a Parallel CORBA Object

The integration of MPI-based parallel codes to CORBA infrastructures, such as Jaco3, is an important issue in several research and development projects [3]. The usual way of encapsulating such codes is to adopt a master/slave approach. In that case, a particular process plays the role of a master that is connected to slave processes through the MPI layer. Only the master process is encapsulated into a CORBA object. Such a simple solution requires some modifications to existing MPI-based codes if they did not already follow a master/slave approach. Moreover, the master might represent an important bottleneck when two MPI codes, encapsulated into CORBA objects, have to communicate with each other. The master process has to gather data from the slave processes, using MPI, and has to send them back to the callee through the ORB. The callee will then call the other CORBA object that in turn will scatter the data to its slave processes. This approach does not offer a scalable solution to the encapsulation of parallel codes. As the number of slave processes or the size of the problem (amount of data transmitted between two parallel codes) increases, it will entail a high overhead. To avoid such undesirable behavior, we advocate the use of new kinds of CORBA objects we call parallel CORBA objects. The concept of a parallel CORBA object (we will use "parallel object" from now on) is simply a collection of identical CORBA objects as shown in Fig. 4.

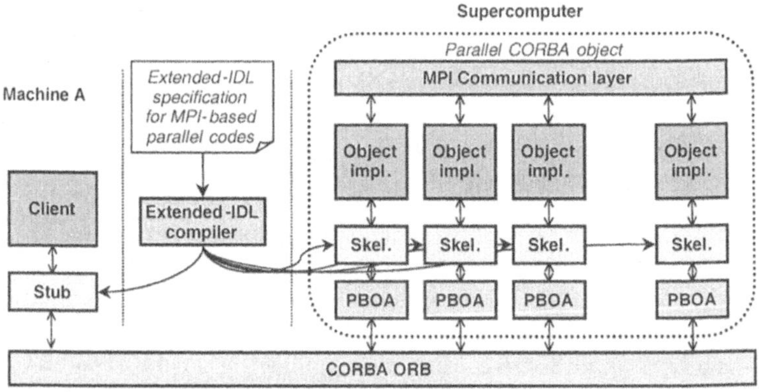

Fig. 4. Encapsulation of MPI-based parallel codes into CORBA objects.

These objects can be manipulated as a single entity by our system. Many problems appear when dealing with coarse-grain parallelism on a distributed memory parallel architecture like a cluster of PCs. Our goal is to hide as many of these problems as possible. First of all, the calling of an operation by a client will result in the execution of the associated method by all objects belonging to the collection at the server side. Execution of parallel objects is based on the SPMD execution model. This parallel activation is done transparently by our system. Data distribution between the objects belonging to a collection is entirely handled by the system. However, to let the system carry out parallel execution and data distribution between the objects of the collection, some specifications have to be added to the object interface. A parallel object interface is thus described by an extended version of IDL, called Extended-IDL. It is a set of new keywords, added to the IDL syntax, to specify both the number of objects within the collection and the data distribution. The following example is an Extended-IDL specification of a service implementing matrix and vector operations.

```
interface[*:2*n] MatrixOperations {
  const long SIZE=100;
  typedef double Vector[SIZE];
  typedef double Matrix[SIZE][SIZE];
  void multiply(in dist[BLOCK][*] Matrix A, in Vector B,
                out dist[BLOCK] Vector C);
  void skal(in dist[BLOCK] Vector C, out csum double skal);
};
```

These extensions allow the specification of the mapping of objects to computing nodes, data distribution among the objects of the collection, and collection operation performed on output parameters. There are described briefly in the following sections. A more complete description is given in [12,13].

Mapping of objects. The number of objects in the collection, as well as the shape of the virtual node array where objects of the collection will be mapped, are specified within the two brackets after the IDL keyword `interface`. The expression may be an integer value, an interval of integer values, a function, or the "*" symbol—which means that the number of objects is chosen at runtime. The shape of the virtual node array is an optional parameter that describes how objects must be organized. A virtual node array is used during data distribution of multidimensional arrays. It indicates the number of objects to be distributed on a given dimension for which a distribution is specified.

Data distribution. Data distribution is specified using the `dist` keyword before the type of each parameter to be distributed. In the previous example, operation `multiply` has two parameters (matrix A and vector C) that are distributed. After the `dist` keyword, the distribution mode for each array dimension is specified. Distribution modes are similar to the ones defined by HPF (High Performance Fortran). The "*" indicates that the corresponding array dimension is not distributed. A non-distributed parameter, as vector B, is replicated among each object of the collection.

Collective operation. Collective operation is a simple way to perform computations on the values returned by the objects belonging to the collection. Operation `skal`, in the previous example, illustrates the use of this new extension. The `csum` keyword indicates that the value returned by the operation is the sum of all the values given by all objects belonging to the collection.

6 Related Works

Several projects deal with software infrastructure for distributed numerical simulation application combining the benefits of distributed and parallel programming. Some projects exploit a client/server approach such as RCS [1], NetSolve [4], and Ninf [14]. One drawback of these environments is the difficulty for the user to add new functions in the libraries. Moreover, they are not compliant to relevant standard such as CORBA. The Legion [8] project aims at creating a worldwide environment for high-performance computing. Many of CORBA's principles (such as heterogeneity management and object location) are provided by the Legion run-time system, although Legion is not

CORBA-compliant. It manipulates parallel objects to obtain high performance. The Globus project [7] is developing basic software infrastructure for computations that integrate geographically distributed computational and information resources. Globus provides a set of basic services, such as resource management, communication, information infrastructure, and remote data access.

The architecture of the Jaco3 software environment adopts some of the key aspects of the Globus architecture related to the resource management, the information infrastructure, and the remote data access. However, we took the CORBA ORB as a communication infrastructure, instead of Nexus that is used in Globus. Concerning the encapsulation of parallel codes into CORBA objects, the PARDIS [9] project proposes a solution very close to our approach, because it extends the CORBA object model to a parallel object model. A new IDL type is added that is a generalization of the CORBA sequence. It describes data type, data size, and how data must be distributed among objects. Comparing to PARDIS, our approach provides a more detailed description of object parallel services.

7 Conclusion and Perspectives

This paper has outlined some of the Jaco3 services. The resulting software infrastructure will increase the accessibility of distributed simulation techniques to a wider range of users. It will make possible the use of a combination of different simulation tools, possibly from different companies, where each tool is optimized for a certain part of a system and can be delivered as a black box by the manufacturer of the part without revealing any details. We are only in the final steps of the architecture design of the Jaco3 environment. To assess the approach, two industrial applications from the aerospace sector will be considered. One will couple thermal, optical, and mechanical simulations for the optimization of the optical payload and precision of a satellite; and the second one will couple electromagnetic simulation codes to study the impact of an antenna installation on an aircraft. It is planned to perform the experiment on various computing resources and networking technologies.

Acknowledgments

This work is funded by the European Commission under the Esprit work programme. The authors of this paper would like to thank all the partners of the Jaco3 project who contributed to the design of the Jaco3 software environment: Aérospatiale Matra (coordinator of the project), Alcatel Space Industries, Allgon, Ericsson-Saab, Intecs, and KTH.

References

1. P. Arbenz, W. Gander, and M. Oettli. The Remote Computation System. In *HPCN Europe '96*, volume 1067 of *LNCS*, pages 662–667, 1996.
2. J. Bester, I. Foster, C. Kesselman, J. Tedesco, and S. Tuecke. GASS: A data movement and access service for wide area computing systems. In *6th Workshop on I/O in Parallel and Distributed Systems (IOPADS-99) Proceedings*, pages 78–88, New York, May 5 1999. ACM Press.
3. T. Breitfeld, S. Kolibal, A. Schreiber, and M. Wagner. Java for controlling and configuring a distributed turbine simulation system. In *First UK Workshop Java for High Performance Network Computing*, 1998.
4. H. Casanova and J. Dongara. NetSolve: A Network Server for Solving Computational Science Problems. *The International Journal of Supercomputer Applications and High Performance Computing*, 11(3):212–223, 1997.
5. D. Ceneda. VisualORB: Visual simulation tool on an ORB. Aérospatiale internal document, August 1999.
6. K. Czajkowski, I. Foster, and C. Kesselman. Resource co-allocation in computational grids. In *Eighth IEEE International Symposium on High Performance Distributed Computing Proceedings*, pages 219–228. IEEE, August 1999.
7. I. Foster and C. Kesselman. Globus: A metacomputing infrastructure toolkit. *The International Journal of Supercomputer Applications and High Performance Computing*, 11(2):115–128, Summer 1997.
8. A. S. Grimshaw, W. A. Wulf, and the Legion team. The Legion vision of a worldwide virtual computer. *Communications of the ACM*, 1(40):39–45, January 1997.
9. K. Keahey and D. Gannon. PARDIS: CORBA-based architecture for application-level parallel distributed computation. In ACM, editor, *1997 ACM/IEEE SC97 Conference Proceedings*. ACM Press and IEEE Computer Society Press, November 1997.
10. T. M. Kurc, A. Sussman, and J. Saltz. Coupling multiple simulations via a high performance customizable database system. In *The Ninth SIAM Conference on Parallel Processing for Scientific Computing*, March 1999.
11. OMG. The Common Object Request Broker: Architecture and Specification. Technical Report PTC/96-03-04, Object Management Group, 1996. Version 2.0.
12. T. Priol and C. René. Cobra: A CORBA-compliant programming environment for high performance computing. In *Euro-Par'98*, pages 1114–1122, Southampton, UK, September 1998.
13. C. René and T. Priol. MPI code encapsulation using parallel CORBA object. In *Eighth IEEE International Symposium on High Performance Distributed Computing Proceedings*, pages 3–10. IEEE, August 1999.
14. M. Sato, H. Nakada, S. Sekiguchi, S. Matsuoka, U. Nagashima, and H. Takagi. Ninf: A network based information library for global world-wide computing infrastructure. In *HPCN Europe '97*, volume 1225 of *LNCS*, pages 491–502, 1997.

New Generalized Data Structures for Matrices Lead to a Variety of High-Performance Algorithms

Fred G. Gustavson

IBM T. J. Watson Research Center; Yorktown Heights, NY 10598; USA;
<gustav@watson.ibm.com>

Abstract. We describe new data structures for full and packed storage of dense symmetric/triangular arrays that generalize both. Using the new data structures, one is led to several new algorithms that save "half" the storage and outperform the current blocked-based level-3 algorithms in LAPACK. We concentrate on the simplest forms of the new algorithms and show for Cholesky factorization they are a direct generalization of LINPACK. This means that level-3 BLAS's are not required to obtain level-3 performance. The replacement for level-3 BLAS are so-called kernel routines, and on IBM platforms they are producible from simple textbook type codes, by the XLF Fortran compiler. In the sequel I will label these "vanilla" codes. The results for Cholesky, on Power3 with a peak performance of 800 Mflop/s at $n \geq 200$ is over 720 MFlop/s and reaches 735 MFlop/s. Using conventional full-format LAPACK DPOTRF with ESSL BLAS's, one first gets 600 MFlop/s at $n \geq 600$ and only reaches a peak of 620 MFlop/s. We have also produced simple square blocked full-matrix data formats where the blocks themselves are stored in column-major (Fortran) order or row-major (C) format. The simple algorithms of LU factorization with partial pivoting for this new data format is a direct generalization of LINPACK algorithm DGEFA. Again, no conventional level-3 BLAS's are required; the replacements are again so-called kernel routines. Programming for squared blocked full-matrix format can be accomplished in standard Fortran through the use of three- and four-dimensional arrays. Thus, no new compiler support is necessary. Finally we mention that other more complicated algorithms are possible, for example, recursive ones. The recursive algorithms are also easily programmed via the use of tables that address where the blocks are stored in the two-dimensional recursive block array.

1 Introduction

Recursive Blocked Data formats have been introduced as a replacement for standard Fortran/C array storage (see [6],[7]). One of the key innovations was to see that storing a matrix as a collection of submatrices (e.g., square blocks of size NB) led to very high performance on today's RISC processors. Let a rectangular matrix A have M rows and N columns. For clarity, assume $M = m * NB$ and $N = n * NB$. A consists of mn submatrices of size $NB \times NB$. In [6] and [7], we demonstrated that recursion (i.e., divide-and-conquer) should be used to order these mn blocks in storage. This storage arrangement leads to L2, L3, and memory blocking automatically. However, the ordering of the blocks is non-linear, and tables are needed to properly address these mn blocks.

A simpler way to order the mn blocks is standard Fortran/C order; that is, store the mn blocks either in column-major or row-major order. If one applies a recursive algorithm to this simpler block data layout, one approximates an automatic blocking for L2, L3, and memory; that is, performance for this layout will approximate recursive blocked data layout [6,7]. The main benefit of the simpler data layout is that addressing of an arbitrary (i, j) element of $A(0 : M - 1, 0 : N - 1)$, namely, $a(i, j)$ can be easily handled by a compiler and/or a programmer. Let $0 \leq i < M$ and $0 \leq j < N$, and write $i = I*NB+i1$ and $j = J*NB+j1$. If A is stored in column-block-major order and all blocks are in column-major order, then $a(i, j)$ is in block $I + m * J$ at location $i1 + NB * j1$ of that block. Also, IBM Power3 processors have three fixed-point units that are mostly idle during floating point computations. Hence the additional index computations above can probably be overlapped, meaning there will be little or no performance loss.

The ESSL (Engineering and Scientific Subroutine Library) has as its basis for level-3 algorithms, kernel routines that achieve peak performance when the underlying arrays fit into L1 cache [1]. If one were to adopt these new simpler data formats, then BLAS and LAPACK algorithms become almost trivial to write. We will discuss it briefly here, and Sect. 2 will show an example.

Take any vanilla code, say Gaussian elimination with partial pivoting or QR factorization of an $M \times N$ matrix A. This code has a block equivalent where the stride one distance is NB^2. The row stride for the vanilla code is LDA. For the block equivalent it is $m*NB^2$. It is quite easy to write the block equivalent code from the vanilla code. In the vanilla code, a floating point operation is usually an FMA ($c = c + ab$); in the block equivalent, this is a call to a DGEMM kernel. Similar analogies exist; for example, for $b = b/a$ or $b = b * a$, we have either a DTRSM or a DTRMM kernel. In the simple block equivalent codes, we are led to one of the variants of IJK order [2]. For these types of algorithms, the BLASes are simply calls to kernel routines. In reality, the BLASes disappear entirely. However, more complicated algorithms can be employed; for example, recursive algorithms. In that case, BLAS's calls access several submatrix blocks [7,6]. So, a blocking routine is required to call the kernel routines multiple times on different sub-matrix operands as the blocking is now variable.

New algorithms can be obtained from simple Dense Linear Algebra vanilla codes if one first introduces the new data formats. One way to do this is to ask users to input their data in standard Fortran or C order with some additional storage appended below their standard array. Then the standard Fortran or C array could be transformed in-place to the block square format. Next, the block equivalent of the vanilla code (with calls to standard ESSL type kernels) could be performed on the transformed square block full-column format. Our recent Power3 result for Cholesky factorization demonstrates that the performance is superb.

Now we turn briefly to SMP algorithms for Dense Linear Algebra Codes. In [3], superb SMP performance was obtained using recursive blocking on matrices stored in standard Fortran order. The new square blocked full-order storage could be used instead of standard Fortran order. In that case, performance should improve even further. Similar statements can be made for other dense linear algebra algorithms, such as Cholesky and general LU factorization.

Recently, other new block data formats [8,5] have been discovered while working on this project, both at Uni-C and IBM Research. These have to do with triangular/symmetric arrays. The major difference between these new array formats is they more closely mirror standard storage formats than the recursive arrays, which store the blocks in a non-linear fashion. To be precise, these symmetric/triangular arrays store the blocks either as block columns or in standard packed format order. We describe these new formats in Sect. 3 and Sect. 4.

2 Simpler Square Blocked Full Data Formats

These new data formats are best described by an example. Let A be an $(M = 11)\times(N = 10)$ matrix with $LDA = 12$. In Fortran, this matrix would be stored as shown in Fig. 1.

$$A = \begin{bmatrix} 1 & 13 & 25 & 37 & 49 & 61 & 73 & 85 & 97 & 109 \\ 2 & 14 & 26 & 38 & 50 & 62 & 74 & 86 & 98 & 110 \\ 3 & 15 & 27 & 39 & 51 & 63 & 75 & 87 & 99 & 111 \\ 4 & 16 & 28 & 40 & 52 & 64 & 76 & 88 & 100 & 112 \\ 5 & 17 & 29 & 41 & 53 & 65 & 77 & 89 & 101 & 113 \\ 6 & 18 & 30 & 42 & 54 & 66 & 78 & 90 & 102 & 114 \\ 7 & 19 & 31 & 43 & 55 & 67 & 79 & 91 & 103 & 115 \\ 8 & 20 & 32 & 44 & 56 & 68 & 80 & 92 & 104 & 116 \\ 9 & 21 & 33 & 45 & 57 & 69 & 81 & 93 & 105 & 117 \\ 10 & 22 & 34 & 46 & 58 & 70 & 82 & 94 & 106 & 118 \\ 11 & 23 & 35 & 47 & 59 & 71 & 83 & 95 & 107 & 119 \\ * & * & * & * & * & * & * & * & * & * \end{bmatrix}$$

Fig. 1. Standard Fortran column-major storage order.

Let $NB = 4$ be the block size, and suppose A is stored in column-major block order. Here $m = n = 3$, and A is a 3×3 block matrix. Each square block is an $NB\times NB$ submatrix of A. In the new data format, A would be stored as in Fig. 2.

$$A = \begin{bmatrix} 1 & 5 & 9 & 13 & 49 & 51 & 57 & 61 & 97 & 101 & * & * \\ 2 & 6 & 10 & 14 & 50 & 52 & 58 & 62 & 98 & 102 & * & * \\ 3 & 7 & 11 & 15 & 51 & 53 & 59 & 63 & 99 & 103 & * & * \\ 4 & 8 & 12 & 16 & 52 & 54 & 60 & 64 & 100 & 104 & * & * \\ 17 & 21 & 25 & 29 & 65 & 69 & 73 & 77 & 113 & 117 & * & * \\ 18 & 22 & 26 & 30 & 66 & 70 & 74 & 78 & 114 & 118 & * & * \\ 19 & 23 & 27 & 31 & 67 & 71 & 75 & 79 & 115 & 119 & * & * \\ 20 & 24 & 28 & 32 & 68 & 72 & 76 & 80 & 116 & 120 & * & * \\ 33 & 37 & 41 & 45 & 81 & 85 & 89 & 93 & 129 & 133 & * & * \\ 34 & 38 & 42 & 46 & 82 & 86 & 90 & 94 & 130 & 134 & * & * \\ 35 & 39 & 43 & 47 & 83 & 87 & 91 & 95 & 131 & 135 & * & * \\ * & * & * & * & * & * & * & * & * & * & * & * \end{bmatrix}$$

Fig. 2. New square blocked full column-major order.

Now suppose a user inputs the matrix A in standard Fortran or C order with some additional space in the array, holding A directly below A. For example, in the 11×10 example above with $LDA = 12$, the minimum storage for A is $LDA * n = 120$ double words. If the user supplied $ns \geq 144$ elements (extra storage of ≥ 24 double words below A), one could transform Fortran storage order to the new square blocked full-column order. After this data transformation is completed, one can execute the block equivalent of a standard vanilla code with calls made to standard kernel routines.

We briefly describe this procedure for Gaussian elimination with partial pivoting. In the vanilla version, the outer loop is on $j = 0, n-1$; and for each j, one finds the pivot in column j and swaps it with $a(j,j)$. Then $a(j+1 : m-1, j)$ is scaled by the reciprocal of the pivot to form column j of L. Then, columns $j+1$ to $n-1$ are processed in two steps. Let k be the generic column. First, a swap of row j and the pivot row is made. Second, a DAXPY update is performed: $a(j+1 : m-1, k) = a(j+1 : m-1, k) - a(j+1 : m-1, j) * a(j, k)$. For the block version, refer to Subroutine 1.

Subroutine 1. `DGEFB`

```
      subroutine dgefb (m,n,a,nb ,ipiv,info)
      implicit none
      integer*4 m,n,nb ,info
      real*8 a(0:nb-1,0:nb-1,0:*)
      integer*4 ipiv(0:1,0:*)
      real*8 one,zero,t
      parameter (one=1.d0,zero=0.d0)
      integer*4 nb2,m1,n1,m2,n2,j,k,i,bj,bk,bi,ms,ks,ns
      integer*4 iti,ibi,iai,ici ! pointers
      info=0
      nb2=nb*nb            ! size of a square block
      m1=(m+nb-1)/nb       ! row order of block matrix
```

```
      n1=(n+nb-1)/nb      ! col order of block matrix
      m2=m+nb-m1*nb       ! row size of last block
      n2=n+nb-n1*nb       ! col size of last block
      bj=0                ! block j index
      do j=0,n-1,nb
*
*     factor column swath A(j:j:m-1,j:j+nb-1)
*
        bk=bj
        iti=bj+m1*bj     ! (0,0)-th element of block iti is A(j,j)
        ns=nb
        if(bj.eq.n1-1)ns=n2
        call rgetf3(m-j,ns,a(0,0,iti),nb,ipiv(0,j),info)
*     globalize pivot indices
        do i=j,j+ns-1
          ipiv(1,i)=ipiv(1,i)+bj
        enddo
        do k=j+nb,n-1,nb
          bk=bk+1         ! block k index
          ks=nb
          if(bk.eq.n1-1)ks=n2
          ibi=bj+m1*bk   ! (0,0)-th element of block ibi is A(j,k)
*     forward pivot A(j:j+m-1,k:k+ks-1) ! a(0,0,bk*m1) -> A(0,k)
          call dlaswpb(ks,a(0,0,bk*m1),nb,j,j+ns-1,ipiv,1)
*     solve L*X = A(j:j+nb-1,k:k+ks-1)
          call dllnu4(nb,ks,a(0,0,iti),nb,a(0,0,ibi),nb)
          bi=bj
          do i=j+nb,m-1,nb
            bi=bi+1
            ms=nb
            if(bi.eq.m1-1)ms=m2
            iai=bi+m1*bj ! (0,0)-th element of block iai is A(i,j)
            ici=bi+m1*bk ! (0,0)-th element of block ici is A(i,k)
*
*     update A(i:i+ms-1,k:k+ks-1) = A(i:i+ms-1,k:k+ks-1)
*     - A(i:i+ms-1,j:j+jb-1)*A(j:j+nb-1,k:k+ks-1)
*
            call dab4(ms,ks,nb,a(0,0,iai),nb,a(0,0,ibi),nb,
     &                 a(0,0,ici),nb)
          enddo
        enddo
        bi=0
        do i=0,bj-1
*     back pivot A(j:j+m-1,i:i+nb-1) ! a(0,0,bi) -> A(0,i*nb)
          call dlaswpb(nb,a(0,0,bi),nb,j,j+ns-1,ipiv,1)
          bi=bi+m1
        enddo
        bj=bj+1           ! next block j col
      enddo
```

```
      return
      end
```

In the blocked version, the outer loop is on $bj = 0, n1-1$ and for each bj, one factors a block column $L*U = P*A(j:m-1, j:n-1)$ by calling kernel routine **RGETF3**. Then columns $j+nb$ to $n-1$ are processed in three steps. Let $k:k+ks-1$ be the generic block column. First, there is a forward pivot step. Next, there is a DTRSM computation done by kernel routine **DLLNU4**. Finally, there is a DGEMM update, which is done by a series of calls to kernel routine **DAB4**. After these three steps, there is a back pivot step. There are three kernel routines in the block equivalent (see Subroutine 1). They are a factor a panel of size $M \times N$, where $N \leq NB$ kernel called **RGETF3**, a DTRSM kernel called **DLLNU4**, and a DGEMM kernel called **DAB4**. We mention that factor part has the same function as LAPACK routine DGETF2. However, it is a level-3 routine, done recursively, as the prefix R and suffix 3 indicates. Note that the vanilla routine, actually the LINPACK routine DGEFA, does not have a back pivot step. So the blocked version does extra work that could be avoided.

DGEFB might be packaged as a subroutine in standard Fortran. Following LAPACK, we suggest using the input format of **DGETRF(M, N, A, LDA, IPIV, INFO)**. The new routine would have a nearly identical calling sequence: **DBGTRF(M, N, NS, A, LDA, IPIV, INFO)**. The new parameter is $NS \geq n * LDA$. If NS is not sufficiently large, **DBGTRF** just returns by placing in NS the amount of storage necessary for good level-3 performance. If NS is sufficiently large, the input storage is rearranged into square block format, and then **DBGTRF** is executed giving level-3 performance. Like the $LWORK$ parm, the value returned in NS will be the value used by **DBGTRF** for good level-3 performance, namely $m1 * n1 * NB^2$.

3 Blocked Packed Formats for Symmetric/Triangular Arrays

These new formats are a generalization of packed format for triangular arrays. They are also a generalization of full format for triangular arrays. The main benefit of the new formats is that they allow for level-3 performance while using about half the storage of the full array cases.

For blocked packed formats, there are two parameters LDA and NB, where $LDA \geq N \geq NB$.

In packed format, the elements of a triangular matrix are laid out in memory as shown in Fig. 3 (the numbers represent the position within a data array).

In the blocked packed format, we first choose a block size NB ($NB = 4$ in this example), and then we lay out the data in stripes NB wide. This allows level-3 BLAS to be used on the stripes. For both $uplo = U$ or L, the $LDA \geq N$. For $uplo = L$, the first vertical stripe is $LDA \times NB$. It holds the

$$
\text{Packed Lower} \qquad\qquad\qquad\qquad\qquad\qquad \text{Packed Upper}
$$

$$
\begin{bmatrix}
1 &&&&&&&&& \\
2 & 11 &&&&&&&& \\
3 & 12 & 20 &&&&&&& \\
4 & 13 & 21 & 28 &&&&&& \\
5 & 14 & 22 & 29 & 35 &&&&& \\
6 & 15 & 23 & 30 & 36 & 41 &&&& \\
7 & 16 & 24 & 31 & 37 & 42 & 46 &&& \\
8 & 17 & 25 & 32 & 38 & 43 & 47 & 50 && \\
9 & 18 & 26 & 33 & 39 & 44 & 48 & 51 & 53 & \\
10 & 19 & 27 & 34 & 40 & 45 & 49 & 52 & 54 & 55
\end{bmatrix}
\quad
\begin{bmatrix}
1 & 2 & 4 & 7 & 11 & 16 & 22 & 29 & 37 & 46 \\
 & 3 & 5 & 8 & 12 & 17 & 23 & 30 & 38 & 47 \\
 & & 6 & 9 & 13 & 18 & 24 & 31 & 39 & 48 \\
 & & & 10 & 14 & 19 & 25 & 32 & 40 & 49 \\
 & & & & 15 & 20 & 26 & 33 & 41 & 50 \\
 & & & & & 21 & 27 & 34 & 42 & 51 \\
 & & & & & & 28 & 35 & 43 & 52 \\
 & & & & & & & 36 & 44 & 53 \\
 & & & & & & & & 45 & 54 \\
 & & & & & & & & & 55
\end{bmatrix}
$$

Fig. 3. Packed format arrays.

first trapezoidal $N \times NB$ part of L. The next stripe has $LDA = LDA - NB$, and it holds the next trapezoidal $(N - NB) \times NB$ part of L. In Fig. 4 for $uplo = L$ with $n = 10$, $NB = 4$, and $LDA = 12$, the numbers represent the position within the array where $a(i,j)$ is stored. Note the missing numbers (e.g., 13, 25, 26, 37, 38, and 39), which correspond to the upper right corner of the first stripe. $LDA \geq N$ is chosen so the first and hence all other stripes map well into L1 cache. An example of blocked lower packed format is given in Fig. 4. The above blocked lower packed array consists of three rectangular

$$
\begin{bmatrix}
1 & * & * & * &&&&&& \\
2 & 14 & * & * &&&&&& \\
3 & 15 & 27 & * &&&&&& \\
4 & 16 & 28 & 40 &&&&&& \\
5 & 17 & 29 & 41 & 49 & * & * & * && \\
6 & 18 & 30 & 42 & 50 & 58 & * & * && \\
7 & 19 & 31 & 43 & 51 & 59 & 67 & * && \\
8 & 20 & 32 & 44 & 52 & 60 & 68 & 76 && \\
9 & 21 & 33 & 45 & 53 & 61 & 69 & 77 & 81 & * \\
10 & 22 & 34 & 46 & 54 & 62 & 70 & 78 & 82 & 86 \\
* & * & * & * & * & * & * & * & * & * \\
* & * & * & * & * & * & * & * & * & *
\end{bmatrix}
$$

Fig. 4. Blocked lower packed format.

arrays: the first is 12×4, the second is 8×4, and the last is 4×2. Addressing this set of three arrays as a composite block array is straightforward. An example of blocked upper packed format is given in Fig. 5. The blocked upper packed array consists of three rectangular arrays: the first is 4×2, the second is 8×4, and the last is 12×4. The LDA's of these blocks are 4, 8, 12. This $n \times n$ upper triangular matrix consist of three rectangles.

$$\left[\begin{array}{cccccccccc}
1 & 5 & 9 & 17 & 25 & 33 & 41 & 53 & 65 & 77 \\
* & 6 & 10 & 18 & 26 & 34 & 42 & 54 & 66 & 78 \\
* & * & 11 & 19 & 27 & 35 & 43 & 55 & 67 & 79 \\
* & * & * & 20 & 28 & 36 & 44 & 56 & 68 & 80 \\
 & & * & * & 29 & 37 & 45 & 57 & 69 & 81 \\
 & & * & * & * & 38 & 46 & 58 & 70 & 82 \\
 & & * & * & * & * & 47 & 49 & 71 & 83 \\
 & & * & * & * & * & * & 60 & 72 & 84 \\
 & & & & & & * & * & 73 & 85 \\
 & & & & & & * & * & * & 86 \\
 & & & & & & * & * & * & * \\
 & & & & & & * & * & * & *
\end{array}\right]$$

Fig. 5. Blocked upper packed format.

With extra storage appended directly below a standard packed array, one can move to these new data formats without extra storage. For the examples above, AP requires 55 storage elements. If there are $88-55=33$ free locations below AP, then one can move the packed array downward into the block packed by starting at the end of AP and moving columns down backward one at a time.

The main innovation in using the blocked packed format is to see that one can use the standard packed algorithms where each i, j element is now a rectangular or square block. Because of the storage layout, the beginning of each block is easily located. Also key is the fact that this format supports level-3 BLAS. Hence old packed code is easily converted into blocked packed level-3 code. In a nutshell, we keep "standard packed" addressing so that the library writers and users can handle their own addressing in a Fortran or C environment.

Now turn to full format. We continue the example with $N = 10$, and $LDA = 12$. Setting $NB = 10$, one obtains full format; that is, block packed gives a single block triangle, which happens to be full format (see Fig. 6).

The unused storage of size $n * (LDA + LDA - n - 1)/2$ consists of n-fragmented vectors of sizes $LDA - n : LDA - 1 : 1$. We use colon notation (see [4]). These vectors are interspersed with $1 : n : 1$ vectors of the symmetric matrix A. In the example above, the symmetric matrix consists of ten vectors of sizes 1 to 10 in steps of 1 (55 elements total). The unused storage consists of ten vectors of sizes 2 to 11 in steps of 1 (65 elements total). Users of dense linear algebra codes do NOT utilize this fragmented storage. If this is so,

$$uplo = L \qquad\qquad\qquad\qquad uplo = U$$

$$\begin{bmatrix}
1 & * & * & * & * & * & * & * & * & * \\
2 & 14 & * & * & * & * & * & * & * & * \\
3 & 15 & 27 & * & * & * & * & * & * & * \\
4 & 16 & 28 & 40 & * & * & * & * & * & * \\
5 & 17 & 29 & 41 & 53 & * & * & * & * & * \\
6 & 18 & 30 & 42 & 54 & 66 & * & * & * & * \\
7 & 19 & 31 & 43 & 55 & 67 & 79 & * & * & * \\
8 & 20 & 32 & 44 & 56 & 68 & 80 & 92 & * & * \\
9 & 21 & 33 & 45 & 57 & 69 & 81 & 93 & 105 & * \\
10 & 22 & 34 & 46 & 58 & 70 & 82 & 94 & 106 & 118 \\
* & * & * & * & * & * & * & * & * & * \\
* & * & * & * & * & * & * & * & * & *
\end{bmatrix}
\begin{bmatrix}
1 & 13 & 25 & 37 & 49 & 61 & 73 & 85 & 97 & 109 \\
* & 14 & 26 & 38 & 50 & 62 & 74 & 86 & 98 & 110 \\
* & * & 27 & 39 & 51 & 63 & 75 & 87 & 99 & 111 \\
* & * & * & 40 & 52 & 64 & 76 & 88 & 100 & 112 \\
* & * & * & * & 53 & 65 & 77 & 89 & 101 & 113 \\
* & * & * & * & * & 66 & 78 & 90 & 102 & 114 \\
* & * & * & * & * & * & 79 & 91 & 103 & 115 \\
* & * & * & * & * & * & * & 92 & 104 & 116 \\
* & * & * & * & * & * & * & * & 105 & 117 \\
* & * & * & * & * & * & * & * & * & 118 \\
* & * & * & * & * & * & * & * & * & * \\
* & * & * & * & * & * & * & * & * & *
\end{bmatrix}$$

Fig. 6. Blocked packed formats when $nb = n$.

one could convert full format to packed blocked format, thereby freeing up a contiguous block of storage.

There are $n1 = \lceil n/NB \rceil$ block rectangles in packed block storage. The first $n1 - 1$ rectangles have NB columns, and the last has $n2 = n + NB - n1 * NB$ columns. Full storage requires $LDA * n$ locations. Packed blocked storage requires $(n1 - 1) * LDA * NB - ((n1 - 1) * (n1 - 2)/2) * NB^2 + (LDA - n + n2) * n2$ locations. The saved storage is the difference, namely $((n1 - 1) * (n1 - 2)/2) * NB^2 + (n - n2) * n2$ locations. When $n = n1 * NB$, namely $n2 = NB$, this expression becomes $n * (n - NB)/2$. So, for practical problems, about half of the storage is saved. To summarize the full format case: With a data movement, one could transform full storage A to packed block storage AB and return to the user contiguous storage of about $n^2/2$ locations.

Now we turn to programming dense linear algorithms in the new formats. As an example, see Subroutine 2, and consider the $uplo = L$ code for `DPBTRF` (positive definite blocked), which produces the lower Cholesky factor for positive definite symmetric matrix AB. We wanted to use pb for positive definite block; however, b is reserved for band. Algorithm `DPBTRF` is a simple right-looking algorithm as the code illustrates. Routine `DPOFN` is an ESSL factor kernel. In ESSL, all BLASes and level-3 factorization routines use kernel routines. For example, in ESSL's DGEMM, a blocking routine is called to partition the matrix operands, A and B, into submatrices (matrix blocks); and then calls are made to kernel routines that operate on the blocks. Data copying of the operands to the kernel routines is decided on by the ESSL DGEMM blocking routine. Now in `DPBTRF` we can call the kernel routine directly, thereby avoiding copying, since NB and LDA were chosen for good L1 cache behavior. We do not give any performance results for this code, but we suspect it will perform very well.

Subroutine 2. `DPBTRF`

```
      subroutine dbptrf(uplo,n,nb,a,lda,info)
      implicit none
      character*1 uplo  ! only uplo ='L' currently implemented
      integer*4 n,nb,lda,info
      real*8 one
      real*8 a(*)
      parameter (one=1.d0)
      integer*4 j,k,i,kb,ib,ldaj,ldak,jj,kj,kk
      info=0
      ldaj=lda                   ! lda of first j block
      jj=1                       ! -> a(j,j), j=1
      do j=1,n-nb,nb
*
*     factor a(j:j+nb-1,j:j+nb-1)
*
        call dpofn(a(jj),ldaj,nb,info,*30)
*
*     solve l(j:n,j:j+nb-1)*a(j:j+nb-1,j:j+nb-1)**T =
*     a(j:n,j:j+nb-1) for l
*
        call dtrsm('R','L','T','N',n-j-nb+1,nb,one,a(jj),ldaj,
     &        a(jj+nb),ldaj)
*
*     initialize pointers for the k loop
*
        kk=jj                    ! -> a(k,k), k=j
        ldak=ldaj                ! lda of k block, k=j
        kj=jj                    ! -> a(k,j), k=j
        do k=j+nb,n,nb
*
*     update pointers for the k loop
*
          kj=kj+nb               ! -> a(k,j)
          kk=kk+ldak*nb          ! -> a(k,k)
          ldak=ldak-nb           ! lda of current k block
          kb=min(n-k+1,nb)
*
*     update a(k:k+kb-1,k:k+kb-1) = a(k:k+kb-1,k:k+kb-1)
*     - a(k:k+kb-1,j:j+nb-1)*a(k:k+kb-1,j:j+nb-1)**T
*
          call dsyrk('L','N',kb,nb,-one,a(kj),ldaj,one,
     &          a(kk),ldak)
          i=k+kb                 ! row i of columns j and k
          ib=n-i+1               ! there are ib rows of a remaining
*
*     update a(i:n,k:k+kb-1) = a(i:n,k:k+kb-1)
*     - a(i:n,j:j+nb-1)*a(k:k+kb-1,j:j+nb-1)**T
```

```
*
            if(ib.gt.0)call dgemm('N','T',ib,kb,nb,-one,
     &           a(kj+kb),ldaj,a(kj),ldaj,one,a(kk+kb),ldak)
         enddo
*
*       update pointers for the j loop
*
         jj=jj+ldaj*nb               ! -> a(j,j), j=j+nb
         ldaj=ldaj-nb                ! lda for next j block
      enddo
*
*       factor a(j:n,j:n)
*
      call dpofn(a(jj),ldaj,n-j+1,info,*30)
      return
 30   info=info+j-1
      return
      end
```

4 Square Blocked Packed Formats for Symmetric/Triangular Arrays

This new format is a generalization of packed format for triangular arrays. It is also a generalization of full format for triangular arrays. The main benefit of the new formats is that they allow for level-3 performance while only using about half the storage of the full array cases.

For square blocked packed formats, there are two parameters, NB and $TRANS$, where usually $N \geq NB$.

Now packed format arrays are stored as in Fig. 3. In square blocked packed format, we first choose a block size NB ($NB = 4$ in this example), and then we lay out the data in squares of size NB^2. This allows level-3 BLAS to be used on the squares. For both *uplo* = U or L, there is $TRANS$. For *uplo* = L, the first vertical stripe is $n \times NB$ and consists of $n1$ square blocks, where $n1 = \lceil n/NB \rceil$. Each square block can be in row-major order ($TRANS = T$) or column-major order ($TRANS = N$). It holds the first trapezoidal $n \times NB$ part of L. The next stripe has $n1 - 1$ square blocks, and it holds the next trapezoidal $(n - NB) \times NB$ part of L. In the following drawing for *uplo* = L with $n = 10$, $NB = 4$ and $TRANS = T$, the numbers represent the position within the array where $a(i,j)$ is stored. Note the missing numbers (e.g., 2, 3, 4, 7, 8, and 12, which correspond to the upper right corner of the first stripe). An example of square blocked lower packed format ($TRANS = T$) is given in Fig. 7. The blocked lower packed array consists of six square block arrays: the first three are 4×4, 4×4, and 2×4. The next two are 4×4 and 2×4. The last square block is 2×2. Note the padding, which is done for ease of addressing. Addressing this set of six square blocks as a composite block array is straightforward. An example of Square Blocked Upper Packed

Format ($TRANS = N$) is given in Fig. 8. The blocked upper packed array consists of six square block arrays: the first is 4×4; the next two are 4×4; and the last three are 4×2, 4×2, and 2×2. Each block is in column-major order. Note the padding, which is done for ease of addressing. Addressing this set of six square blocks as a composite block array is straightforward.

$$\begin{bmatrix}
1 & * & * & * & & & & & & & & \\
5 & 6 & * & * & & & & & & & & \\
9 & 10 & 11 & * & & & & & & & & \\
13 & 14 & 15 & 16 & & & & & & & & \\
17 & 18 & 19 & 20 & 49 & * & * & * & & & & \\
21 & 22 & 23 & 24 & 53 & 54 & * & * & & & & \\
25 & 26 & 27 & 28 & 57 & 58 & 59 & * & & & & \\
29 & 30 & 31 & 32 & 61 & 62 & 63 & 64 & & & & \\
33 & 34 & 35 & 36 & 65 & 66 & 67 & 68 & 81 & * & * & * \\
37 & 38 & 39 & 40 & 69 & 70 & 71 & 72 & 85 & 86 & * & * \\
* & * & * & * & * & * & * & * & * & * & * & * \\
* & * & * & * & * & * & * & * & * & * & * & *
\end{bmatrix}$$

Fig. 7. Square blocked lower packed format.

$$\begin{bmatrix}
1 & 5 & 9 & 13 & 17 & 21 & 25 & 29 & 49 & 53 & * & * \\
* & 6 & 10 & 14 & 18 & 22 & 26 & 30 & 50 & 54 & * & * \\
* & * & 11 & 15 & 19 & 23 & 27 & 31 & 51 & 55 & * & * \\
* & * & * & 16 & 20 & 24 & 28 & 32 & 52 & 56 & * & * \\
 & & & & 33 & 37 & 41 & 45 & 65 & 61 & * & * \\
 & & & & * & 38 & 42 & 46 & 66 & 70 & * & * \\
 & & & & * & * & 43 & 47 & 67 & 71 & * & * \\
 & & & & * & * & * & 48 & 68 & 72 & * & * \\
 & & & & & & & & 81 & 85 & * & * \\
 & & & & & & & & * & 86 & * & * \\
 & & & & & & & & * & * & * & * \\
 & & & & & & & & * & * & * & *
\end{bmatrix}$$

Fig. 8. Square blocked upper packed format.

With extra storage appended directly below a standard packed array, one can move to these new data formats without extra storage. For the examples above, AP requires 55 storage elements. If there are $96 - 55 = 41$ free locations below AP, then one can move the packed array downward into the block packed by starting at the end of AP and moving the square blocks in a block column into a buffer of of size $n1 * NB^2$, either in row-major or column-major order. The entire buffer is then copied back over the vacated block column.

One of the main innovations in using the square blocked packed format is to see that one can use the standard packed algorithms where each i, j element is now a square block. Because of the storage layout, the beginning of each block is easily located. Also key is the fact that this format supports level-3 BLAS. Hence old packed code is easily converted into square blocked packed level-3 code. We keep "standard packed" addressing so the library writers and users can handle their own addressing in a Fortran or C environment.

Now turn to full format. Please refer to Sect. 3 on packed blocked format as the ideas there carry over here.

Now we turn to programming dense linear algorithms in the new formats. As an example, see Subroutine 3. The *uplo* $= L$ code for `DPSTRF` (positive definite square) produces the lower Cholesky factor transposed for positive definite symmetric AB. Algorithm `DPSTRF` is a simple right-looking algorithm as the code illustrates. Routine `DPOFU` is a ESSL factor kernel. In ESSL, all BLASes and level-3 factorization routines use kernel routines. For example, in ESSL's DGEMM, a blocking routine is called to partition the matrix operands, A and B, into submatrices (matrix blocks), and then calls are made to kernel routines that operate on the blocks. Data copying of the operands to the kernel routines is decided on by the ESSL DGEMM blocking routine. Now in `DPSTRF`, we can call the kernel routine directly, thereby avoiding copying, because NB was chosen for good L1 cache behavior.

Subroutine 3. `DPSTRF`

```
*       Square Blocked Lower Packed Transposed Format Cholesky Factor
        subroutine dpstrf(uplo,n,nb,a,info)
        implicit none
        character*1 uplo      ! only uplo='L' and trans='T' handled
        integer*4 n,nb,info   ! mod(n,4) = 0 is assumed
        real*8 one,a(*)
        integer*4 j,k,i,kb,ib,jj,jk,kk,ji,ki,nb2,n1,n2,nn
        parameter (one=1.d0)
        info=0
        nb2=nb*nb                  ! size of a square block
        n2=(n+nb-1)/nb             ! order of block matrix
        n1=n2*nb2                  ! block lda
        jj=1                       ! -> a(j,j), j=1
        do j=1,n-nb,nb
*       factor a(j:j+nb-1,j:j+nb-1)
          call dpofu4(a(jj),nb,nb,info) ! factor kernel
          if(info.gt.0)goto 30
          ji=jj
          do i=j+nb,n,nb
            ji=ji+nb2
            ib=min(n-i+1,nb)
```

```
*     solve a(j:j+nb-1,j:j+nb-1)**T*u(j:j+nb-1,i:i:ib-1) =
*     a(j:j+nb-1,i:i+ib-1) for u
          call dslvl4(a(ji),nb,ib,a(jj),nb,nb) ! trsm kernel
        enddo
*     initialize pointers for the k loop
        kk=jj                     ! -> a(k,k), k=j
        nn=n1                     ! lda of k block, k=j
        jk=jj                     ! -> a(j,k), k=j
        do k=j+nb,n,nb
*     update pointers for the k loop
          jk=jk+nb2               ! -> a(j,k)
          kk=kk+nn                ! -> a(k,k)
          kb=min(n-k+1,nb)
*     update a(k:k+kb-1,k:k+kb-1) = a(k:k+kb-1,k:k+kb-1)
*     - a(j:j+nb-1,k:k+kb-1)**T*a(j:j+nb-1,k:k+kb-1)
          call dtata4(kb,nb,a(jk),nb,a(kk),nb) ! syrk kernel
          ji=jk                   ! -> a(j,i) i=k
          ki=kk                   ! -> a(k,i) i=k
          do i=k+nb,n,nb
            ji=ji+nb2             ! -> a(j,i)
            ki=ki+nb2             ! -> a(k,i)
            ib=min(n-i+1,nb)      ! block a(k,i) has size nb by ib
*     update a(k:k+kb-1,i:i+ib-1) = a(k:k+kb-1,i:i+ib-1)
*     - a(j:j+nb-1,k:k+kb-1)**T*a(j:j+nb-1,i:i+ib-1)
*     next, call gemm kernel
            call datb4(kb,ib,nb,a(jk),nb,a(ji),nb,a(ki),nb)
          enddo
          nn=nn-nb2               ! lda for next time through loop
        enddo
*     update pointers for the j loop
        jj=jj+n1                  ! -> a(j,j), j=j+nb
        n1=n1-nb2                 ! lda for next j block
      enddo
*     factor a(j:n,j:n)
      call dpofu4(a(jj),nb,n-j+1,info) ! factor kernel
      if(info.gt.0)goto 30
      return
 30   info=info+j-1
      return
      end
```

We suggest a way to package this data storage in LAPACK. We continue with the current routine for Cholesky factorization, *uplo* = *L*. For *uplo* = *U* a similar procedure would be followed. Define a new LAPACK routine called subroutine `DPSTRF(UPLO, N, AP, NS, INFO)`. Input parameters $UPLO$, N, AP, $INFO$ have the same meaning as the corresponding parameters of LAPACK routine `DPPTRF` and hence do not need any description. The new parm NS stands for the storage the user inputs for AP: $NS \geq n*(n+1)/2$. If NS is not sufficiently large, `DPSTRF` just returns in NS the amount of storage necessary for good level-3 performance. If NS is sufficiently large, the input storage is rearranged into square block format and then DPSTRF is executed giving level-3 performance. Like the $LWORK$ parm, the value returned in NS will be the value used by DPSTRF for good level-3 performance.

5 Vectors

Vectors generalize as follows. Each vector is a collection of subvectors. The subvectors have the same format as Fortran 77 vectors: $X(0 : NB-1 : INCX)$. The "stride" between two subvectors (either constant or variable) has to be determined. This vital information becomes part of the definition of the vector as a collection of subvectors.

6 Recursion

We now have block-based data structures stored in the conventional column-major or row-major order. However, recursion requires a new way to store the blocks. To be able to do this, we can address the blocks through the use of integer tables. Because the blocks contain NB^2 elements, the number of blocks will be relatively small. This means that the additional storage for the tables will be tiny.

7 How the BLASes Change

The main thing to note is that data copying has been removed. A new set of BLASes, namely factor kernels, must be defined. However, the conventional BLASes become simpler to write, because neither data copying nor data allocation need to be considered.

8 Summary and Conclusions

We have described several novel data formats for dense linear algebra and have described some novel simple algorithms that use these new data structures. We have relied on a heuristic that is the key factor governing performance on processors with deep memory hierarchies, namely, blocking or tiling. To be able to use this heuristic we have made use of the following fact from linear algebra: "Every point algorithm has a submatrix block formulation." Lacking are many concrete demonstrations that these data formats improve the performance of standard linear algebra software that are typified by ESSL and LAPACK. This paper shows that the result is true for Cholesky factorization. Also, we presented a simple algorithm for LU factorization with partial pivoting without any performance results. Finally, there is agreement that the new software being developed can become part of LAPACK and ESSL if sufficient gains in performance and or storage utilization are demonstrated. Preliminary results indicate that this will happen.

Acknowledgments

The work described here is the outcome of my collaboration with HPCN at the University of Umeå, Sweden, and Uni-C in Lyngby, Denmark. At Umeå my collaborators are Erik Elmroth, Andre Henriksson, Isak Jonsson, Bo Kågström, and Per Ling. At Uni-C they are Bjarne-Stig Andersen, Alex Karaivanov, Minka Marinova, Jerzy Waśniewski, and Plamen Yalamov. In some sense they should be co-authors. Certainly, this work would not have progressed to this point without our collaboration. A catalyst for our progress is Carl Tengwall of IBM, who believed from the beginning that these collaborations would be an excellent match between academia and industry.

References

1. R. C. Agarwal, F. G. Gustavson, and M. Zubair. Recursion leads to automatic variable blocking for dense linear-algebra algorithms. *IBM Journal of Research and Development*, 38(5):563–576, September 1994.
2. J. J. Dongarra, F. G. Gustavson, and A. Karp. Implementing linear algebra algorithms for dense matrices on a vector pipeline machine. *SIAM Review*, 26(1):91–112, January 1984.
3. E. W. Elmroth and F. G. Gustavson. Applying recursion to serial and parallel QR factorization leads to better performance. Submitted to *IBM Journal of Research and Development*.
4. G. Golub and C. Van Loan. *Matrix Computations*. Johns Hopkins Press, Baltimore and London, second edition, 1989.
5. F. G. Gustavson. Notes on blocked packed format and square blocked format for symmetric/triangular arrays.
6. F. G. Gustavson. Recursion leads to automatic variable blocking for dense linear-algebra algorithms. *IBM Journal of Research and Development*, 41(6):737–755, November 1997.
7. F. G. Gustavson, A. Henriksson, I. Jonsson, B. Kågström, and P. Ling. Recursive blocked data formats and BLAS's dense linear algebra algorithms. In B. Kågström et al., editors, *Applied Parallel Computing: Large Scale Scientific and Industrial Problems*, volume 1541 of *Lecture Notes in Computer Science*, pages 195–206, 1998.
8. F. G. Gustavson, A. Karaivanov, J. Wasniewski, and P. Yalamov. The efficiency of a new packed storage for symmetric indefinite matrices. Working draft, September 1999.

Technologies for High-Performance Computing in the Next Millennium

Dave Turek

Technical Strategy and Business Opportunity; IBM Server Group IBM; USA

Abstract. The environment for high-performance computing is changing rapidly. On the hardware side, chip-related advances such as copper metallization, silicon-on-insulator, and high-resolution lithography will continue to contribute to rapid increases in device speed. Although the investments needed continue to grow, we don't see any fundamental roadblocks during the next ten years. With higher levels of integration, the importance of packaging grows, and IBM has paid quite a bit of attention to this. On the software side, the changes brought about by an increased interest in Open Source promises to fundamentally lower the capital requirements for the introduction of new platforms and to affect the computer industry in many other ways as well.

The evolution of the IBM RS/6000 SP System continues—after it proved its ability to play chess successfully against the reigning World Champion Gary Kasparov in 1997. IBM is on track to deliver in 2000 the ASCI White 10-Tflop/s system to the Department of Energy's Lawrence Livermore Laboratory. Equally interesting is the fact that the whole area that IBM calls Deep Computing is growing dramatically. Commercial use of HPC Technologies is reaching new highs as exemplified by the high placement on the recent Top500 list of an IBM RS/6000 SP System at Charles Schwab, a financial services company offering stock trading on the Web.

IBM follows up on these successes by continued investments. Most recently announced is the Blue Gene project, which aims to have a Petaflop/s system ready by 2004 to tackle the protein-folding problem. This challenging problem is of great importance to better understand, among other things, how proteins actually interact with the living organism.

Deep Computing promises to affect our lives in fundamental and interesting ways. More information on these activities is available on the IBM Web pages, at `<http://www.research.ibm.com/dci/>`.

Global Tele-Immersion: Working in Cyberspace

Maxine D. Brown

Electronic Visualization Laboratory/University of Illinois at Chicago; USA

Abstract. Tele-immersion is defined as collaborative VR over networks, which extends the human/computer interaction paradigm to human/computer/human collaboration. The computer provides real-time data in shared, collaborative environments, which enables computational science and engineering researchers to interact with each other as well as their computational models, over distance.

Tele-immersion combines face-to-face meetings with audio- and video-conferencing in ways that involve data mining and heavy computation. When participants are tele-immersed, not only can they virtually see and interact with each other, but when they leave, their environment can continue to evolve, autonomously controlling computation and data access.

The shared virtual environment can be many things–part of a car design; the interior of an ancient cultural shrine that has fallen into ruin; the eye of a storm; or a three-dimensional environment that does not physically exist. Collaborators enter the environment as avatars, which are lifelike computer-generated representations. The environment transmits gestures as well as audio and video, so users have a greater sense of presence in the shared space than they would with other collaborative media.

This adds a new dimension to virtual collaboration. In tele-immersion, participants are not talking about a molecule; they are standing inside it. They are not looking at a scale model of a car design; they are sitting in the driver's seat. They are not using telescopes to observe astronomical phenomena; they become bigger than our galaxy. Participants can change their size, can interactively change parameters on a supercomputing calculation, and can get "inside" their data. With tele-immersion, they can then work with colleagues to collectively study the impact of those changes. Enhanced with these capabilities, collaborators may choose to work virtually even if more traditional face-to-face meetings are possible.

The University of Illinois at Chicago's Electronic Visualization Laboratory (EVL) has been involved with several applications that demonstrate tele-immersion. Several applications will be presented as part of this presentation.

ActiveSpaces on the Grid: The Construction of Advanced Visualization and Interaction Environments

Lisa Childers[1], Terry Disz[1], Mark Hereld[1,3], Randy Hudson[1,2], Ivan Judson[1], Robert Olson[1], Michael E. Papka[1,3,4], Joe Paris[1], and Rick Stevens[1,2,3,4]

[1] Futures Laboratory, Mathematics and Computer Science Division; Argonne National Laboratory; Argonne, IL 60439; USA
[2] ASCI FLASH Center; University of Chicago; Chicago, IL 60637; USA
[3] Computation Institute; University of Chicago; Chicago, IL 60637; USA
[4] Department of Computer Science; University of Chicago; Chicago, IL 60637; USA

Abstract. The Futures Lab group at Argonne National Laboratory and the University of Chicago are designing, building, and evaluating a new type of interactive computing environment that couples in a deep way the concepts of direct manipulation found in virtual reality with the richness and variety of interactive devices found in ubiquitous computing. This environment provides the interactivity and collaboration support of teleimmersive environments with the flexibility and availability of desktop collaboration tools. We call these environments ActiveSpaces. An ActiveSpace is a physical domain that has been augmented with multiscale multiscreen displays, environment-specific and device-specific sensors, body and object trackers, human-input and instrument-input interfaces, streaming audio and video capture devices, and force feedback devices—and has then been connected to other such spaces via the Grid.

1 Toward the Evolution of ActiveSpaces

The Futures Lab group at Argonne National Laboratory and the University of Chicago is developing prototype collaboration and visualization environments that we call *ActiveSpaces.* ActiveSpaces are the workspaces of the future, places that combine existing workspace infrastructure with high-tech information technology. The goal is to construct a workspace that enhances the work experience, enables the user to be more productive, and does not intimidate the user. The construction of ActiveSpaces is a cross-cutting research project that combines research in display technology, collaboration environments, networking, and many other areas into a seamless environment. Other groups are working on similar cross-cutting problems [6,10,14,15] or similar focused problems addressing a specific area [8,11].

The key point of ActiveSpaces is not the construction or integration of any one technology into the workspace, but the integration of many diverse components. These components are individually developed to address a certain research goal or problem, and they are then incorporated into the larger picture of the workspace to become an ActiveSpace. Our premise is simple: in order for advanced visualization and collaboration technology to succeed, it must be widely used and used in a variety of application and interaction domains; and this wide use will happen only when the technology is empowering and the environments are compelling. The technology needs to be integrated into the places and modalities that characterize modern intellectual work.

Our experience has indicated that a considerable number of research activities require people to work in loosely coupled collections of small groups. This multiple-level group structure is the target of our technology work. We are interested in developing working environments that enable groups to visually and interactively investigate large scientific datasets using large-format and immersive visualization technologies in the context of shared collaborative spaces. These shared collaborative spaces are characterized by a number of attributes that distinguish them from both current desktop-oriented IP-based teleconferencing systems and from traditional low-bandwidth videoconferencing. They create the illusion of being in a shared workspace that is permanently connected to other workspaces. The illusion is supported by using multiple cameras, large-format displays, and full-duplex ambient audio, all of which enable natural conversations between participants as if they were in the same work room.

Our goal in this work is to understand how to engineer into future workspaces the technology that can support high-performance collaboration and scientific visualization. We believe that the following areas of work are needed to pursue this goal.

- Group workspaces need to be designed to be comfortable, flexible, attractive, and compelling environments that encourage users to congregate and naturally express themselves.
- Computing and communications technologies that are needed to support collaboration (e.g., cameras, displays, microphones, interfaces, and controls) should be highly integrated into the physical design of the space.
- The user interface should support a natural set of interaction modalities (for collaboration, this could mean hands-free full-duplex high-quality audio, multiperspective video streams; for visualization this could mean that 3D datasets should be experienced in 3D, with direct interaction interfaces of virtual reality devices).
- User-owned and user-managed resources should be easily and temporarily integratable into the active scope of the workspace. For example, a user's laptop, phone, and personal digital assistant [PDA] should include a simple mechanism that enables it to become part of the collective resource set for the time the user is participating in the activity of the workspace.
- Interconnecting collaboration and visualization provides a uniquely challenging set of requirements for the systems design, the physical and soft-

ware integration, and the use of high-performance networking and communications for enabling shared interactions.

- Comprehensive sets of middleware services are needed to provide a high-level set of abstractions for multimodal communication, security, scheduling, and resource management on which the ActiveSpace environments can be layered.

Below we expand on each of these areas. We will then discuss the progress in our collaboration and advanced display technology research.

1.1 Creating Compelling Group Workspaces

A major challenge for designers of modern workspaces—rich in information technology—is to design something that people want to use. We believe that while fully immersive VR environments can be technologically exciting and highly attractive for occasional use, they do not (yet?) replace a well-designed physical space that has comfortable seating, excellent lighting, variable lines of sight, multiple work surfaces, and a flexible layout. One can create images of similar environments in a VR system, but there are many problems involved. Supporting multiple users in the same proximity is difficult. Many physical constraints rapidly tire the user, including lack of mobility, occluded views, the persistent use of low-resolution displays, and constant-focus display planes. In addition, the amount of computing and graphics power needed to provide just the ambient environment is often many times more than what is currently available and focused on the task (e.g., design model or scientific visualization dataset).

Thus we believe that we need to design spaces that work first as physical spaces; then, with sufficient introduction of information technology, they will become truly compelling spaces. We show in Fig. 1 and Fig. 2 how we have modified a small meeting room/library and a workshop room with communications technology that in our opinion creates a compelling space to work, particularly with the displays enabled. The implication of our analysis and experiments is that we need to return to the more basic thinking of how IT should be built into the workspace. It would not be a stretch to say that these ideas have been influenced by the Arts and Crafts style of interior architecture and the realization that display and communications technology is poised to leap off the desktop into the very fabric of our work and living spaces.

1.2 Integrating Technology into the Structures

Soon we will have a vast assortment of computing and communications device, form, and function factors to use in developing new environments. As a starting point for our work, we have made the following assumptions.

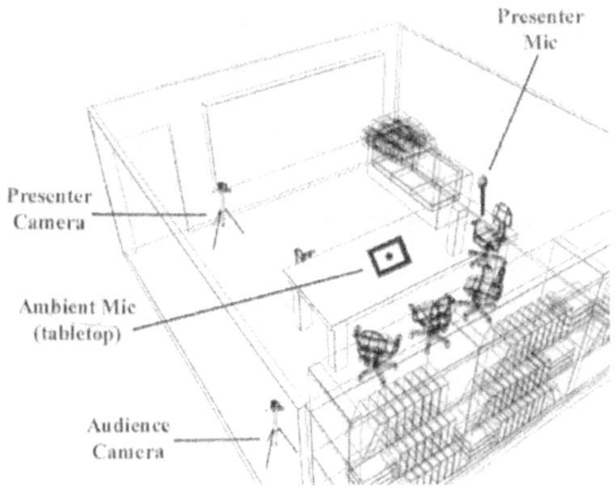

Fig. 1. AccessGrid node in a small meeting room setting.

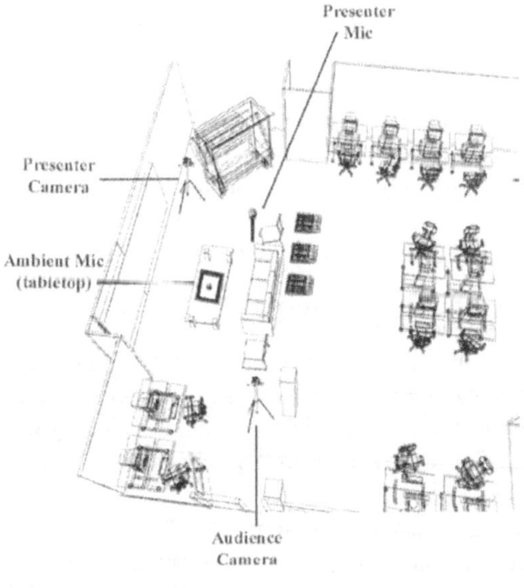

Fig. 2. AcessGrid node in a larger workspace setting.

- Within a few years, the functionality of the desktop PC will be available in a variety of physical form factors ranging from handheld and wearable devices to embedded systems.
- PC-based computer systems will become increasingly componentized, enabling one to assemble new types of ensemble and clustered devices that may share subsystems with desktop computers, but that will not be recognizable as desktop systems per se.
- New classes of devices (e.g., sensors, lights, controllers, switches, transducers) will become available as computer peripherals.
- Existing analog and digital peripherals will continue to decrease in price and increase in quality as chip technologies and economies are more fully used.
- Building, room, and personal-area networking will explode with new types of low-power and low-cost wireless capabilities.
- Moving off the desktop will create a huge new number of application opportunities, as well as the need for new user interaction mechanisms and organizing metaphors.

These six factors will drive computers into the environment in a way that is only partially understood at present. We certainly believe that rooms of all types will become one of the next application environments to be explored (along with automobiles, aircraft interiors, luggage, and the personal environments people carry with them). An essential point here is that we recognize that it will be possible to integrate substantial technology into the physical environment; the important questions are what type of technology it will be and how the integration will work. In our current and planned experiments, we are focusing on integrating cameras, projectors, microphones, speakers, screens, and tracking systems. However, we expect in the future to increase the number of environmental sensors and controls; we also expect to include lighting, seating, interactive work surfaces, and boundary (door and window) sensors. These devices will be built into the space and become a permanent part of the environment.

1.3 Natural Interaction Modalities

After the room and related structures become the application environment, it becomes immediately obvious that we need new user interfaces. Interaction modalities that these interfaces might use include voice, gesture via camera images, intentful motion via trackers, and handwriting or touch via embedded sensor mats or tracked pens.

In the collaboration application, natural interaction can mean a number of things. It might mean using full-duplex always-open voice channels to have a natural conversation with a remote participant. It might mean using perspective video to support a casual nonverbal visual cue. When these environments are used for scientific visualization, it can mean having a variety of

3D direct manipulation interfaces, each permitting the user to interact, manipulate, and modify the data without breaking the visualization metaphor in use. Workspaces of the future will include haptics, audio feedback, and audio commands.

1.4 User Resource Docking

It is very common today for users to have one or more personal computing and communication devices that they routinely take with them throughout the workplace throughout the workday. These devices include laptop computers, cellular telephones, two-way pagers, and PDAs. It is becoming clear that making all these tools available for use in a group activity would be an improvement over the current situation. A common example of this is a user who brings a presentation to a meeting on a laptop and needs to make it available to the group.

In more complex examples, the user's personal resources can augment the group environment by providing private channels of communication, or by providing local user interfaces to group- or room-oriented tools. Exploiting this capability requires software interfaces in the ActiveSpace that can "dock" with a personal resource, as well as common-room networking or personal-area networking (like Bluetooth) to provide the seamless physical interface for the duration of the session [4]. We are investigating the types of group applications that this capability would enable.

1.5 Collaboration and Visualization

Of particular interest to our group is the merging of collaboration technology with high-end scientific visualization systems. Part of this vision is the notion of people working together over large distances while engaged in close cooperative visualization of a large dataset that might be part of a computational science exploration or a complex design project. One approach to this problem is to understand how people use advanced visualization tools when they are physically co-located and to approximate that experience via networking by combining teleconference and tele-immersion capabilities with the visualization tools.

Another approach is to create a virtual shared space in which multiple types of visualization might be pursued simultaneously (e.g., immersive visualization and large-format visualization), and use collaboration technology to provide multiple foreground and background channels to link the groups together. A third approach is to focus on techniques that scale the human interaction beyond what is possible in the physical world (i.e., multicasting a visualization session to hundreds of people, giving them each a large degree of freedom to investigate the data, and synchronized to one or more lead sites). The goal here is to create an environment that is intrinsically more powerful than any single system. An example of this would be to link multiple displays and groups of people that can explore a single dataset in multiple modalities simultaneously or at multiple space- and time-scales simultaneously.

These latter cases are our primary motivation; however, we recognize the need to support a wide variety of collaborative approaches to visualization. We are also interested in how collaboration technologies can be scaled. Currently our focus is on small group collaboration, with on the order of five connected sites, with up to six people per site. This appears to be a reasonable goal for the near future because in our experience, this is the most common scale of scientific and engineering collaborations.

1.6 ActiveSpace Middleware Needs

As we develop these systems, we have identified a need for a common set of system infrastructure capabilities. This set includes the following.

- Widespread and robust support of IP-based multicast, and tools for converting multicast to unicast and unicast to multicast.
- A common set of authentication tools that provide a one-time-per-session logon to the collection of ActiveSpace resources.
- High-performance communications libraries that support both best-effort service as well as emerging differentiated service.
- Resource discovery and management tools, which enable sites to advertise and discover resources that can be allocated as part of a distributed ActiveSpace session.
- Comprehensive scheduling tools that can integrate both human and non-human resource availability into coherently managed schedulable resource sets.
- Database-driven software distribution, configuration, updating, and monitoring capability that will enable a small group to manage a large number of distributed resources nearly automatically.

The importance of middleware cannot be overstated. It is critical to enable the rapid experimentation and development of ActiveSpace-like systems. By building on this layered structure, we not only have the ability to avoid overly specific implementations but also can leverage the development resources of a number of groups that are developing middleware-related tools and libraries.

1.7 Our Testbeds

The prototype environments we have built include a number of "AccessGrid Nodes" aimed at exploring group-oriented room-scale collaboration mechanisms. For visualization research we have developed software systems that use the CAVE and the ImmersaDesk, and we have been actively involved in the development of CAVE-related technology since 1992. In the past year, we have developed a large-format tiled display system we call the *ActiveMural* and a smaller version known at the *μMural*. We have initiated several long-term projects to further develop the technology and software systems for these devices, and we have a number of collaborative projects that are aimed at networking together multiple sites that have similar environments [1,3,2].

2 The Grid

A new concept is shaping the way the research community is thinking about the evolution of the Internet. We call this concept the Computational Grid or simply the Grid. The name Grid comes from the analogy of the electric power grid, which provides a pervasive service via a complex network of providers, and yet has functional interoperability of devices and services. Perhaps we stretch the analogy a bit far, but the term Grid is short and works.

The concept of a Computational Grid has emerged in the university and national laboratory research communities in the past few years. Its origin can be traced to early (i.e., early in Internet time: 1991) metacomputing experiments, where multiple supercomputers were linked together to create a single large virtual machine that could be used to attack problems larger than any single machine could handle. In these early efforts to build distributed virtual machines, the supercomputer sites were interconnected by high-speed networks—the modern Internet precursor NSFnet being one of them. However, there were no additional capabilities available beyond the high-speed transport of data.

The early metacomputing experiments were sometimes successful, and they led to the thinking that interconnecting additional high-performance devices (e.g., virtual reality systems, telescopes, electron microscopes, terabyte-scale data archives) via common high-performance Internet-based networks would be even more interesting. After prototyping dozens of distributed high-performance applications on these ad hoc testbeds, it became apparent that many of the same software services and functions were needed by multiple applications. In each case, the application was having to solve the same problems; among these were authentication in a distributed multidomain environment, remote access to data without the benefit of a common name space, lack of network performance interfaces, and lack of high-performance wide area data transport interfaces. It soon became clear that these and some additional services formed a common collection of services that would benefit many applications. It is this layer of services that has formed the basis for the middleware layer of the Grid.

By deeply incorporating into the architecture this notion of common middleware services, the Grid represents a much more comprehensive view of one possible future direction for the Internet. Central to this view is the idea that the network can be the conduit of advanced applications delivery beyond the near-term evolution of the Web and desktop-to-server–based applications. In one type of Grid application, the user is presented with the capability of harnessing a diverse set of computational, informational, collaborative, and possibly remotely controlled systems to build an application. These applications could range from simulation-based design of products, to distributed data mining, to inclusion of networked virtual realty demonstrations in a sales seminar.

The Computational Grid extends the current notion of the Internet to include a variety of integrated services, including computational servers, remote data services, collaboration infrastructure, remote instrumentation control, distributed computation, and tele-immersive visualization (distributed virtual reality). These high-level services are implemented on a common software substrate of middleware that provides application-independent services, such as data security, use authentication using public key infrastructure, distributed resource management, directory services, resource brokering services, and distributed resource scheduling.

The key advantage of the Grid is that through these advanced middleware services, applications that are much more powerful than today's Internet applications can be developed and broadly deployed. Applications developers will be able to develop new types of applications (e.g., distributed data mining with integrated collaborative information visualization) that can include high-performance access to data and devices in generalized ways. To build the Grid, we need to add new middleware capabilities to the existing suite of IETF protocols that support the Internet, and we need to begin to build large-scale testbeds that can validate the Grid concept on real applications with real users. Fortunately, this is happening. The federal Next Generation Internet program is funding a significant amount of work in this area, the university-based Internet2 initiative is building testbeds that support further exploration of the Grid concept, and the commercial sector is starting to view the Grid as a new modality of the Internet that may open up significant business applications and markets for advanced networking services.

3 Advanced Display Environments

An advanced display environment takes one beyond the traditional desktop display system, beyond a single monitor-based system. As part of an active space, the advanced display environments work to integrate the physical space, the user, and the computing environment. We have been investigating advanced display environments from two directions: immersive stereo-based environments, such as the CAVE and ImmersaDesk, and high-resolution tiled environments.

3.1 Immersive Stereo Environments

Immersive stereo displays, such as the CAVE and ImmersaDesk, provide a rich environment for the development of interactive 3D applications. The immersive environments couple real-time tracking to the software-generated images for correct viewer-centered rendering of a given scene. The CAVE environment uses a single projector to display surface configuration, with a single desktop resolution projected onto a 10×10-foot surface. Our work in immersive environments has focused on developing immersive scientific visualization applications and on building these applications to work within the real-time requirements of VR [13]. We have also worked on understanding what value is added by the use of immersive environments and how people work within them.

3.2 Tiled Display Environments

Tiled display environments allow for the construction of arbitrary resolution surfaces. The display surfaces are constructed by tiling display devices together to form a single display. Several groups are working on the design and development of tiled displays that are being driven by a variety of different computing platforms [8,10,11]. We have constructed two such devices: the ActiveMural, a fifteen-projector system; and the μMural, a portable six-projector version. Both systems can be driven by either a high-end SGI machine or a Linux cluster.

ActiveMural. The ActiveMural (AM) is a rear-projected large-tiled display with a 16×8-feet visible projection surface (see Color Plate 2 on page 285). The current configuration is driven by either a Linux cluster, with each computer having a graphics accelerator card, or an SGI Onyx2 with eight Infinite Reality2 graphics cards, with two-channel output. Table 1 shows the various resolutions that the ActiveMural is capable of based on the current configuration. The screen material for the AM is JenMar Visual Systems BlackScreen, with a resolution greater than 200 lines per inch. The screen material is extremely good at the rejection of ambient light, allowing the AM to be used in normal room-lighting conditions, unlike the immersive environments described above. Interaction with the AM is currently limited to keyboard/mouse input; we have experimented with a 3D tracked joystick, which operates much like the CAVE wand, with mixed results. A promising input research path we have taken is the use of the Fujitsu Stylistic 2300, running WindowsNT and using a wireless networking card as a mobile interface to the AM.

μMural. The μMural is a smaller portable version of the AM described above. The μMural currently uses six projectors and can also be driven by either a Linux cluster or the SGI Onyx2. When being driven by the Onyx2, the μMural can use either one IR (Infinite Reality2) board with the eight-channel option or three IRs using the dual-channel option. Interaction with the μMural is the same as interaction with the AM. The major difference

Table 1. Different Resolutions of ActiveMural and μMural.

Device	Compute Engine	Graphics Cards	Total Resolution
ActiveMural	Linux	15	5120 × 2304
ActiveMural	SGI	8	5120 × 2304
μMural	Linux	6	3072 × 1536
μMural	SGI	3	3072 × 1536
μMural	SGI	1	1948 × 1140

between the two, other than size, is the inclusion of hardware shadow masks to help with edge blending. Figure 3 shows a head-on picture of the μMural using hardware-based edge blending.

Fig. 3. Image of a computer model rendered at full μMural resolution.

3.3 Visualization Systems

The advanced display environments discussed above are output devices with no knowledge of the Grid. It is the software infrastructure that makes these devices endpoints on the Grid. Using the advanced display systems, we are constructing visualization systems to integrate the advanced display software environments onto the Grid and into ActiveSpaces. For immersive display systems, we built on the CAVE library [5], layering on top of it additional tools such as the Visualization Toolkit [12]. In addition, we have worked to enable the CAVE library to support more than eight display surfaces for use on tiled displays. Color Plate 3 on page 286 shows the same application running in both a CAVE and on the ActiveMural. The CAVE library has been ported to a variety of different compute platforms and provides a mechanism for rapidly deploying applications on a variety of different display environments.

In addition to developing tools for the CAVE library, we are constructing a set of open source tools and libraries for the tiled display environments. We have begun by building a set of tools for maintaining these environments, which includes alignment, color correction, and blending for the tiles.

4 Advanced Collaboration Spaces: AccessGrid

Today, the most common vision of computer-based collaboration tools is one of people sitting at their computer terminals, trying to look at their little cameras while also looking at a small, grainy video image and all the while saying, "Can you hear me? How do I run this software? Can you hear me?" This low image is the result of a lofty goal: to use the Internet to provide access to and collaborate with people in other places without having to travel.

Our own unsatisfactory experiences with desktop collaborative technology caused us to rethink what was really required to enable wide-area collaboration. First, we realized that we most often worked with colleagues as small groups, and so we began to think in terms of wide-area group collaboration. Second, although we attend structured meetings, workshops, and conferences, we often tend to be more productive in an unstructured manner with a lot of brainstorming, problem solving, casual conversation, and spontaneous idea generation. From this, we realized the need to support multiple modes of interaction, from very structured to completely casual. Third, we usually have our computers with us and often want to share with others some information stored in our computer, be it a visualization, a spreadsheet, a presentation, a Web site, a document, or a movie. Finally, we realized that one of the problems plaguing existing efforts was the perceived need to accommodate wide ranges of pre-existing equipment, software, and capabilities. We could see there would be significant advantages to be gained from having all participants use exactly the same gear and software.

We envision our ideal collaborative environment as an intentionally designed space: one that is rewarding to be in, and one that provides a sense of co-presence with other groups using similar spaces. We envision a space with ambient video and audio; large-scale displays; and software to enable the relatively transparent sharing of ideas, thoughts, experiments, applications, and conversation. We envision a space where we can "hang out" comfortably with colleagues at other places, and also use that same space to attend and participate in site visits, remote conferences, tutorials, lectures, and other structured meetings. We imagine the space will support the same capabilities, through remote interaction, that we have now in face-to-face meetings: subconscious floor control through social conventions; the ability to have private, one-on-one, whispered conversations; the ability to gather a small group in a corner and caucus; and all the other things we take for granted when we are a group in the same physical place. In addition, we envision that the space will be "smart" enough to recognize that someone has brought personal computing resources to it and will allow the export of items from one computing device to other individuals or groups.

The challenges this vision presents are many and varied; some are easily addressed, while others will require groundbreaking research efforts.

In realizing the first AccessGrid, we focused on basic enabling infrastructure for groups of people to find, talk to, see, and share ideas with other groups. Our philosophy is to use open source software wherever possible. First, this avoids forcing participants to purchase from and be slave to a particular vendor. Second, this allows every AG organization an equal chance to participate fully in research and development in AccessGrid technology.

4.1 Display

An AccessGrid Node, as we call a single room outfitted for AG use, requires a tiled display of sufficient physical size to comfortably accommodate a small group of people, up to a dozen or so, sitting around the display, all with good sight lines to the display. Second, the display must have sufficient resolution and size to accommodate the projection of multiple video streams from multiple sites, projecting near–life-size images of people at other sites. Solutions to this vary, but we are most satisfied with a three-projector, front-projection wall. The projected area is about 18 feet by 6 feet, with a seating area about 25 feet wide and about 20 feet deep. The projectors are mounted in the ceiling and are of sufficient brightness that the room can operate in normal light, allowing people to read and interact. We are experimenting with a corner display that uses four projectors, two per wall, to see if this enhances the sense of co-presence.

4.2 Video

An AccessGrid Node must generate multiple video streams from different perspectives in the room in order for people at other sites to get a feel for the room and its occupants. We specify four video streams: a wide audience shot, a close-up shot of the presenter or main speaker, a wide-area shot of the display screen (it is important for remote sites to be able to see what the originating site sees), and a roving audience and room camera. We use remote control pan-tilt-zoom cameras for maximum flexibility. They are placed to be unobtrusive and to facilitate the feeling of eye contact. We place cameras just below the projected screen area and place video images of people with whom we are conversing just above one of the cameras.

4.3 Audio

Being able to converse freely with people at other sites, unencumbered by microphones, wires, floor control protocols, or gadgets, is a cornerstone of AG usability. We achieve this ability by placing sufficient numbers and types of microphones and speakers within the space. We make sure there is adequate pickup everywhere in the room that there are likely to be people. We also use professional-quality echo cancellation gear by Gentner Corp. to ensure full-duplex audio. We currently place two speakers strategically in the front of the room to project good-quality audio into the space.

4.4 Computing

An AccessGrid Node uses four computers. The Display Computer runs windows NT and has a multiheaded video card. This is the machine that manages the tiled display and allows us to treat the multiple projectors as a single desktop. This machine is decoding all of the video streams, which can be several dozen, so it needs to be as robust as possible. The Video Capture Computer runs Linux and has four video capture cards. This machine encodes all the video streams captured at a node and then broadcasts those streams to the AccessGrid. It too must be a robust configuration to keep up with encoding demands placed on it. The Audio Capture Computer also runs Linux and performs the audio encoding and broadcasting as well as the audio decoding of the multiple streams being sent from other AG Nodes. The Control Computer runs Windows 98 and is used to run control software for the audio gear. This separation of function allows us to optimize each piece of gear for its intended purpose.

4.5 Software

Aside from the operating system mentioned above, a compliant AG Node requires several pieces of software developed by AccessGrid partners.

A distributed PowerPoint master and server Software that allows one node to control the flow of PowerPoint presentations at all participating nodes.

A status beacon Software that runs on AG Nodes and regularly reports the status of the node to an AG database.

Reflector A piece of software that reflects network packets back to the node used for audio debugging.

Vic Network video capture and display software, originally written at LBL, modified and distributed by University College London, and further modified by AG developers.

Rat Network audio capture and playback software, written and distributed by University College London.

Virtual Venue software A room-based metaphor to control the scope of interaction on the AccessGrid. The Virtual Venue software contains a set of rooms in which AG node participants can interact. This is a method of allocating, controlling, and automatically assigning multicast addresses. This software allows users to leave one group and join another with simple clicks on a Web-based map interface. The software automatically tears down existing connections and builds new ones as dictated by the addresses related to each room.

Auto-layout Software that automatically lays out windows across the screen real estate, based on preselected preferences. It eases managing the placement of windows on a large display.

tkMOO A text-based virtual space client used as a reliable communications back channel during live events and as a virtual community meeting place at all other times.

4.6 Network

The AccessGrid tools depend on network multicast to work well. Sites without multicast capability are forced to use some kind of traffic reflector. We use the Fermi National Accelerator Laboratory's MultiSession Bridge. Use of the bridge introduces delay, complexity, and significantly increases network load. Sites wishing to become AccessGrid Nodes should see that multicast capability is supplied to their site. The other practical network consideration is available bandwidth. A full AG session can deliver many dozens of video streams to a Node, typically four from each participant as well as the originating one. The bandwidth required by each stream depends on settings at the origin and can vary from 128KB/s to 512KB/s or more. The effect of inadequate bandwidth on the AG Node is dropped packets, which results in unintelligible audio and jerky-motion video. Other effects can be detrimental performance for, and possible hostility from, other users on the local network.

4.7 Operations

The AccessGrid has been used in several major events in 1999: the Access DC grand opening event in April; the three Alliance Chautauquas in the summer; and SC99, where several sites brought Nodes to the floor while others participated from their home sites. From these events, we have learned a great deal about operating an AG node and conducting a live event using AG technology. An operator's manual is being developed, which encapsulates and codifies the practices we have learned. Figure 4 shows a diagram of a basic AccessGrid node.

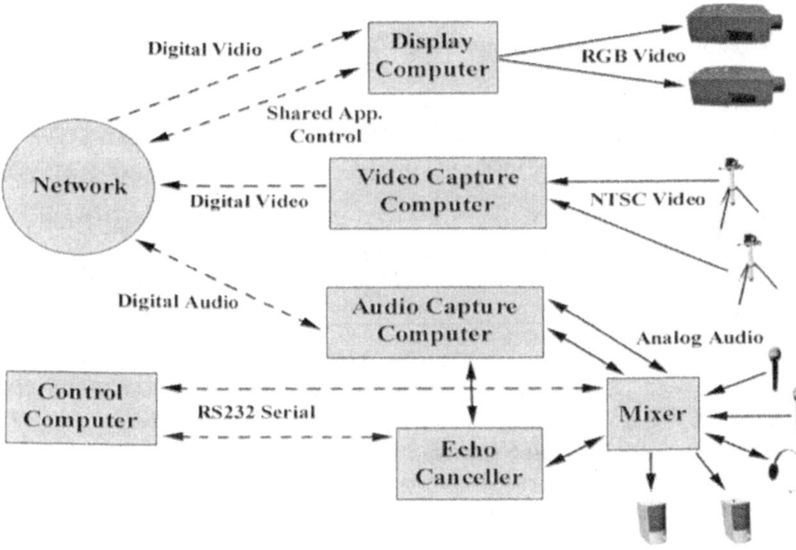

Fig. 4. Conceptual diagram of an AccessGrid node.

5 Conclusion

We have initiated several long-term projects [1,3,2] to develop the technology and software systems for ActiveSpaces. ActiveSpaces represent a convergence of collaboration, advanced visualization, and smart spaces concepts and technology. We have identified six trends in the development of information technology that will provide the driving forces for ActiveSpace environments. We are actively prototyping testbed systems to support exploration of the software, user interfaces, and applications that ActiveSpaces will enable. We firmly believe that our software must be designed and developed in a way that can leverage the emerging Grid Middleware infrastructure [7], and that these concepts need to be evaluated with real applications. Our applications partners include dozens of researchers from academia and national laboratories. Our goal is to release our prototype software to the community via open source software [9].

Acknowledgments

This work was supported by the Mathematical, Information, and Computational Sciences Division subprogram of the Office of Advanced Scientific Computing Research, U.S. Department of Energy, under Contract W-31-109-Eng-38.

References

1. Argonne National Laboratory. *AccessGrid.* `<http://www.mcs.anl.gov/fl/accessgrid/>`.
2. Argonne National Laboratory, Lawrence Berkeley National Laboratory, Electronic Visualization Laboratory, Los Alamos National Laboratory, Princeton University, University of Utah. *CorridorOne.* `<http://www.mcs.anl.gov/fl/corridorone/>`.
3. Argonne National Laboratory, Los Alamos National Laboratory, University of Utah. *Advanced Visualization Technology Center.* `<http://www.avtc.org/>`.
4. C. Bisdikian, S. Bouet, J. Inouye, R. Mettala, B. Miller, K. Morley, T. Muller, M. Roter, and E. Solboom. Bluetooth protocol architecture: Version 1.0. White Paper 1.C.120/1.0, Bluetooth, August 1999. `<http://www.bluetooth.com/>`.
5. C. Cruz-Neira, D. J. Sandin, and T. A. DeFanti. Surround-screen projection-based virtual reality: The design and implementation of the CAVE. In *SIGGRAPH '93 Annual Conference Proceedings*, pages 135–142, 1993.
6. T. L. Disz, M. E. Papka, and R. Stevens. Ubiworld: An environment integrating virtual reality, supercomputing, and design. In *Heterogeneous Computing Workshop Proceedings.* International Parallel Processing Symposium, April 1997.
7. I. Foster and C. Kesselman, editors. *The Grid: Blueprint for a New Computing Infrastructure.* Morgan Kaufmann, 1998.
8. G. Hunphreys and P. Hanrahan. A distributed graphics system for large tiled displays. In *IEEE Visualization '99 Proceedings*, October 1999.
9. Mathematics and Computer Science Division, Argonne National Laboratory. *Futures Laboratory Website.* `<http://www.mcs.anl.gov/fl>`.
10. R. Raskar, M. S. Brown, R. Yang, W.-C. Chen, G. Welch, H. Towles, B. Seales, and H. Fuchs. Multi-projector displays using camera-based registration. In *Visualization '99 Proceedings*, October 1999.
11. R. Samanta, J. Zheng, T. Funkhouser, K. Li, and J. P. Singh. Load balancing for multi-projector rendering systems. In *Eurographics '99 Proceedings*, 1999.
12. W. Schroeder, K. Martin, and B. Lorensen. *The Visualization Toolkit: An Object-Oriented Approach to 3D Graphics.* Prentice Hall PTR, 1998.
13. H. Tufo, P. Fischer, M. E. Papka, and K. Blom. Numerical simulation and immersive visualization of hairpin vortices. In *International Conference of High Performance Computing and Communications Proceedings*, Portland, Oregon, November 1999.
14. M. Weiser. The computer for the 21st century. *Scientific American*, 265(3), September 1991.
15. T. Winograd. Towards a human-centered interaction architecture. Unpublished draft available at `<http://graphics.stanford.EDU/projects/iwork/papers/humcent/>`, April 1999.

The Global Technology Grid: Its Role in Virtual Reality

Tom DeFanti

851 South Morgan St., Room 1120, M/C 154; Electronic Visualization Laboratory/University of Illinois at Chicago; Chicago, IL 60607-7053; USA

Abstract. The Global Technology Grid is the interlinking of simulation computers, large databases, and high-end visualization environments like CAVEs and multi-screen high-resolution displays over high-speed networks. Achieving this Grid is a massive undertaking, requiring everything from networking quality of service, to authenticated reservations of resources, to new levels of international cooperation. This talk will cover these topics from a visualization point of view.

Steering and Visualization of Electromagnetic Simulations Using Globus

Erik Engquist

Department of Numerical Analysis and Computing Science; Royal Institute of Technology; SE–100 44; Sweden

Abstract. A framework for computational steering of a finite difference code for electromagnetic simulation has been developed and implemented. In computational steering, we need to develop software that allows the user to enter an interactive visualization or VR environment and from there control the computation.

A proof of concept implementation has been carried out using an existing code for 3D finite difference time domain approximation of Maxwell's equations. Large parts of the computational steering software are general, but details in the choice of control variables and visualization is specialized to the electromagnetics code.

To handle the large computational requirements of both simulation and visualization, the system can be distributed across multiple machines. This is possible through the use of the Globus toolkit for communication, data handling, and resource co-allocation. This program also makes use of VTK for data filtering and the generation of visualization elements, and IRIS Performer with pfCAVELib for 3D interactive rendering on CAVE compatible devices.

Two test cases are presented. In one example with a smaller number of computational cells, full computational steering with recomputation is possible. In another with a large number of computational cells, the solution is precomputed and only the visualization is interactive. The scalability of the computational code is tested for different computers in order to determine the size of the problem that can be handled with full computational steering on the available local hardware.

1 Introduction

The objective of this project is to produce a framework for a distributed computational steering code. This framework will be implemented with a computational code that is used for large-scale Finite Difference Time Domain (FDTD) electromagnetic calculations. In electromagnetic simulations, the electric fields, magnetic fields, and surface currents are computed. Typical applications are in electromagnetic compatibility, radar cross section research, and antenna design.

To enable visualization and interaction in a 3D immersive environment, the code made use of the Visualization Toolkit (VTK) [9] in combination with IRIS Performer [2] and pfCAVELib [7]. The graphics facilities on which this project has been run include an ImmersaDesk and the VR-Cube at PDC. In one application, the program was used for the interactive visualization of the results of a very large FDTD computation. The geometry contained two million polygons, and the domain of computation consisted of more than one billion cells. This was made possible through the use of data decomposition, parallelization, and variable levels of detail.

A user interface was also created as part of the project. Interaction is carried out through the use of a tracked wand and floating menu buttons. The user may choose to display different features of the computed solution. Such features include surface currents, electric or magnetic fields in cutting planes, and electric or magnetic field strength isosurfaces. Regarding interaction with the computational process, the user can stop and start the computation as well as adjust all the simulation parameters.

The Nexus [5] communication library from the Globus toolkit is used to connect the flow of control information from the visualization process to the computation process and the flow of computed data in return. Nexus is a multi-threaded communication library that is especially well suited for a heterogeneous computing environment, as well as the event-driven nature of the communication. Via this computational Grid implementation, the application has made use of computational resources ranging from a Linux cluster to an IBM SP.

The final product consists of approximately 4,000 lines of Fortran90, C, and C++ code. There are also additional shell script routines used in organizing and configuring the package. The computational code has been edited to communicate with the visualization code as well as be interactively controlled by the user. Data transfer and translation routines have been written for the exchange of information between separate processes. Synchronization structures and routines have been created to allow multi-threaded communication. A visualization code was written that handles both geometric and volume data with multiple methods of presentation. The displayed information can be manipulated in an immersive environment using a wand and menu interface.

2 Problem

For the implementation of the steering and visualization package, we chose an existing computational code. Because the code is a part of a much larger project, we needed to minimize the changes required to the preexisting code.

2.1 The Numerical Scheme

The electromagnetic calculations are based on Maxwell's equations,

$$\frac{\partial E}{\partial t} = \frac{1}{\varepsilon}\nabla \times H - J \tag{1}$$

$$\frac{\partial H}{\partial t} = -\frac{1}{\mu}\nabla \times E \tag{2}$$

$$\nabla \cdot E = 0 \tag{3}$$

$$\nabla \cdot H = 0 \tag{4}$$

The 3D electric, $E(x,t)$, and magnetic, $H(x,t)$, fields are defined as functions of space, $(x \in \Omega)$, and time, $(\mathbb{R}^4 \to \mathbb{R}^6)$. The current $J(x,t)$ is regarded as a forcing term; and the material parameters—permittivity, $\varepsilon(x)$, and permeability, $\mu(x)$—could be space dependent. The electric and magnetic fields are divergence free.

The most common numerical approximation of (1) and (2) is the Yee scheme [11], which is based on centered differences in space and time. This means it is a leap-frog scheme applied to Maxwell's equations on a staggered grid. The Yee scheme has the advantage that very few operations are required per grid point, the E evolution depends on H and vice versa. Additionally, the discrete forms of (3) and (4) are preserved by the scheme. A typical difference step for the x-component of E looks like the following.

$$E^{n+1}_{x,i,j,k} = E^{n}_{x,i,j,k} + \frac{\Delta t}{\varepsilon_{i,j,k}} \left(\frac{H^{n+\frac{1}{2}}_{z,i,j+\frac{1}{2},k} - H^{n+\frac{1}{2}}_{z,i,j-\frac{1}{2},k}}{\Delta y} - \frac{H^{n+\frac{1}{2}}_{y,i,j,k+\frac{1}{2}} - H^{n+\frac{1}{2}}_{y,i,j,k-\frac{1}{2}}}{\Delta z} \right) \tag{5}$$

Absorbing boundary conditions of different types are possible at open boundaries, which limits the computational domain [10].

2.2 The FDTD Code

This code was a part of GEMS [1], the Large Scale FDTD project within the Parallel and Scientific Computing Institute (PSCI). The programming language is Fortran90 with MPI message passing. Domain decomposition is used as a basis for parallelization. The input to the code consists of an initialization file that describes simulation parameters and defines the names of files that describe geometry, sources, and output. These output files contain the electromagnetic fields and surface currents. Modifications to the code have been limited to rewriting the initialization and output routines. One goal of the project was to require minimal changes to the computational code in order to simplify its integration into existing projects.

The solver originally consisted of almost 5,000 lines of code. Our changes amounted to the removal of an initialization file read-in function and two result write-out commands. These were replaced by corresponding data I/O routines in the communication code. Inside the main computational loop, we also added a call to a state update function in the communication code.

The number of flop/s required each time-step of the computational code is approximately $N_x \times N_y \times N_z \times 36$ for each field, where N_x, N_y, and N_z are the number of grid points in each dimension of the computational domain.

Unfortunately, (5) shows that there are six unknowns for each component update. On scalar architecture machines, this will mean a high ratio of memory access to computation, which results in the inherent memory fetch penalties.

The performance of this code on the Linux cluster was around 80-90 MFlop/s per node, as seen in Fig. 1. On the four-way 333MHz PowerPC 604e SMP nodes on the SP, performance rose to 220–240 MFlop/s per node, while the Power2 nodes performed at 180–190 MFlop/s. This type of information is useful for gauging the feasibility of steering a particular computation. One can look at the performance requirements of a particular case and then demand the necessary hardware. More likely, one can see the limits imposed by a given hardware and can then scale the size of the computation appropriately.

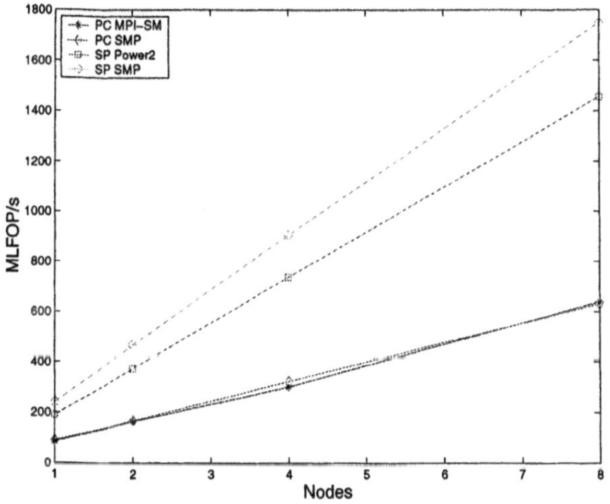

	MFLOP/s			
Nodes	PC MPI-SM	PC SMP	SP Power2	SP SMP
1	85	89	190	240
2	164	167	373	467
4	301	323	737	906
8	639	631	1455	1749

Fig. 1. Performance of FDTD code on PC cluster using MPI with shared memory (PC MPI-SM), threaded code (PC SMP), on IBM Power2 nodes (SP Power2), and IBM SMP nodes (SP SMP).

The basic performance requirements show the necessity of distributed computing for this case. Because the visualization machine is burdened by transforming the field data and generating geometric elements, it cannot simultaneously support large-scale numerical computations.

3 Resources

At the Center for Parallel Computers at KTH (PDC), there exists a broad range of hardware resources. We made use of hardware systems for visualization, computation, and communication.

3.1 Visualization Resources

Of special interest to this project were PDC's visualization facilities, including the VR-Cube, a six-surface immersive environment [6]. The Cube is driven by a 12-processor SGI Onyx2 with three InfiniteReality2 pipes. As a development platform there exists an ImmersaDesk, which is connected to a dual-processor SGI Octane with EMXI graphics. Both of these facilities support CAVELib and IRIS Performer, making development for both completely identical. A tracking system determines the location of the viewer and adjusts the 3D images accordingly.

The VR-Cube is a projection-based immersive environment with six surfaces (floor, ceiling, and four walls). The interior of the Cube measures $3 \times 3 \times 2.5$ meters. The rear wall is hinged, and the Cube sits on a 2.5-meter-high platform above the floor in order to allow rear projection on all six sides. This removes any shadows due to users and allows for a completely immersive environment. The depth effect is achieved through the use of liquid crystal shuttered glasses. Images for the left and right eye are alternated as an IR transmitter synchronizes the glasses. The wall resolutions are 1152×960, while the floor and ceiling are 1152×1152. Each graphics pipe drives two screens at a frequency of 96Hz, which in turn means the user experiences stereo images at a frequency of 48Hz per eye.

3.2 Computation Resources

The computational resources are even more varied. For initial development, the computation was carried out on the visualization machine itself. Once distributed computation was implemented, the development continued on a Linux cluster. The cluster consists of eight dual Intel Pentium II nodes. The processors run at 350MHz, and the nodes have 256MB of memory. Communication between the nodes consists of fast Ethernet NICs connected to a central switch. This machine was lightly loaded and well suited for development purposes. The heavier computational runs were carried out on an IBM SP. Some SP nodes have 160MHz Power2 processors and others have 4-way PowerPC processors operating at 333MHz. The Power2 nodes have 256MB memory, while the SMP nodes have 512MB. The SP machine is one of the main computational resources at PDC and is therefore heavily loaded and not as well suited for development work.

4 Implementation

In our program, we make use of a wide range of specialized software tools. Aside from the CAVE library and the rendering libraries, which are tied to the SGI machine, these tools are freely available and quite portable. The extensibility and openness of the tools we used simplified the integration process and also adds to the extensibility of the final code.

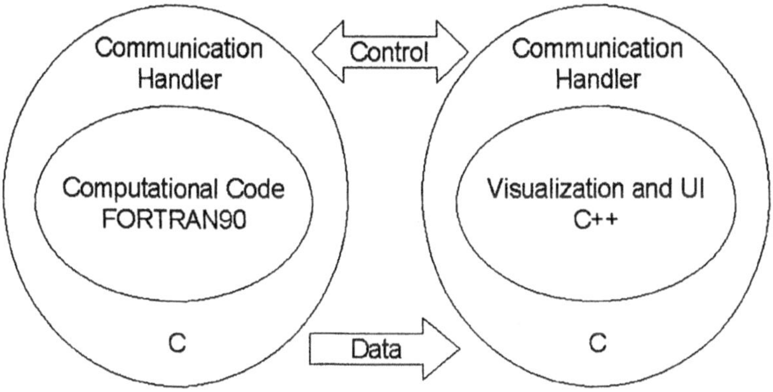

Fig. 2. Program structure.

4.1 Program Structure

The choice of different tools for each aspect of the project resulted in the need to program in several different languages. As one can see in Fig. 2, there are three main parts of the code (calculation, visualization, and communication) and three respective languages (Fortran90, C++, and C). The function-calling interface between C and C++ is similarly supported by most compilers, after one takes steps to avoid C++ name mangling. Unfortunately, this is not the case with C and Fortran; the few compilers we tested each used different naming conventions.

When assembling requirements for our software, we identified a visualization hierarchy, ranging from concurrent computation and visualization to pure visualization of a static dataset. We begin with fully interactive computation and visualization as shown in Fig. 3. At this level the control data, consisting of user input and information about the state of the computational code, are exchanged between the visualization and computation hardware. These data are asynchronous and low volume. The computed data consist of much more information and move only from the calculating machine to the visualization machine.

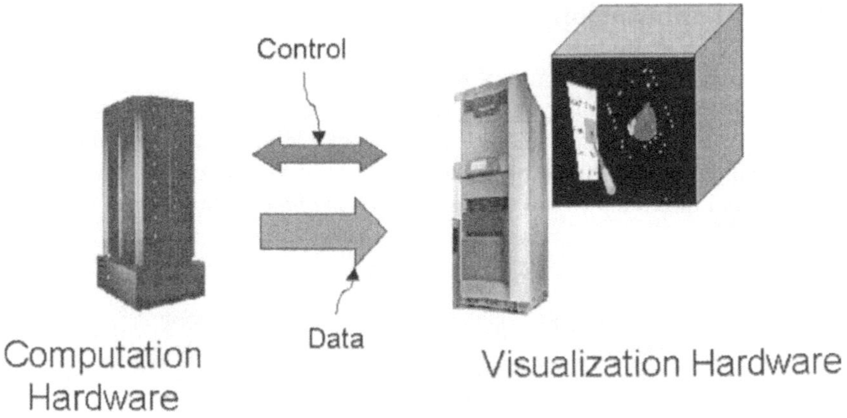

Fig. 3. Computational steering.

The final level of this visualization hierarchy can be seen in Fig. 4. This is simply post-processing data visualization. After a computational run has been completed, the results are saved to disk, and visualization work can begin on the dataset. In this case the user does not have interactive control of the computation, but the calculations might be so demanding that interactivity is not a reasonable possibility.

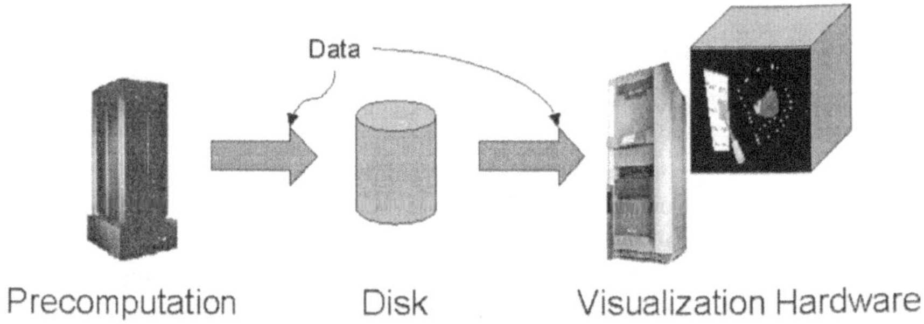

Fig. 4. Interactive visualization.

4.2 Visualization

The data flow structure of our visualization code is shown in Fig. 5. The initial visualization efforts were put into a stand-alone visualization program.

The program needed to function in 3D environments and handle generic volume data as well as the geometries used by the FDTD code. To make this prototype implementation as simple as possible, it read data files written in VTK internal formats.

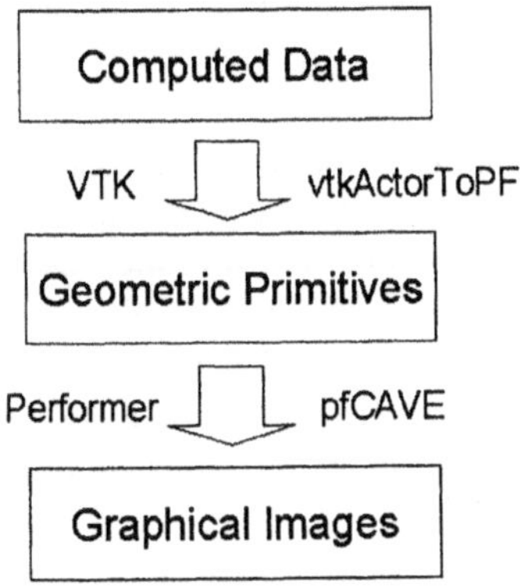

Fig. 5. Visualization structure.

This initial work developed into a simple viewer/interactor, where one could write a VTK pipeline and have the resulting geometry viewable in the Cube. The code became very useful later on as a test bed for viewing new visualization features. We are only making use of the data-processing capabilities of VTK. The products of our VTK pipeline, `vtkActors`, are translated into Performer `pfGeodes` by the `vtkActorToPF` [8] code. The actor information is translated to geodes, and by using `vtkActorToPFTranslator` the connection is updated as the actor is modified as well.

The VTK-generated structures are loaded into a Performer scene graph. Performer uses a tree structure to track the relative transformations and other connections of geometric objects. By traversing this graph, Performer can find collisions, invisible objects, and other information valuable to the final rendering of a scene.

Visualization elements. The first element is the conducting geometry. The geometry consists of a coordinate list, a connectivity list, and a value list. The connectivity information describes the ordering of coordinates into squares. Because most surfaces have two sides, there is connectivity information for each normal direction. The data values corresponding to the squares describe the surface current induced on the conductor. An example is the blue-green object in Color Plate 4 on page 287. Because most VTK filters originally handled only data per vertex and not data per cell, the data are mapped to the vertices of the square. For performance and storage reasons, these squares are then triangulated and connected into triangle strip structures. To make the current information clearer, a logarithmic color scale is used.

The volume data consist of E_x, E_y, E_z, H_x, H_y, and H_z values at each point. There are almost endless ways in which one can display these data; here, users can take advantage of VTK's versatility and add whatever objects they would want. Currently we have implemented isosurfaces, cutting planes, and a vector probe.

The isosurface is implemented using the `vtkContourFilter` class. A user can view the strength of a particular component of the electric or magnetic fields, or the total field strength. The user interface described in Sect. 4.3, can set the value at which the isosurface should be displayed. The isosurface is represented as triangle strips just like the conducting geometry. The surface is also made transparent so that it does not hide any other structures that are being displayed. For example, the transparent red object in Color Plate 4 on page 287 is an isosurface.

Cutting planes can be used to display individual field component strengths as well as total field strength. In this case, two different methods have been used. For smaller datasets, there is a general plane that the user can move with the wand. This implementation uses `vtkClipVolume` and `vtkPlane`. Unfortunately, the implicit function clipping in VTK is not a particularly fast operation. Therefore, we also created a cutting plane for the large dataset using `vtkExtractVOI`. This plane is restricted to remain orthogonal to the x, y, or z axis. Such a cutting plane is visible in Color Plate 5 on page 287.

Finally, the vector probe is used to inspect data at a given point. At the end of the wand, a vector will point in the direction of the electric and/or magnetic field. This vector is scaled by the magnitude of the vector. There is also an option for the separate component vectors to be shown. The numerical information at a point, as well as the global maximum and minimum values used in scaling the vector, come from the VTK data pipeline. The actual vector arrow is a Performer object. While this might not be the cleanest implementation, it was done mainly for performance reasons.

Presentation of geometries. VTK data consist of several parts. One part consists of the scalar values—in our case, the results of the computations. Another element is the coordinate points of the data. In the case of the free fields, these data are very structured, while the conducting geometry is not as regular. Finally, the points are organized in terms of connectivity describing cells. Cells are the basic geometric elements: points, lines, triangles, etc. VTK currently contains twelve different cell types, ranging from 1D to 3D cells. The data consisting of points, cells, and values are manipulated by VTK to be rendered on screen.

Presentation of volume data. Computational codes such as the CEM code used for this project produce data for the whole domain of computation. How to view these data in an efficient and understandable manner is an important question. While volume rendering can be defined generally as the rendering of volumetric data, it is more common to use the term in reference to specific volumetric display methods. These are the methods that directly use 3D primitives, such as tetrahedra and voxels. VTK contains `vtkVolumeRayCast` classes for volume rendering that implements ray-casting techniques. There is also some progress in using hardware-object-ordering techniques, and one can also look at SGI's Volumizer library. Unfortunately this type of visualization is very demanding and not well suited for larger volumes, especially in interactive applications. There are many methods one can use to produce the effect of true volume rendering or to present volume data in a form that is less taxing on the machine.

Less-demanding rendering methods involve the use of 2D objects to present volume data. Several transparent cutting planes placed in parallel, for example, can be combined to present information about the volume at several different levels. Isosurfaces calculated pointwise can present information similarly to such a surface generated through a ray-casting method. Due to the need for interactivity in the application, we decided to use these simpler methods to display data.

Large datasets. As one can see in Sect. 2.2, the problem size is limited by the interactivity of the computation. However, one could imagine the case where the computation produced more data than can be interactively displayed. A test of this case presented itself at a meeting of the General ElectroMagnetic Solvers (GEMS) project in June of 1999. Here, the data from a large run were to be shown in the VR-Cube. The geometry consisted of two million polygons, and the volume of computation consisted of $1260 \times 1260 \times 635$ cells.

To approximate the limits of the available hardware, we made some simple calculations. SGI specifications state a peak polygon rate for an InfiniteReality2 pipe around 13M polygons/s. This is split over two screens, leaving 6.5M polygons/s if the walls are equally loaded. In stereo, one needs a separate image per eye, which leaves 3.25M polygons/s. For the appearance of motion, one needs $\approx$ 20 frames per second, which brings us to 163K polygons/frame. Of course all things are seldom optimal, and a more reasonable count is around 100K polygons. Again, the hardware performance sets a limit for the complexity of the problem to be displayed. In this case, it is not the computational load, but rather the number of conducting objects, surfaces, and menus, that must react to the user in an immersive environment.

The raw geometric data were not suitable to be displayed in the Cube. To minimize the changes in a relatively working code, we felt the best approach was to create a dataset that was better suited for interactivity. Initially, all information not visible from the inside of the plane was clipped. This was done rather simply by including only the geometry that fit inside a cylinder with the dimensions of the cabin. In this way, the wings and tail section were removed from the dataset. The next step was removing information in the shell of the plane that faced outwards. The structure of the data produced by the CEM code is such that two identical cells are produced for each wall. These cells differ in scalar values as well as normal directions, thus producing an inside and outside wall. Taking the cross product $\bar{n} \times \bar{c}$, where $\bar{n}$ is the normal of a cell and $\bar{c}$ faces from the centroid of the plane to the cell, gives us -1 for outward-facing cells. At this point the data consisted of 300 thousand polynomials.

A useful technique when dealing with large datasets is to use multiple levels of detail (LOD). Objects that are closer to the viewer should be fully resolved, while faraway details need not be shown as clearly. Optimally one would want this loss of detail to be continuous and relatively unnoticeable to the viewer. Both Performer and VTK have facilities for the handling of LODs. The `vtkLODActor` in VTK is a simple implementation wherein an object is replaced by progressively simpler objects depending on the amount of CPU time allocated to the renderer. The default implementation starts with the full object, then steps to a point cloud, and finally to an outline box. Performer, on the other hand, has very advanced LOD capabilities, such as `pfASD` or active surface definition.

The most straightforward LOD method is similar to the one that exists in VTK. Here a `pfLOD` is created with different levels for each object. The user can then switch between levels depending on the machine load or distance from the observer. To minimize the "blinking" effect of one object being replaced by another, the user can define a blending zone. In the blend zone, both objects are displayed with varying opacity. The result is that one object ghosts into another without the complexity of `pfASD`.

For our LOD work, we chose to use the Performer facilities over those in VTK. The geometric objects were created by a VTK pipeline and then translated into `pfLODs`. The replacement threshold was set beyond the three-meter width of the Cube, the reasoning being that other users in the Cube should not be too affected by the location of the tracked glasses. Past this first threshold, the walls were replaced by the smooth rings, though the internal structures remained. These structures faded out of view after another three meters or so, with the cabin capped at both ends by projections of the cockpit and tail onto 2D surfaces. As the user approaches the cockpit, it is automatically replaced by a fully detailed version. Under these conditions, the object required around 90,000 polygons. When combined with cutting planes from the volume data, the model behaved quite well.

The volume data were clipped by a large rectangular box in order to contain the volume visible in the cabin. These data were also transposed to make the slices seen in Color Plate 5 on page 287 contiguous in memory. To color the cutting plane, we used a `vtkAtanLookupTable`, which we wrote for

VTK to highlight the variation in field strength. While constraining, these optimizations made the interactive visualization possible.

Due to our lack of access to a smooth triangulated model of the plane, we were not able to test the `pfASD` structures. It would be interesting to project the calculated data onto a nicer model and let there exist a continuous variation in the level of detail. Currently, this is being developed as parts of the software package are integrated into the GEMS project.

4.3 User Interface

The design of a user interface for a 3D environment is itself a sufficiently difficult undertaking. A great deal of research exists on the subject of interaction and manipulation of immersive applications. It is still a rather open area and as such is lacking in any form of standardization. Common sets of widgets and interaction frameworks such as those found for 2D windowing environments do not exist to the same extent.

There are two sources of information about the user in the immersive environment, a set of tracked glasses and a tracked three-button wand. The glasses can be used to follow the user's movements, and information from the glasses is used by CAVELib for perspective correction as well.

The design goal for the user interface was to keep it simple and easy to modify. The fundamental concept is borrowed from windowing user interfaces, where the user chooses to apply focus using the pointer. The wand code allows the user to carry out a picking operation on the displayed structures. The chosen structure is sent a message to activate while the previously activated object is deactivated. The activated object is now allowed to intercept information from the user. This decentralization of user interaction code greatly reduces the complexity of the core visualization code and puts no restrictions on an individual module's use of interaction information.

The wand. A schematic of the wand (Fig. 6) shows three buttons and a small joystick. In the initial code described in Sect. 4.2, the user interface was limited to toggling the display state by pressing button combinations. The joystick was used to manipulate the visualization elements, such as changing an isovalue or moving a cutting plane.

As the GUI evolved, the position and heading information that could be gathered from the wand became important. In order to give the user feedback, a 3D wand widget shown in Fig. 6 was rendered to follow the movement of the real wand in the virtual environment. The location of the tip of the wand was used for display elements, such as the vector probes and the freely moving cutting plane. In other cases, the wand is used as an infinite pointer. When pressing buttons on the menu panel, for example, only the direction of the wand is considered. The rendered image does not actually need to coincide with the menu. Another element of interaction is moving objects around in

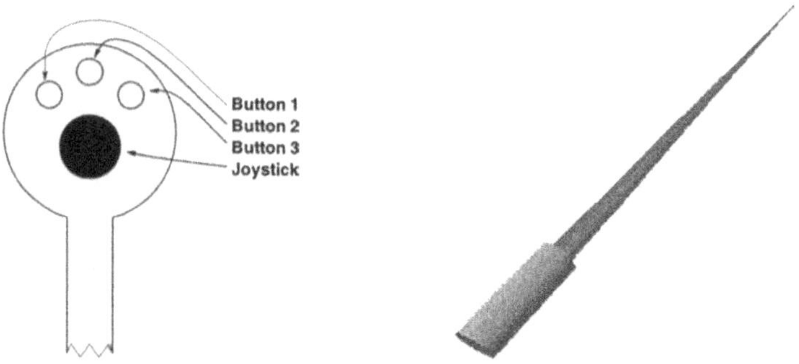

Fig. 6. Diagram of wand and wand widget.

space. We found the most natural effect to be produced by "grabbing" and moving objects relative to the wand. To grab an object, one finds the wand's transformation matrix relative to the world and combines the inverse with the object's transformation matrix. To keep from having one complex interaction code, each object handles its own movement depending on whether or not it has been selected by the user.

The menu box. A method for changing the display state became necessary when we were able to visualize more than one piece of information. As the user's choices grew, the easiest way to allow a user to keep track of the current state while also displaying other choices was to use a panel of buttons. As a model for the behavior of our menu box, we implemented a tool similar to the one found in the EnVis [3] program produced at PDC. The buttons on the floating menu simply consist of text that has been texture mapped onto the menu object. When the menu is active, the user can point at the menu and press a button on the wand. In the menu code, the pointing coordinates are used to identify which button has been pressed, and then the button object bound to this button is sent an event message. As with the wand interaction code, the menu code is greatly simplified by placing the activity routines in the object.

4.4 Communication

The communication between processors in the computational code is MPI. There are MPI libraries on all the big machines at PDC, and using MPICH-G [4] one can even use MPI over distributed resources. The information exchange between the computation and visualization code seen in Fig. 2 uses the Nexus communication library.

The thread interface routines are used to query the state of communicated information. After each time step, the computational code checks to see if it should continue. When initializing, the code checks what control information is dirty and needs to be updated. There are also flags that inform the visualization process to update itself with new data. On the visualization side, the code can access the state information directly, which removes the need for any such interface routines.

In the FDTD code. When dealing with parallel code, one has to deal with the problem of synchronization. In this project there are not only separate processes running on different machines, but also separate threads running within some processors. This situation is further complicated by the different programming languages involved, for example, Fortran90 does not have a standard threading interface. Threading exists in Fortran through the use of APIs or compiler directives; an example is OpenMP, which is used on the SMP machines described in Sect. 3.2 and Sect. 2.2. The interface with the computational code consists of small C functions wrapped around the shared memory structures.

The communication routines handle the exchange of data between different processes, as well as the allocation of memory when setting up the computation and presenting the results. When adding interactivity to the computational code, control of the program parameters was moved to the UI process. Adjusting computational parameters can be described as follows.

1. The user commits a parameter adjustment.
2. The new parameter is received by the computational process and a dirty flag is set for the parameter.
3. The Fortran code queries the states of the communication flags and halts on the dirty parameter.
4. Adjustments are made to the computational process, and calculation can resume.

For the FDTD code, the computation is usually restarted after a parameter change, especially when changing computational domains or parallel data distribution.

In the visualization code. The integration of the message handlers and the C++ visualization code was more straightforward. The UI was built with the unpredictable behavior of the user in mind, so special synchronization routines were not necessary for the control information exchange. The VTK routines were able to handle externally allocated memory, which simplified receiving the computational data, but we had to coordinate the arrival of new data with the drawing processes.

To avoid illegal memory accesses, the computational result arrays were protected by locks. This way, the communication handler threads and visualization processes could safely share the primary data without necessitating any duplication. After the VTK and Performer pipelines are reinitialized, such precautions are not needed because there are facilities for internal updates of data. The resulting steps in the visualization process when data are being received is to halt any data-processing tasks, receive the data array, and update the visualization pipeline.

5 Conclusion

A new system for steering and visualization of an FDTD code for electromagnetic simulations has been developed and implemented. The system is based on public domain communication and visualization software.

The user can interactively control computation as well as the visualization environment. The user can also display information elements, such as cutting planes, isosurfaces, and vector probes by selecting from a floating menu. This menu gives the user control over such computation elements as the domain of computation and computational cell sizes.

Even with powerful hardware consisting of an IBM SP for computation and a SGI Onyx2 for visualization, speed is still a limitation. For a fully interactive session of computation and visualization, the maximum number of discretization cells is about 3,000 running on a few SP nodes. For interactive visualization, a geometry with about 100,000 polygons is the limit for PDC's Onyx2 installation. With more complex objects, compression and levels of detail must be introduced. This has been carried out successfully on a billion-cell computation.

6 Acknowledgments

I would like to thank my advisor, Dr. Per Öster, for his guidance and knowledgeable input into the project. I would also like to thank Prof. Jesper Oppelstrup for many interesting discussions. Also, I extend my gratitude to PDC and its staff for support, ideas, and use of the systems. Continuing, I would like to thank Ulf Andersson and Gunnar Ledfelt for their help in understanding their FDTD code, as well as advice on visualization.

References

1. U. Andersson and G. Ledfelt. Large-scale FD-TD: A billion cells. *15th Annual Review of Progress in Applied Computational Electromagnetics*, 1:572–577, March 1999.
2. G. Eckel. *IRIS Performer Programmer's Guide.* Silicon Graphics, 1997.
3. J. Engström. Visualization of CFD computations. Master's thesis, Center for Parallel Computers, Royal Institute of Technology, Stockholm, Sweden, 1998.
4. I. Foster and N. Karonis. A grid-enabled MPI: Message passing in heterogeneous distributed computing systems. In *SC'98 Proceedings.* ACM Press, 1998.

5. I. Foster, C. Kesselman, R. Olson, and S. Tuecke. Nexus: An interoperability toolkit for parallel and distributed computer systems. Technical report, Mathematics and Computer Science Division, Argonne National Laboratory, 1993.
6. J. Ihrén and K. Frisch. The fully immersive cave. In *3rd International Immersive Projection Technology Workshop*, pages 59–64, May 1999.
7. D. Pape. *pfCAVE CAVE/Performer Library (CAVELib Version 2.6).* Electronic Visualization Laboratory, University of Illinois at Chicago, March 1997.
8. P. Rajlich, R. Stein, and R. Heiland. *vtkActorToPF.* `<http://hoback.ncsa.uiuc.edu/~prajlich/vtkActorToPF/>`.
9. W. Schroeder, K. Martin, and B. Lorensen. *The Visualization Toolkit: An Object-Oriented Approach To 3D Graphics.* Prentice Hall, 1997.
10. A. Taflove. *Computational Electromagnetics: The Finite Difference Time-Domain Metod.* Artech House, Norwood, MA, 1995.
11. K. S. Yee. Numerical solution of initial boundary value problems involving Maxwell's equations in isotropic media. *IEEE Trans. Antennas Propagat.*, 14(3):302–307, March 1966.

Immersive Displays for the Individual, the Group, and for Networked Collaboration

Henry Fuchs

Department of Computer Science; University of North Carolina at Chapel Hill; USA

Abstract. Despite unprecedented progress in information technology over the past quarter century, the fundamental interface between users and their personal computers remains the same as that first demonstrated by the XPARC Alto in 1973. That system was a brilliant advance over the then-current state of the art. However, its vision—Alan Kay's notion of a computer as a personal "dynamic book"—is closer to the ultimate portable computer than to the ultimate office environment. We are developing an approach based on several beliefs: 1) the user will find useful computer-generated imagery over a much larger field of view, and with many more pixels, than provided by today's computer screens; 2) users would rather work in their daily office environment that is always available than in a specialized facility that has to be shared and scheduled; 4) the size of the user's office is not likely to increase significantly in the future; 5) much of the work in the future, as today, is likely to be a mixture of solo personal work and interaction with one or two people either locally or at a distance.

Guided by these beliefs, we are developing a sequence of increasingly capable environments that we are starting to use ourselves.[1] This development began within the National Science Foundation's Science and Technology Center in Graphics and Visualization, and more recently it has been focused within the National Tele-Immersion Initiative sponsored by Advanced Network and Services. Our systems consist of multiple video projectors and cameras, most of them mounted on the ceiling, like track lights. (We also occasionally use flat panel displays.) The aim is to have every millimeter of visible surface in the office both lighted by at least one projector and observed by at least one camera. In this way, we can show a wide variety of integrated imagery onto all of the room's surfaces. We have also experimented with head tracking of the user and with stereo display (achieving left right separation either through selective polarization or time-sequential presentation), in order to deliver effective 3D imagery within this kind of immersive environment. The user should be able to view and move (2D or 3D) imagery around the office in an intuitive way that feels similar to working with "real" papers and other physical artifacts in that same office.

Although many years of work remain, we are increasingly optimistic that an immersive environment with many of these capabilities will be a very compelling and useful interface for the increasing variety of daily computer-related activities.

[1] R. Raskar, G. Welch, M. Cutts, A. Lake, L. Stesin, and H. Fuchs. The office of the future: A unified approach to image-based modeling and spatially immersive displays. In *ACM SIGGRAPH'98 Conference Proceedings*, pages 179–188, 1998.

Distributed Visualization and the Grid

Carl Kesselman

University of Southern California, Information Sciences Institute; USA

Abstract. It has been often said that the purpose of computing is insight, not numbers. However, with increases in computing capability, our growing ability to perform detailed simulations and the development of more and more sophisticated scientific instruments, the task of extracting insight from the increasing volume of data becomes increasingly difficult. As the volume of data to be visualized increases, we exceed the limits of current visualization systems to access, process, and display the data.

One proposed solution to these problems is to distribute the visualization process across the network. In doing so, we can minimize the need to send all of the data to be visualized from where it is being stored to where it is being displayed, we can exploit high-end processing resources to perform visualization processing, and we can support collaborative and interactive analysis and display. However, distributed visualization introduces a new set of problems into the visualization process. Specifically, we must now manage collections of resources and configure them to form a visualization-processing pipeline. This is exactly the set of capabilities that are provided as part of the emerging computational infrastructure known as the Grid.

In this talk we introduce the distributed visualization problem and give an overview of Grids and Grid technology. We describe how many of the problems associated with distributed visualization are addressed by Grid services, and we illustrate with examples how the Globus Grid toolkit can be used to construct a variety of distributed visualization systems.

Acceleration of a Formfactor Calculation through the Use of the 2D Tree

Sungye Kim, Hyekyung Ko, and Kyunghyun Yoon

Department of Image Engineering; Graduate School of Advanced Imaging Science, Multimedia, and Film; Chung-Ang University 221; HuksukDong, DongJakKu; Seoul; South Korea

Abstract. Radiosity is a rendering technique that requires a lot of CPU time. The formfactor calculation is the most time-consuming part of the radiosity flow. As a result, accelerating the formfactor calculation becomes an important factor in the improvement of the algorithm. In this paper, we use a geodesic dome to approximate the hemisphere; and to simplify the computation, we represent the dome with a set of points. A kD Tree has been applied to select the clipping area of the projected region. This method is simpler and faster than the hemicube approach, because it avoids time-consuming processing such as that associated with polygon clipping.

1 Introduction

One of the goals of computer graphics is to generate photo-realistic images by properly controlling the effect of light. The illumination model explains such control of light. Illumination models can be classified as either local or global. With the global illumination model, it is possible to express indirect lighting through the diffusion and reciprocal reaction of light by taking into account the characteristics of the objects that exist in the environment. Ray tracing and radiosity are two examples of global illumination models. Radiosity produces a more realistic image by capturing the diffuse reflection of light. Radiosity was introduced by Goral et al.[8] and Nishita et al.[11], and it has been studied to simulate the diffuse reflected light. Radiosity manages diffuse reflection of light in the enclosure. Consequently, soft shadows and more natural color bleeding effects are possible.

One of the advantages of radiosity is its view-independent property. Because a formfactor is a strictly geometric quantity, its value depends only on the shape and relative location of surfaces in the scene. This property enhances the speed of walkthrough because it doesn't require the formfactor to be re-calculated under different illumination conditions. The computation of formfactors is generally the most time-consuming part of the radiosity algorithm. Therefore, fast and accurate calculation of the formfactor is one of the main ways to improve the radiosity algorithm. A new algorithm to speed up the formfactor calculation is proposed in this paper, and its result is compared with those of previous algorithms.

2 Related Works

2.1 Radiosity

Radiosity, calculated in (1), is defined by a scalar value according to each position of an object[1]. The value of radiosity at position i is the sum of the emitted energy($Emission_i$) from the object, and reflected energy after the radiated energy ($Radiosity_j$) from all other objects is absorbed into the object. The total energy radiated from position i is influenced by the geometric relation between other objects. Using this model, $Formfactor_{ij}$ is computed for the emitting patch and each of the other patches in the environment.

$$Radiosity_i = Emission_i + Reflectivity_i \int_{env} Radiosity_j Formfactor_{ij} \quad (1)$$

The calculation of the radiosity matrix is needed in order to figure out the radiosity equation. There are various methods for solving the matrix. The most common one is the progressive refinement method. This method, proposed by Cohen[3], modifies the Southwell algorithm, which enables the radiosity matrix to be calculated repeatedly. The radiosity computing process is achieved when all of the sources of the scene's energy have been distributed. It is also possible to stop the radiosity process if a useful image is obtained.

2.2 Formfactor

Formfactor calculation is a very important but time-consuming step when determining a radiosity solution. Calculating the formfactors takes up to 90% of the time of a radiosity program[12]. So, it is important to calculate the formfactor more rapidly. The formfactor is the physical property that represents heat transfer between two patches. Formfactor, F_{ij}, is represented by the ratio of two energies, one of which is radiated from the Lambertian surface, E_i, the other of which is absorbed into patch, E_j. The reflectance or emittance properties of two patches don't influence the setting of the formfactor because they are determined only by the geometric relationship of the two patches. So, the formfactor between two patches is defined by (2), where the function V_{ij} represents the visibility between the patches E_i and E_j.

$$F_{dA_{ij}} = \frac{1}{A_i} \int_{A_i} \int_{A_j} \frac{\cos\theta_i \cos\theta_j}{s^2} V_{ij} dA_j dA_i \ , \quad (2)$$

where

dA_i is the differential area of the E_i
s is the distance between the two patches
dA_j is the differential area of the E_j
V_{ij} is the visibility function

2.3 Recent Research

The formfactor calculation algorithms can be divided into an analytic method [2], a ray-tracing method[17], and a projection method using hemi-space[4]. The analytic method uses a contour integral to calculate the formfactor. Goral [8] used a contour integral using the Stoke's law. However, this method can only be guaranteed for environments with no occlusion. The algorithm using the ray-tracing method[17], introduced by Wallace, uses sampling to calculate the formfactor. Although this method can be applied to curved environments, it takes a large amount of processing time. The two methods mentioned above guarantee the accuracy of the calculation, but they use a lot of processing time and are therefore not suitable to apply to a real-time application. Many different methods using the hemi-space have been suggested and each of them uses mainly the hemicube, cubic tetrahedron, and hemisphere.

To subdivide the hemisphere, Spencer[15] used the traditional spherical polar split. But he immediately rejected the method because of the wildly varying areas of the surface elements of the hemisphere. Subsequently, he found equi-area regions by using unevenly spaced values of Φ in the discretization. Foley et al.[6] used a modified method that had originally been proposed by Renka[13]. Renka's algorithm shows that the surface of a unit sphere is approximated by a triangular mesh of points based on the regular tetrahedron. This is one of the platonic solids. Renka created a discretized hemisphere by finding the midpoint of each of the geodesic arcs that correspond to the tetrahedron's edges. However, Foley used an alternative method to Renka's algorithm, because it does not allow the user to select the number of split triangles. For example, if one requires a triangle to be split into 3,000 triangles, then the specific options are either 1,024 or 4,096.

Neil Gatenby et al.[7] introduced an alternative to the hemicube algorithm that used a unit hemisphere as its view volume. To make a hemisphere, they used non-platonic tetrahedron, because it results in a less-regular distribution of areas than would have been achieved with the platonic tetrahedron suggested by Foley et al. In their algorithm, each direction formed by the discretization was considered the formfactor. They used Renka's algorithm and the method of Sillion et al.[14] to calculate the formfactor.

Eric Languenou et al.[9] suggested a method of formfactor calculation combining hemisphere and ray tracing. In their paper, the hemisphere is discretized into surface elements by the spherical coordinate, and each element corresponds to a small formfactor called Delta formfactor. This method requires the $N \times M$ intersection test, where N is the number of discretized elements of the hemisphere and M the number of elements within the environment. It fires a ray of light from the center of the hemisphere to the selected points on the surface of the hemisphere. Consequently, the number of pixels on the hemisphere greatly affects the quality of the final image. However, the accelerating method from the ray tracing must be included to improve the speed in a complex environment. Languenou et al. used the acceleration method using spatial subdivision and a subdivision tree to overcome this problem.

Akio Doi and Itoh Takayuki[5] proposed an improved version of the methods suggested by S. N. Spencer[15] and Van Wyk Jr.[16]. Spencer and Van Wyk proposed hemisphere base methods with edge subdivision. They calculated the degree of the arc between pairs of points projected onto a hemisphere base. This value was used to determine a number of intermediate points along the element's edge. However, the method using only edge subdivision failed to remove hidden surfaces. Their method subdivided patches in the scene into small triangles before projecting them onto the hemisphere. The formfactor was calculated by projecting each triangle onto the base of the hemisphere. Akio also used a parallel processor that divided the base of the hemisphere into regions and assigned a region to each processor to obtain a high performance.

3 The Formfactor Calculation

3.1 Approximating the Hemisphere Using a Geodesic Dome

The word "geodesic" (earth-divide) means to divide the shape of a sphere such as the earth into exact points. Generally, a geodesic dome is a hemisphere composed of many triangles. Various types of geodesic domes can be created according to the initial model. Tetrahedrons, octahedrons, and icosahedrons are used mainly as a base model. Figure 1 shows a geodesic dome being created with various basic units. The icosahedron approximates the sphere most closely, as shown in Fig. 1.

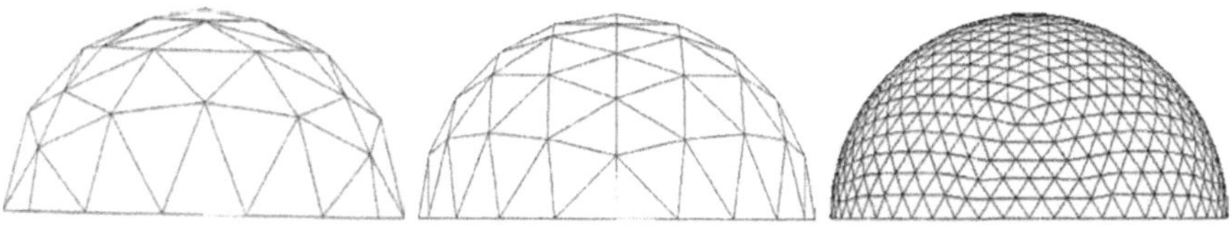

(a) by Tetrahedron (b) by Octahedron (c) by Icosahedron

Fig. 1. The geodesic domes.

3.2 kD Tree (2D Tree)

The factor k in the kD tree denotes the dimensional view of the tree. That is, if we use two dimensions, the kD tree becomes a 2D tree. A 2D tree is a binary tree that recursively subdivides the plane by horizontal and vertical cut lines[10]. As it descends down the tree, the cut lines alternate in direction. Even levels contain vertical cut lines, and odd levels contain horizontal cut lines. This suggests that a set of points on a 2D plane is a set S, and then a horizontal or vertical line that goes through a point in the S set becomes a value of a node in the tree. Figures 2a and 2b represent a subdivision by a 2D tree algorithm and tree corresponding to it. Figure 2b shows the nodes (a, d, f, j) that are points contained in a rectangle query range like Fig. 2a.

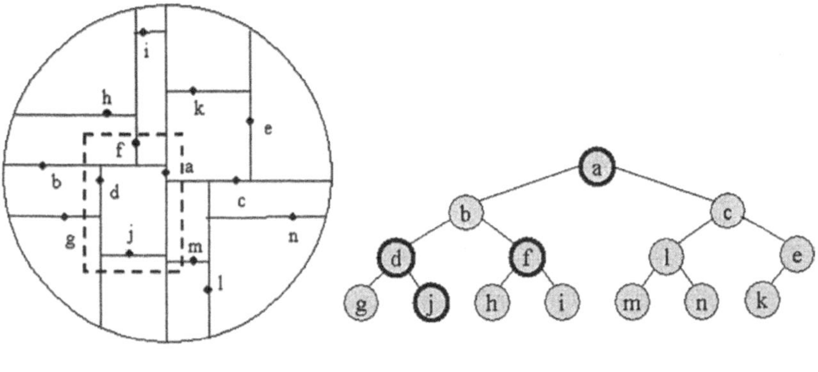

(a) subdivision of the plane (b) search of the 2D tree

Fig. 2. 2D Tree.

A 2D tree is constructed over a set of n points in $T(n)$ time, where $T(n)$ is expressed by the recurrence, (3), for suitable constants a and b. Hence $T(n) \in O(nlogn)$.

$$T(n) = \begin{cases} 2T(\frac{n}{2}) + an & if n > 1 \\ b & if n = 1 \end{cases} \tag{3}$$

3.3 Formfactor Calculation Using the Geodesic dome

In this paper, we use the icosahedron as the basic unit that forms the geodesic dome. And a geodesic subdividing algorithm generates the triangle cells that form 20 surfaces of icosahedron. New edges are made from the normalization of vectors from the center of the dome to the midpoints of three edges of each cell. The process is shown in Fig. 3. As the subdivision repeats, a dome covered with triangular cells will form close to the hemisphere.

Figure 4 shows the overall flow of the system. Initially, information about the 3D model and the geodesic dome is created. Next, we configure the 2D tree using information about the dome from the preprocessing step. And then, among the patches in the environment, we select the one with the

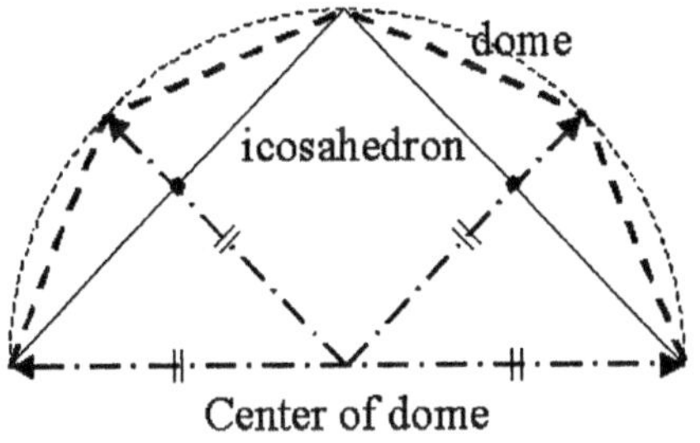

Fig. 3. Creating a dome by geodesic subdivision algorithm.

highest energy and set it as an energy-emitting patch. The formfactor is calculated between the patch chosen and all the other patches. Because we use a progressive refinement method, we can create intermediate images at each step. After each step arrives at a satisfied convergence, we can get result images.

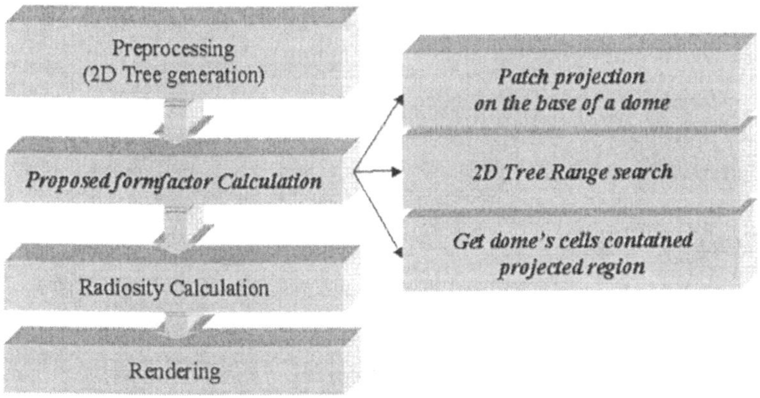

Fig. 4. System flow.

To calculate a formfactor, we use a Delta formfactor notion of a hemicube algorithm. And to simplify the clipping process, we apply a 2D tree to determine the projected region. In preprocessing, the data structure about the dome is created. The geodesic dome is moved to the center of the emitter. This emitter is a patch that has the highest energy. For calculating the formfactor of patches in the environment and the emitter, each patch projects to the center of the dome. Formfactor calculation of each dome cell in the projection region moves on to the formfactor calculation of patches in the environments and the emitter. That is, each cell of the geodesic dome corresponds to a grid cell of the hemicube. The Delta formfactor of each cell can be easily obtained by examining the cells in the projected region, because flat triangles cover the surface of the dome (see Fig. 5).

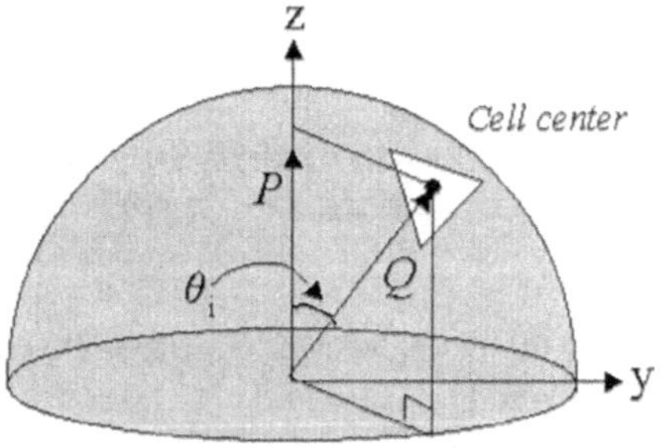

Fig. 5. Delta formfactor for cell of dome.

In this paper, the formfactor of each cell in the dome is represented in (8), which combines equations (4) through (7).

$$P = emitter.normal \tag{4}$$

$$Q = domecell.center - emitter.center \tag{5}$$

$$\cos\theta_i = \frac{P \cdot Q}{|P||Q|} \tag{6}$$

$$\cos\theta_j = \frac{(-Q) \cdot (-domecell.normal)}{|-Q|| - domecell.normal|} \approx 1 \tag{7}$$

$$\begin{aligned} F_{dE_i dE_j} &\approx \frac{\cos\theta_i \cos\theta_j}{\pi radius_{dome}{}^2 Area_{cell}} \\ &= \frac{\cos\theta_i}{\pi} Area_{cell} \\ &= \frac{1}{\pi} \frac{P \cdot Q}{|P|\,|Q|} Area_{cell} \ , \end{aligned} \tag{8}$$

where

emitter is the emitter patch radiating energy

domecell is the cell of the dome located in the projected area
P is the normal vector of the emitter
Q is the vector from the center of the emitter to the center of the dome cell

The value of $\cos\theta_j$ in equation (7) is near 1, because the angle between the normal vector of each dome cell, P, and vector Q to the center of each dome cell is almost 0 degrees. Finally, the formfactor of each cell within the dome is represented by (8). The radius of the dome, $radius_{dome}$, is 1, and the area of each cell, $Area_{cell}$, is easily determined by calculating the area of the cell.

3.4 Determining cells in the projected region

A 2D tree is used for performing a range search about the projected region from the base of the geodesic dome. A typical 2D tree search uses a rectangular area for searching, but in this paper, we applied an arbitrary quadrilateral to search a query range. Points shown in Fig. 6 are represented in 2D coordinates, (x, y) by projecting the center coordinates, which is representative of each cell consisting of a surface that is on the base of the dome.

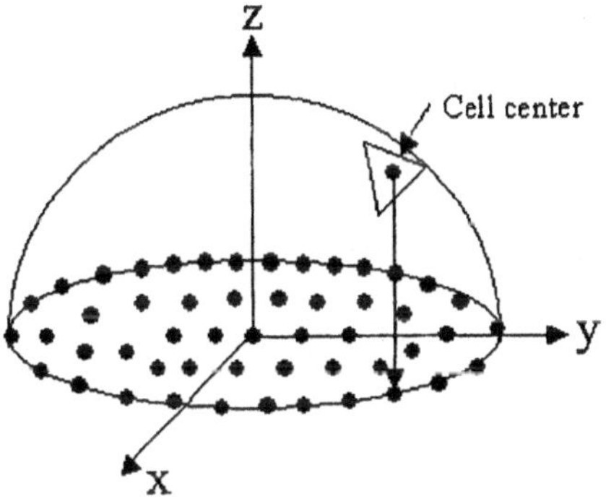

Fig. 6. Dome's 2D representation.

When an arbitrarily shaped polygon, as shown in Fig. 7a, is given as a query range, a minimum-sized bounding rectangle surrounding the polygon is created (see Fig. 7b). Candidate points for the range are selected with the typical 2D tree search method based on the bounding rectangle (see Fig. 7b). As shown in Fig. 7c, we select points included in a projection field of real elements among all of the selected candidate points.

Figure 8 summarizes the total process. Using the information about the cell in Fig. 7c, we can select dome surface cells as shown in Fig. 8b. Finally, selected cells on the dome surface are included in an inner range created by

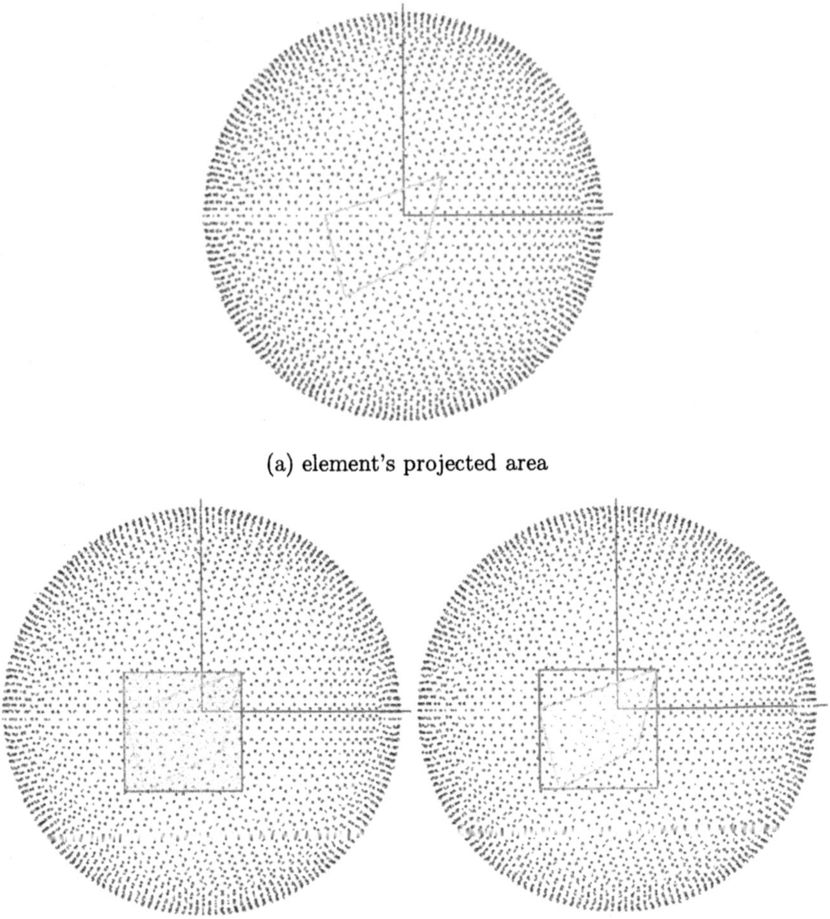

(a) element's projected area

(b) searching for bounding rectangle, (c) searching for element's projected area

Fig. 7. Determination points contained in a projected area.

the first projection, labeled ①, in Fig. 8a. In conclusion, we determine the total sum of the Delta formfactor of these cells between the emitter and the element.

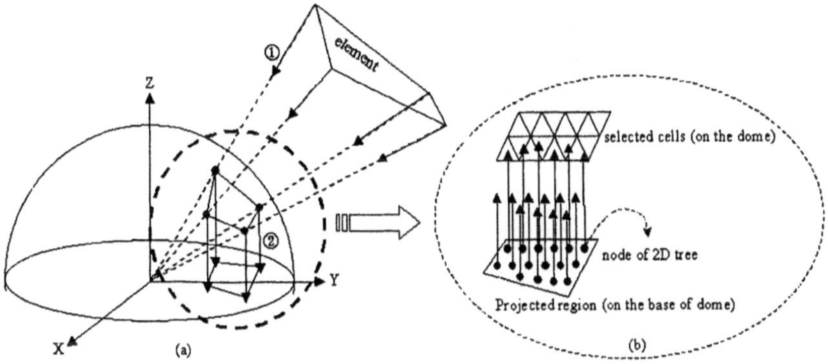

Fig. 8. Determination dome cells contained in a projected area.

4 Implementation & Conclusion

Color Plate 6 on page 288 and Color Plate 7 on page 288 show the images generated by using the proposed method. To evaluate the results, we measured elapsed time between the hemicube algorithm and the proposed geodesic algorithm according to the model in Table 1.

Table 1. Information of test models.

Models	Instance	Surface	Patch	Element
floor and wall	3	3	51	51
cornell room	7	7	127	151
cornell box	9	10	137	191
bedroom	10	18	193	681
TV and sofa	11	15	237	806
office	13	36	340	848
meeting room	12	29	389	945

Figure 9 shows the comparative results over an elapsed time period for hemicube and geodesic methods, using test models. The resolution of the hemicube is 400. In Fig. 9, the horizontal axis shows the increasing complexity of the models, and the vertical axis indicates the elapsed time for the test. The graph shows that the geodesic method consumes less time than the hemicube method. With convergence obtained through processing of the hemicube method, whose iteration step is 100, we measured the elapsed time of a proposed algorithm until the distribution of energy comes close to convergence.

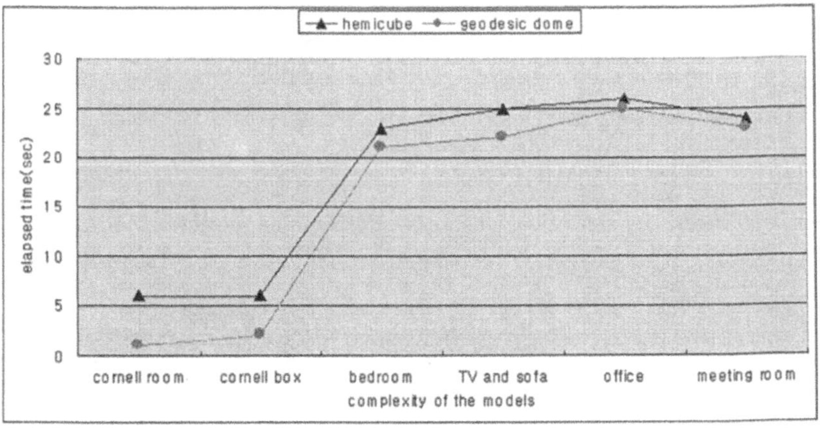

Fig. 9. Elapsed time comparison for each model.

The current radiosity rendering system demands fast processing of complex environments. To meet such demands, a fast calculation of the formfactor is required. The method described in this paper calculates the formfactor by projecting the elements in an environment on the dome. The dome is composed of many triangular cells that are almost equal in area. The formfactors of the elements can be obtained by determining the cells within the region of the projected element. We used a 2D search tree for selecting the cells of the dome contained in the projection area of each element. With further research, geodesic adaptive subdivision algorithms can be applied to create a dome for the complex environment.

Acknowledgments

This research was supported by the Chung-Ang University special research Grants in 1996.

References

1. I. Ashdown. *Radiosity: A Programmer's Perspective.* John Wiley & Sons, Inc., 1994.
2. D. R. Baum, H. E. Rushmeier, and J. M. Winget. Improving radiosity solutions through the use of analytically determined form-factors. *Computer Graphics*, 23(3):325–334, 1989.
3. M. Cohen, S. E. Chen, J. R. Wallace, and D. P. Greenberg. A progressive refinement approach to fast radiosity image generation. *Computer Graphics (SIGGRAPH '88 Proceedings)*, 22(4):75–84, August 1988.
4. M. F. Cohen and D. P. Greenberg. The hemicube radiosity solution for complex environments. *Computer Graphics (SIGGRAPH '85 Proceedings)*, 19(3):31–40, August 1985.
5. A. Doi and I. Takayuki. Acceleration radiosity solutions through the use of hemisphere-base formfactor calculation. *The Journal of Visualization and Computer Animation*, 9(1):3–15, 1998.
6. T. A. Foley, D. A. Lane, G. M. Nielson, and R. Ramaraj. Visualizing functions over a sphere. *IEEE Computer Graphics & Application*, pages 32–40, January 1990.
7. N. Gatenby and T. Hewitt. Radiosity in computer graphics: A proposed alternative to the hemi-cube algorithm. *Second Eurographics Workshop on Rendering*, April 1991.
8. C. M. Goral, K. E. Torrance, D. P. Greenberg, and Bennett Battaile. Modelling the interaction of light between diffuse surfaces. *Computer Graphics (SIGGRAPH '84 Proceedings)*, 18(3):212–222, July 1984.
9. E. Languenou et al. An adaptive discretization method for radiosity. *Eurographics*, 11(3):C–205–216, 1992.
10. M. J. Laszlo. *Computational Geometry and Computer Graphics in C++*. Prentice Hall, 1996.
11. T. Nishita and E. Nakamae. Continuous tone representation of three-dimensional objects taking account of shadows and interreflection. *Computer Graphics (SIGGRAPH '85 Proceedings)*, 19(3):23–30, July 1985.
12. G. Pietrek. Fast calculation of accurate formfactors. *Fourth Eurographics Workshop on Rendering*, pages 201–220, June 1993.
13. R. J. Renka. Interpolation of data on the surface of a sphere. *ACM Transactions on Mathematical Software*, 10(4):417–436, December 1984.
14. F. Sillion and C. Puech. A general two-pass method intgerating specular and diffuse reflection. *ACM SIGGRAPH Computer Graphics*, 23(3):335–344, July 1989.
15. S. N. Spencer. *Photorealism in Computer Graphics*, chapter The hemisphere radiosity method: A tale of two algorithms, pages 127–135. Springer-Verlag, 1992.
16. G. C. Van Wyk, Jr. *A geometry-based insolation model for computer-aided design.* PhD thesis, The University of Michigan, 1988.
17. J. R. Wallace, Elmquist, K. A., and E. A. Haines. A ray tracing algorithm for progressive radiosity. *Computer Graphics*, 23(3):315–324, 1989.

Applications of Volume Rendering in the CAVE

Anton H. J. Koning

Academic Computing Services Amsterdam (SARA); Amsterdam, The Netherlands; `<anton@sara.nl>`

Abstract. This paper presents the work that has been carried out at SARA on various projects that use volume rendering techniques to explore 3D (i.e., volumetric) datasets in the CAVE virtual reality environment. The stereoscopic, large-format images of the CAVE offer insights into the data that a normal rendering viewed on a monitor does not. After evaluating the few existing volume rendering programs available for the CAVE, we concluded that we needed to develop our own software, if we wanted an up-to-date, robust, and extensible application. Because our CAVE is driven by a high-end Silicon Graphics computer, the use of a (hardware assisted) 3D texture-mapping algorithm is the most obvious choice. With this algorithm, interactive framerates can be obtained, which is essential for use in a VR environment. We have used the resulting software, dubbed CAVORE for CAve VOlume REndering, to explore both medical and biological datasets and are extending it with new options as required by the ongoing projects.

1 Introduction

Volume rendering [3] is the preferred visualization algorithm for many types of 3D data sets, when it is not possible or not desirable to extract polygonized surfaces from the data. This volumetric data can come from a variety of fields, including medical (CT, MRI, SPECT, PET), geological (seismic), and biological (CLSM, EM-tomography). The size of these data sets varies from relatively small (e.g., $64 \times 64 \times 32$) to huge (e.g., $2048 \times 2048 \times 1024$). Usually the data consist of a single scalar value per voxel, but vector and tensor datasets are also possible.

In the medical field it is still common for a radiologist to review 3D datasets in a slice-by-slice manner, by printing them on film and viewing them on a lightbox. While radiologists are trained in mentally reconstructing the 3D structures, surgeons, radiotherapists and other medical professionals who might need to look at the images are not. Especially in complicated cases, renderings of 3D views of the relevant structures can greatly enhance their understanding. Nowadays medical workstations are frequently used to review datasets and these often incorporate some form of volume rendering, but these workstations cannot be used to drive VR equipment. In addition, hospitals are still a long way away from having their own VR facility, like a CAVE or an ImmersaDesk. This means that the use of VR is still limited to research projects and to education.

In the Virtual Radiology Explorer (VRE) project, on-site (i.e., in the hospital) display of volumes using VR techniques is combined with remote (i.e., at SARA) data storage and image processing on a supercomputer. In this project, SARA collaborates with the Leiden University Medical Centre and IBM, among others. Various datasets (head, torso, sinus) in different modalities (MRI, MRA, and CT) were visualized with a CAVORE-derived application in the CAVE and on an ImmersaDesk.

Another project in this field has already been defined and will explore the possibilities of using the CAVE to give trainee-neurosurgeons better insight into the structures of the central nervous system. In addition, the usability to visualize cerebral aneurysms will be investigated. This project is a collaboration between SARA and the Academic Medical Centre in Amsterdam.

The oil industry is another field in which volume rendering techniques are frequently used to explore seismic datasets. Several oil companies have invested in various types of VR facilities. A Norwegian company called Norsk Hydro, for which SARA did the software configuration of their CAVE, developed their own volume rendering software. While we have a demonstration of a seismic volume rendering in our CAVE, we are not currently involved in any projects specifically aimed at this field.

The first request we got for a project involving volume rendering was from the Department of Electron Microscopy of the University of Amsterdam. They were interested in viewing Confocal Laser Scanning Microscopy (CLSM) datasets. Two different kinds of datasets were used. The first were a number of images of yeast cells, in which the peroxisomes–organelles involved in the fat metabolism of the cells–were stained. The research question was whether the peroxisomes were recycled or remade. The second series of datasets were of the chromosomes in the cell nucleus of rat testes, during different phases of cell division. In this case the researchers were interested in how a cell manages to avoid knots in the DNA strands when they are condensed into chromosomes prior to cell division.

Although some new insights were gained from the volume visualization in our CAVE, in both cases nothing could be proven conclusively. One of the problems with the CLSM datasets appeared to be the limited resolution in the vertical (or Z) direction, which led to artifacts in the visualization, even though the Huygens image restoration system [12] was used to remove blur and noise from the images. Nevertheless the results were impressive enough for one of the researchers involved to use the CAVE for one of his other projects, titled "Chromosomal structures involved in genetic recombination."

The datasets for this project are Electron Microscopy Tomographs of the synaptonemal complex. This is a sub-cellular structure that is thought to act as a "zipper" during recombination of genetic material during cell division. The datasets have been characterized by one of the researchers involved as resembling "chicken shit," and this person was rather skeptical about using the CAVE to visualize the data. We were told that SARA had done some previous work, trying to visualize the datasets, but with disappointing results.

Upon further investigation, it became clear that surface extraction techniques were attempted, but they failed to yield anything usable. As we shall show later, interactive exploration with volume rendering techniques in the CAVE actually allowed a researcher with very little computer experience to gain insight into the 3D structure of the synaptonemal complex in just a few hours.

In the next section, we will present the application we developed to allow us to explore these various datasets in the CAVE. After that, we will look at the separate projects and review the results. Finally, we will discuss the work so far and look into the future for further developments.

2 Volume Rendering for Virtual Environments

A number of rendering algorithms to do volume visualization exist. Ray casting [8], where rays are traced from the eyepoint through the volume, is the most frequently used algorithm, because it allows many optimizations [10,9] and because all rendering (e.g., shading) parameters can be specified at run time. Other algorithms, like splatting [13], shear-warping [7], and 3D texture mapping [2,11,14], require preprocessing of the data to change most of the rendering parameters. However, for use in a VR environment, high frame-rates are essential to allow a user to interactively explore the data, and ray casting is computationally too expensive to achieve these frame-rates without the use of either special-purpose hardware or a large parallel computer [6].

Because a CAVE requires four high-resolution images to be rendered at a minimum of ten frames per second, the eight-processor Onyx2 we use to drive our CAVE is not powerful enough to use a ray-casting algorithm. Fortunately, the four Infinite Reality graphics pipelines implement 3D texture mapping in hardware, which allows us to use 3D texture-mapping algorithms to do real-time volume rendering at the price of sacrificing some flexibility. Both of the existing CAVE-capable volume rendering programs (Crumbs [1] and VRen [4]) use this approach.

Three-dimensional texture mapping allows a rectilinear volume to be mapped onto polygons that are rendered to the display. By positioning the bounding box of the volume in the 3D world and slicing it with a number of planes, a series of polygons is created. When these polygons are textured with the volume data and drawn partially transparent, back-to-front, with alpha blending enabled, the resulting image is a volume rendering.

Recently SGI released a product called OpenGL Volumizer, a C++ library containing a number of classes and functions that implement this process. When we started this project, OpenGL Volumizer was not yet available, so we had to implement our own functions, based on the many examples available. Currently we are in the process of adapting the software to use Volumizer.

Because the CAVE library takes care of the necessary transformations, it was relatively simple to put up an image in the CAVE. Unlike many other applications though, volume visualization is a highly interactive process, which requires a great deal of user input to be used effectively. The default interaction device in the CAVE is the EVL wand, developed at the Electronic Visualization Laboratory, at the University of Illinois at Chicago. This wand has three buttons and a joystick, as well as six tracked degrees of freedom. While this offers enough possibilities to implement simple forms of interaction in review and navigation applications, our application needed some form of graphical user interface (GUI).

For SARANAV, a Performer-based CAVE application, we developed an object-oriented menu system, which the user can operate with the wand. We ported this menu system to OpenGL and used it to add a menu to the volume rendering application (Fig. 1). The menu systems offers the programmer a choice of four types of menu items: action items, toggle items, value items, and radio items. Action items only cause a callback function to be called when they are selected and have no inherent state. Toggle items, in addition to the callback, have an on/off state, which is shown by a red bulb. Value items have a value associated with them, which can be incremented and decremented, in small and large steps, as implemented by the callback. Finally, radio items consist of a list of options, one of which is always selected, and the callback is called when a new item is chosen.

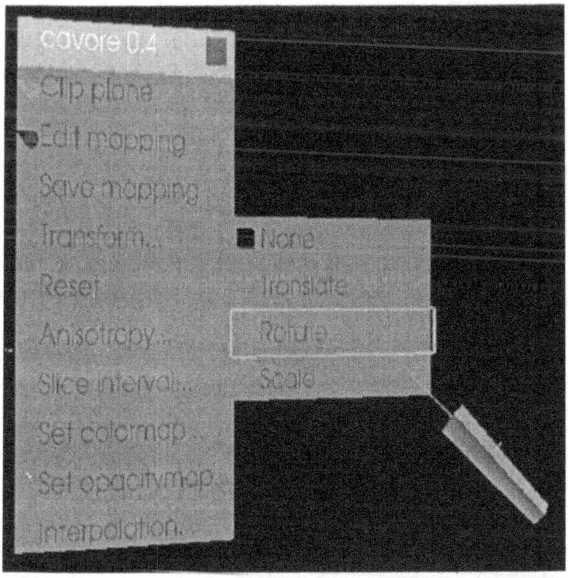

Fig. 1. The menu system.

While the drop-down menu is traditionally associated with 2D GUIs, we found that it also works well in a virtual environment. Purists might argue that the virtual environment is a representation of the real world and should therefore have a more "realistic" interaction model. While this might be true if you are trying to do something virtually that has a real-world equivalent (e.g., operating a car or machine), it is very difficult to present something intuitive for actions that don't have a real-world equivalent. Because this is the case for most interactions with the volume rendering and because many people are familiar with a 2D GUI, the menu system works well. It also presents the possible operations in clear English (or any other language), instead of using icons, which can be hard to interpret.

Although the menu system allows the user to change many parameters of the visualization, it cannot be used for one important volume-visualization tool: the editing of color and opacity lookup tables. These tables, which typically map the original eight-bit gray value to an RGBA (red, green, blue, alpha) value frequently need to be modified interactively to obtain a good visualization of the dataset. Because it is impractical to modify single entries, we need a way to specify and modify a so-called transfer function.

We therefore implemented a widget that allows the user to add, delete, and move control points of a function (Color Plate 8 on page 289), which at the moment is just a linear interpolation between the control points. Separate functions for R, G, B, and A allow the user full control over the color and opacity mapping. Often just a gray-to-gray mapping is required, so we implemented a separate function that sets the R, G, and B components simultaneously. The user can save these mappings for later use, as the program also allows the loading of predefined tables. This mechanism is quite powerful and easy to use. The image in the CAVE is updated continuously, which means the user gets immediate feedback when moving the control points.

3 Chromosomal Structures Involved in Genetic Recombination

The genetic material of a cell, that is the DNA molecules, is usually stored in a structured fashion in the chromosomes. The disturbance of this structure, for instance by rearrangement of the material in the chromosomes (called recombination), can have severe consequences. Many tumors are the result of this recombination. Little is known about the mechanism of recombination.

There is, however, one type of cell division in which chromosomal recombination is normal, and even essential. This is the meiosis, which is responsible for the creation of sperm cells and egg cells. During the prophase of the first meiotic division, homologous chromosomes are paired, followed by recombination by crossing-over of the genes. The chromosomal structure called the Synaptonemal Complex (SC) plays a major role in this process.

This SC consists of a number of substructures: Lateral Elements (LEs), Transversal Elements (TEs), the Central Element (CE), and the Recombination Nodule (RN). This recombination nodule is most probably the substructure of the SC that is responsible for the meiotic recombination. The RN is a little ball, measuring about 80 nm and lying between the LEs in the plane of the CE.

The researchers of the AMC study the assembly, functioning, and destruction of the SC and RN in the cell nucleus during meiosis, using a number of light and electron microscopy techniques. Normal 2D microscope images do not allow the 3D organization of these substructures to be studied, as they tend to be obscured by surrounding or interfering structures. By using EM tomography, we can study them in situ without being disturbed by extraction or isolation techniques.

The 3D datasets for this project are reconstructed from tilt series of transmission electron microscopy images. This means that a number of images are recorded at different viewing angles. The tilt series span -60 to +60 degrees at 5-degree intervals and are reconstructed using CART, or Controlled Algebraic Reconstruction Technique [5]. The reconstruction algorithm is computationally very expensive, due to the limited amount of data. The algorithm has been parallelized and is carried out off-line on the IBM RS6000/SP parallel supercomputer at SARA. It takes about an hour to do a single reconstruction on 13 nodes of the IBM.

Unlike radiologists, who are trained to mentally reconstruct the 3D aspect from the separate slices of volume datasets, biologists often have trouble interpreting this type of data. A demonstration of the volume rendering application in SARA's CAVE convinced the researchers involved to try and use it to gain better insight into these data.

As a single voxel in the dataset measures about a cubic nanometer and the datasets measure about 200 by 200 by 100 voxels, a full-screen image in the CAVE represents a magnification of fifteen million times. This magnification, combined with the stereoscopic images, made it very easy to get an idea of the spatial organization of the substructures of the SC. The most important tool in this case proved to be the interactive manipulation of the gray value and opacity transfer function.

Experimenting with the settings allowed the biologist, with little previous computer experience, to isolate and study the different substructures on his own, after just half an hour of hands-on training in the CAVE. Using CAVORE, it was discovered that the RN consists of several parts, that are connected to the LEs of the SC that lies between the paired chromosomes. It is suspected that the DNA is transferred via a number of bridges to the parts of the RN, where the recombination takes place.

Because it is common to include stereo pairs of microscope images in publications, we added an option to CAVORE that allows the user to save stereo pairs of the image in the CAVE. Figure 2 is an example of such a stereo pair, showing the recombination nodule.

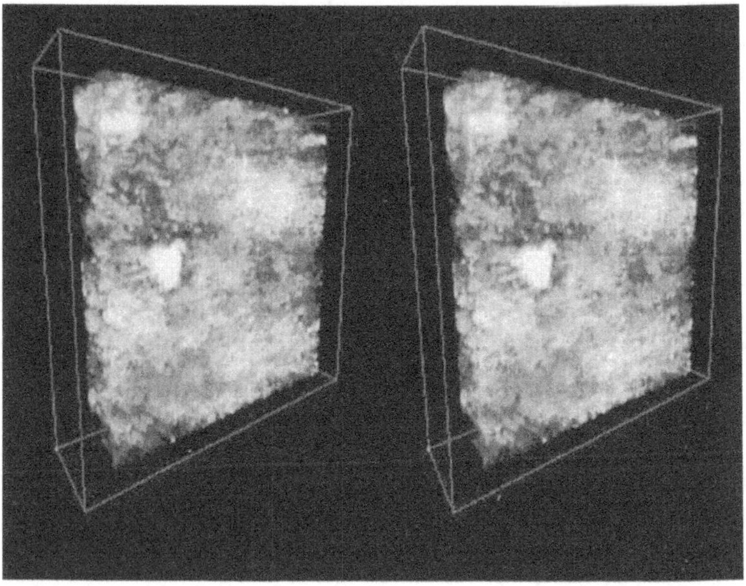

Fig. 2. A stereo pair showing the recombination nodule (RN).

4 The Virtual Radiology Explorer

The Virtual Radiology Explorer (VRE) is a package, based on existing technology previously developed separately by all parties involved in the project, that enables radiologists, radiotherapists, and surgeons to link their personal computers in the hospital to remote high-performance computing (HPC) platforms (i.e., supercomputers) via high-speed network connections. After the datasets are acquired by scanners in the hospital, they are stored in a database running on a computer at the HPC-center. This database can be a simple temporary storage solution or a fully featured Picture Archiving and Communication System (PACS).

On request, the datasets can be retrieved from the database and used in computationally expensive operations, such as image processing, data fusion, and surface extraction. These operations are carried out on an HPC platform, and the results are sent back to the hospital. There, they are presented either on a local VR system, such as an ImmersaDesk, or on the physician's PC using a simulated virtual exploration environment, enhanced with an accelerated graphics engine.

The CAVE was used as a prototyping tool in the VRE project. The reason for this is four-fold. First, we had a volume rendering tool available that allowed interactive exploration of volumetric datasets in a virtual environment. Second, the projection-based virtual environment that the CAVE provides can be seen as the future equivalent of the light box that radiologists are so familiar with. Third, it allows multiple users to investigate the same visualization together and discuss their findings with each other. Finally, SARA's CAVE is connected to several HPC platforms via high-bandwidth networks.

In mid-June of 1999, effectively three months after the start of the project, the prototype that was based on the CAVORE volume rendering package was successfully demonstrated to the partners and the financers. The tool found to be most useful by the physicians was the interactive clipping plane attached to the CAVE wand, which allowed the user to cut away a portion of the dataset at an arbitrary angle (Fig. 3).

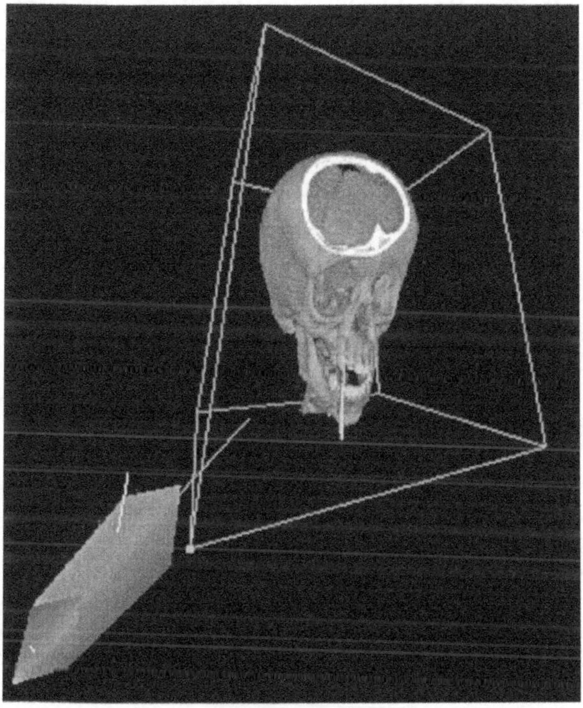

Fig. 3. Clipping plane cutting away part of the skull in a CT image.

We are currently in the process of demonstrating the next version of the software, running the visualization on-site on an SGI Octane workstation driving an ImmersaDesk semi-immersive VR display and retrieving the data from the IBM RS6000/SP parallel supercomputer at SARA. Although the next-generation Internet (Internet-2, Gigaport) has not yet been implemented in the Netherlands, we were able to use the current Surfnet connection to achieve transfer speeds of up to 70 Mbit/s. While this is enough for simple demonstrations, large-scale clinical use of these kinds of systems will require gigabits of bandwidth.

5 Conclusions and Future Work

Volume visualization in virtual environments has proved to be a valuable tool in exploring medical, geological, and biological datasets. Two tools were found to be essential in exploring the volumes: the gray-value and opacity-transfer function editor and the interactive clipping plane allow users with little or no experience with either volume visualization or virtual environments to investigate their data and make new discoveries.

Future extensions that are planned include replacement of low-level rendering routines with SGI's OpenGL Volumizer classes, implementation of parallelized pre-calculation of either gradients or lighting, multimodal visualization, interactive segmentation, measuring tools, and the combination of surface and volume visualization.

References

1. R. Brady, J. Pixton, G. Baxter, P. Moran, C. S. Potter, B. Carragher, and A. Belmont. Crumbs: A virtual environment tracking tool for biological imaging. In *IEEE Symposium on Frontiers in Biomedical Visualization Proceedings*, pages 18–25, 1995.
2. U. Cullip and U. Neumann. Accelerating volume reconstruction with 3D texture hardware. Technical Report TR93-0027, University of North Carolina, 1993.
3. R. A. Drebin, L. Carpenter, and P. Hanrahan. Volume rendering. *Computer Graphics (SIGGRAPH '88 Proceedings)*, 22(4):65–74, 1988.
4. R. Hudson. Vren: A C++ class for VR volume rendering. Final Project Repor EECS590, University of Illinois at Chicago, 1996.
5. R. Jongoo, P. N. M. Duun, J. van Marle, A. J. J. Dietrich, and C. A. Grimbergen. CART: A controlled algebraic reconstruction technique for electron microscope tomography of embedded, sectioned specimen. *Ultramicroscopy*, 76:2003–219, 1999.
6. A. H. J. Koning. *Parallel Volume Visualization*. PhD thesis, University of Utrecht, 1996.
7. P. Lacroute and M. Levoy. Fast volume rendering using a shear-warp factorization of the viewing transformation. In *Computer Graphics (SIGGRAPH '94 Proceedings)*, pages 451–458, 1994.
8. M. Levoy. Rendering of surfaces from volume data. *IEEE Computer Graphics and Applications*, 8(3):29–37, 1988.
9. M. Levoy. Efficient ray tracing of volume data. *ACM Transactions on Graphics*, 9(3):245–261, July 1990.
10. M. Levoy. Volume rendering by adaptive refinement. *The Visual Computer*, 6:2–7, 1990.
11. M. Teschner. Texture mapping. *IRIS Universe*, 29:8–11, Fall 1994.

12. H. van der Voort. *The Huygens Image Restoration System.* Scientific Volume Imaging B.V. `<http://www.svi.nl>`.
13. L. Westover. Interactive volume rendering. In *CH Volume Visualization Workshop*, pages 9–16, New York, 1989. ACM SIGGRAPH.
14. O. Wilson, A. van Gelder, and J. Wilhelms. Direct volume rendering via 3D textures. Technical Report UCSC-CRL-94-19, University of California, Santa Cruz, 1994.

Scalable Visualization of Galaxies, Oceans, and Brains

Bernard A. Pailthorpe[1,2] and Nicole Bordes[1]

[1] SDSC and NPACI; UCSD; La Jolla, CA; USA
[2] School of Physics; University of Sydney; Australia

Abstract. One of the challenges in scientific visualization is to generate software libraries suitable for the large-scale data emerging from tera-scale simulations and instruments. We describe the efforts currently under way at SDSC and NPACI to address these challenges.

The scope of the SDSC project spans data handling, graphics, visualization, and scientific application domains. Components of the research focus on the following areas: intelligent data storage, layout and handling, using an associated "Floor-Plan" (meta data); performance optimization on parallel architectures; extension of SDSC's scalable, parallel, direct volume renderer to allow perspective viewing; and interactive rendering of fractional images ("imagelets"), which facilitates the examination of large datasets.

These concepts are coordinated within a data-visualization pipeline, which operates on component data blocks sized to fit within the available computing resources. A key feature of the scheme is that the meta data, which tag the data blocks, can be propagated and applied consistently. This is possible at the disk level, in distributing the computations across parallel processors; in "imagelet" composition; and in feature tagging.

The work reflects the emerging challenges and opportunities presented by the ongoing progress in high-performance computing (HPC) and the deployment of the data, computational, and visualization Grids.

1 Introduction

Scientific visualization forms one of the cornerstones of HPC, along with computing, networking, and data archival. Two decades of sustained effort are delivering many of the promises of HPC: the realities of teraflop machines, gigabit networks, and petabyte archives are with us already. By contrast the tools, methods, and practices of visualization lag in development and in capability. As dataset sizes move into the 100 GB–1 TB realm, scalable visualization methodologies are required to allow scientists to analyze their data in a timely manner. One of the challenges of scientific visualization is to devise new methods that scale with the data as they grow in size and complexity.

This paper describes an approach to this challenge, under way at SDSC and NPACI. Scientific applications are driving this research: the data streams are generated on-the-fly from a large (teraflop-scale) simulation or instrument, or they are extracted from archival storage. A scientist undertaking the visualization generates data queries, such as selection of a region in geometric or parameter space, data fetch, data filtering, image generation, analysis, and reiteration. Scientists want to compare several datasets, or multiple derived quantities. Often, the sought-after signal is buried in background clutter, presenting further difficulties in finding features. The traditional approach, with all data in core, fails for large datasets.

This work aims to optimize the visualization queries across large volume datasets, utilizing knowledge of data layout in storage, available parallel processing architectures, and graphical displays. The motivating applications are drawn from astrophysics, planetary-scale oceanographic simulations, and human brain mapping. Visualization of large-scale data requires organized data movement as well as efficient computation. Relevant blocks of data must be imported from storage and distributed across the memory of an array of processors; and computations must be performed to yield results for analysis or images for viewing. This suggests a data-visualization processing pipeline, which we describe below.

This chapter is organized as follows: first we introduce the concept of a data-visualization pipeline; then describe its components; finally discuss the motivating applications.

2 The Data-Visualization Pipeline

Visualization of large data (i.e., larger than available core or local disk) requires intelligent orchestration of data movement as well as efficient computation and visualization. These processes can be staged, leading to the notion of a scalable visualization pipeline, as shown in Fig. 1.

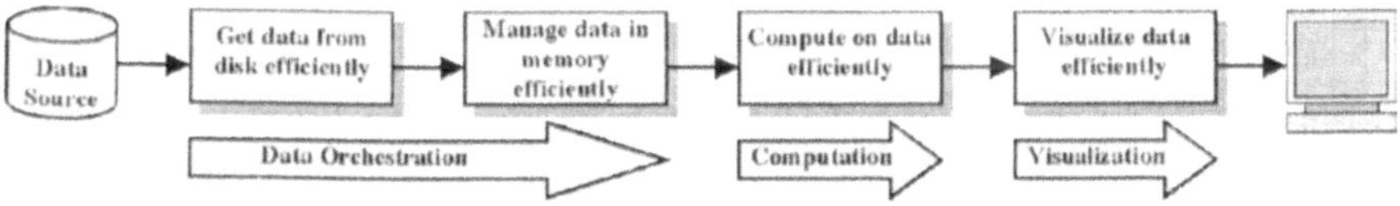

Fig. 1. Data Visualization Processing Pipeline.

- Data selection: selects one or more regions of data on a storage medium, possibly at remote archives and data collections; then restructures the data, breaking them into segments (blocks) to enable fast delivery to a processor array. This is being achieved by extensions to Saltz's Active Data Repository (ADR) and DataCutter [11,18,17].
- Data orchestration: manages the segments while in memory, distributes the data efficiently across multiple processors with appropriate load balancing, based upon current and expected usage patterns. This part of the pipeline uses Baden's Kernel Lattice Parallelization (KeLP) [8,16,2].

- Computation: uses data from multiple sources and undertakes tasks such as calculation of derived quantities, data correlation, filtering, compositing, and structure and patterns extraction. A space-time expression tree abstraction that captures an application's computational intent is being developed at SDSC. It can perform optimizations in a data-aware strategy that manages both computation and data orchestration via KeLP and ADR.
- Visualization: converts the data into displayable representations. SDSC is extending its scalable, parallel, direct volume renderer, Massively Parallel Rendering Environment, MPIRE [10] and investigating pre-processing to segment the volume data into a form suitable for feature detection and tagging.

Filling in the specific proposed technologies under development within NPACI yields the system shown in Fig. 2.

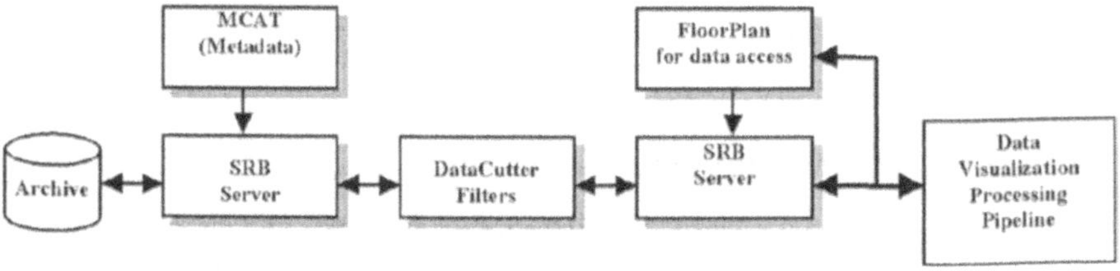

Fig. 2. Data-Handling System for the Pipeline.

The key feature here is the development and integration of these components to generate a higher-order, combined functionality that can be used for large-data visualization. The data-handling system, illustrated in Fig. 2, includes the ability to filter remote datasets to minimize network traffic. The Storage Resource Broker (SRB) [5] provides access to distributed datasets stored in heterogeneous archives, file systems, or network caches. By filtering at the data source, it is possible to interact with datasets that are larger than the local disk. The data source on the left of Fig. 3 represents interactions with that data-handling system and connects the visualization pipeline to the data collections.

3 Components of the Data-Visualization Pipeline

In the sections that follow, we discuss each stage of the visualization pipeline, from data orchestration through computation and visualization algorithms. It is an integrated research program spanning performance optimization for

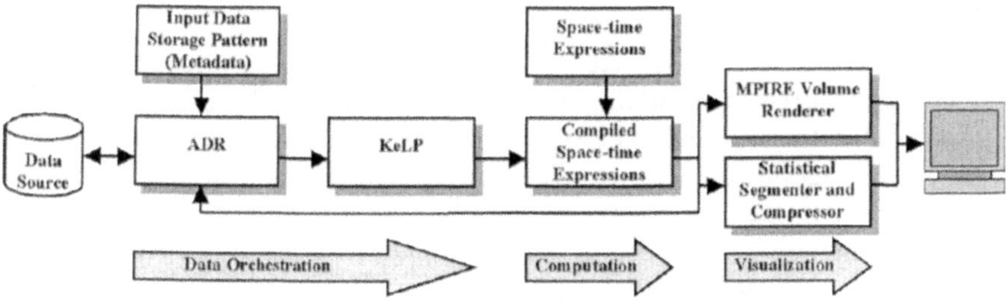

Fig. 3. NPACI Technologies for the Visualization Portion of the Pipeline.

parallel computing architectures; efficient data layout and retrieval; scalable, direct volume rendering; and feature detection and tagging.

This research leverages ongoing work by participating NPACI partners, including ADR for efficient spatial data queries and data movement between disk and memory [11,18]; KeLP, which provides efficient data movement among processor memories based upon a data "FloorPlan" [8]; space-time expression trees and their optimization; direct volume rendering with a perspective viewing model; and feature recognition and tagging.

Throughout we are motivated by scientific applications in modeling galactic and astronomical objects, planetary-scale oceanographic simulations, fluid turbulence studies, and multi-modal brain mapping.

3.1 Data Movement and Orchestration

The visualization pipeline starts with the movement of data from a storage medium into memory. Dataset sizes of megabytes-gigabytes can be read entirely in memory, however this may not be possible for 100 GB–1 TB data. Computation at this scale requires staging of the data, streaming relevant portions of data from archives into memory, distributing it among multiple processors for efficient computation, and later streaming the computed results back to storage. ADR and KeLP manage these data orchestration activities.

For both ADR and KeLP, access to data items starts with a range query. Such a query is expressed as a series of multi-dimensional bounding boxes defined within the underlying multi-dimensional attribute space of a dataset. ADR iterates over the dataset, retrieving only those data associated with coordinates that fall within a range query's bounding boxes. The query bounding boxes are generated by KeLP's geometric calculus, which decomposes irregular regions of data (see below). Through their integration, ADR and KeLP enable efficient movement of data in and out of memory in support of the computations required for visualization.

Active Data Repository (ADR) and DataCutter. Several runtime support libraries and file systems address aspects of the problem of performing efficient I/O in a parallel environment [6]. ADR [7] differs from these systems in several ways. First, ADR carries out range queries directed at both regular and irregular spatially indexed datasets. Second, computation over those datasets is an integral part of the ADR framework: it enables ADR to perform useful computations on data as they stream from disk into the system. In contrast, other systems usually delay computation until the entire I/O operation completes. A third distinguishing feature of ADR is the integration of algorithms that optimize the arrangement of data on disk to enable efficient range queries.

The basic processing steps in ADR consist of (1) retrieving input and output data items that intersect the range query, (2) mapping the coordinates of the retrieved input items to the corresponding output items, and (3) aggregating all the retrieved input items mapped to the same output data items. An intermediate accumulator data structure can hold intermediate results during processing, such as a running sum for an averaging operation.

A crucial optimization carried out by ADR is to orchestrate the flow of data between multiple disks and multiprocessor memory, determining where computations should be carried out and how intermediate values should be cached. ADR determines where the final output is computed and manages the order in which chunks of data are retrieved and processed through the system.

Planning of data movement involves two steps: tiling, and workload partitioning. In the tiling step, if a dataset is too large to fit entirely in memory, it is split up into tiles. Each tile contains multiple chunks of the data of interest, arranged so that its total size is less than the available memory. Since ADR's mapping function can map a single input data item to multiple output tiles, the streaming of data into and out of memory is not necessarily a lock-step operation. ADR arranges data movement so that the right data chunks are in memory when needed. In workload distribution, the overall performance depends on a good distribution of data across processors and avoids the high communications costs of shuffling data back and forth between processors. However, with prior knowledge of the operations to be executed on the input data, the data may be efficiently declustered, spreading it among multiple processors to avoid access bottlenecks, reducing data shuffling, and improving the overall performance. Nadeau's space-time expression mechanism ([19], cf. 3.2) provides such knowledge, indicating what operations to perform on which datasets.

In summary, scalable visualization can utilize an extended ADR by capturing the knowledge embedded in space-time expressions to drive spatial indexing for common queries and declustering [17] based upon operator usage patterns.

Kernel Lattice Parallelism (KeLP). Datasets for many problems are not presented in simple rectangular blocks but within an irregular structure. A fluid simulation of Lake Superior, for instance, produces results only for the water portions of the lake region. Since the lake is not a rectangle, the resulting dataset has an irregular structure. The manipulation of irregular data, both in-memory and out-of-core, is fundamental to all stages of the SDSC visualization pipeline. The KeLP infrastructure [8], developed by Scott Baden at UCSD, supports this kind of irregular data organization. KeLP automatically and transparently decomposes the domain into non-overlapping blocks of data that form a "FloorPlan" [22], as illustrated in panel (a) of Fig. 4.

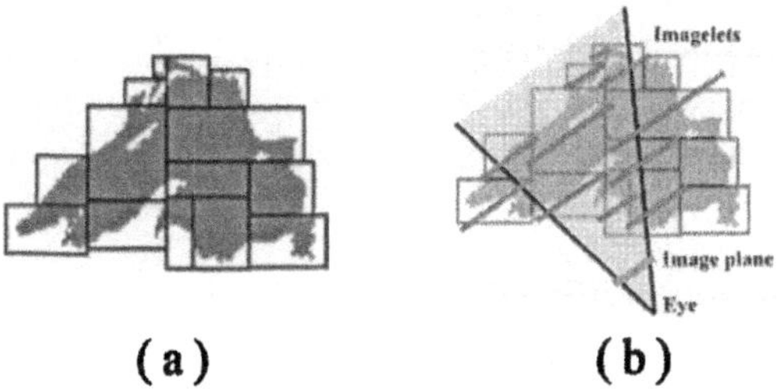

Fig. 4. (a) Typical FloorPlan of data blocks with irregular layout; and (b) image composited by rendering "imagelets" for each data block.

The FloorPlan is a meta-data type for specifying table-based data distributions, which describes irregular data layouts across distributed memories. KeLP's Mover abstraction manages the motion of data between multiprocessor memories for rectangular sections of data described by the FloorPlan. Such an abstraction has been applied to multiblock and adaptive mesh computations [8], which have two characteristics relevant to the visualization pipeline: they employ block sparse data and multi-resolution representations, or both, and hence they provide careful attention to load balancing and communication.

The KeLP FloorPlan and Mover abstractions were applied previously only to data transfers between processor memories. In this new work, the model is being generalized to provide a unified programming model across the extended memory hierarchy from disk to memory. This enables a clean programming interface for expressing out-of-core visualization algorithms. ADR and KeLP dovetail for this purpose: each operates at a different, but related, stage of the visualization pipeline.

Operation of the pipeline starts with a query for data via KeLP, which itself queries ADR to retrieve a FloorPlan of the needed data (FloorPlans are terse and can be fully read into memory). Guided by the FloorPlan, KeLP issues further range queries to ADR, pulling in the data needed by the current computation. As data results arrive and further computations are executed, KeLP works with ADR to shift the data among the processors or back to disk. KeLP's FloorPlan, then, provides the high-level plan of operations, while ADR's spatial indexing and range query abilities enable efficient streaming of data into multiprocessor memories on demand. The demand-driven semantics of this system avoid unnecessary data motion and the proliferation of large temporary files or in-memory data structures. To keep track of the orchestration of disk-memory transfers, KeLP employs both in-memory and on-disk FloorPlans.

In summary, this work is extending KeLP to support in-memory and on-disk data FloorPlans, 3D mesh-covering algorithms, non-rectangular data decomposition schemes for irregular datasets, and application controls to influence data movement and operation execution order. Additionally, science codes (e.g. Noruma's CFD code, KDISTUF [13]) have been parallelized using KeLP, ensuring closer coupling of efficient data handling and computation.

3.2 Optimization of Visualization Queries

The ability to compare and/or correlate multiple large datasets is a challenging problem in visualization. For instance, one might want to overlay one irregular dataset with another regular one, or superimpose datasets with different resolutions. These manipulations can be viewed as operators with datasets as their operands, yielding a traditional expression tree. This concept provides a simple mechanism to describe a complex set of operations dealing with arrangement, computation, and overlay of data in multi-dimensional space and time. Operators in this tree may include traditional mathematical operations or those for compositing, filtering, feature extraction, and domain-specific data computation; or operators to position, orient, and scale datasets. The resulting "data scene" is structurally akin to the scene graphs of current high-level graphics APIs, including SGI OpenInventor and Sun Java3D.

Nadeau [19] is investigating the use of space-time expressions that can be rearranged and "compiled" into a form that enables efficient parallel computation and staging of large-scale data in the visualization pipeline. By drawing analogies between expression trees used in visualization and query trees used in database systems, there is the opportunity to leverage related work in database query optimization and query processing [9,26].

Traditionally, visualization is performed within a data-flow paradigm, rather than one using expression trees. The AVS and DX (data push) and VTK (data pull) products are well-known examples. We note, however, that data-flow networks and expression trees operate on the same data structures, but in opposite directions: data flow is read from the bottom up, starting with the data source; while expressions are read from the top down, starting from operators. Viewing data flow in traditional visualization as expression trees leverages the body of knowledge about expression optimization, which might lead to more efficient yielding of implementations of the visualization pipeline.

3.3 Volume Rendering

Why Volume Visualization? The ability to explore large datasets is important because it enables scientists to investigate complex 3D structures and processes. Many methods of interacting with data volumes use geometric representations of the data, such as isosurfaces or streamlines. The use of surface-fitting algorithms have proved to be a workhorse of scientific visualization. Within NPACI, Bajaj has developed scalable, rapid iso-contouring tools[4] to extend that capability. However, isosurfaces threshold the data and might miss some features or correlations. A comparison of isosurfaces and a volume visualization of Nomura's fluid turbulence simulations (see 4.4.2 from [20]) illustrates this distinction (see Color Plate 9 on page 289).

Interactive Direct Volume Rendering. Direct volume-rendering techniques enable the generation of images from 3D data without first converting the data into intermediate geometric forms, such as isosurfaces. Current volume-rendering techniques arrange volume data into slices held entirely in memory. The slices are walked in a front-to-back order, projecting their voxels onto the image plane to build up an image. SDSC's MPIRE [10,15] provides a suite of such rendering operations, optimized for a variety of parallel architectures. MPIRE implements the ray-casting and splatting algorithms, in both orthographic and perspective viewing models. It has been applied to problem domains including medical imaging and rendering of galactic nebulae. Having demonstrated MPIRE for data sizes of 10s of GB, SDSC is now extending MPIRE to support rendering of datasets larger than 100 GB, which might be too large to fit in memory.

MPIRE can take advantage of the data blocks created by KeLP and ADR because their data decomposition provides a natural subdivision of the rendering task. This extension of MPIRE provides a data-segmented approach to volume rendering that enables the renderer to compute partial images of data blocks as they stream through the system. Each data block is projected to the image plane and rendered into a partial image, or "imagelet" for that block.

Data blocks in KeLP's FloorPlan are convex and non-overlapping, so there is a well-defined front-to-back composite order for assembling rendered imagelets into a final image. Panel (b) of Fig. 4 illustrates a view frustum intersecting multiple blocks within a KeLP FloorPlan of the Lake Superior dataset and compositing of the imagelets. In the simplest case, all n imagelets for n blocks can be rendered first, then composited in a final pass. Alternately, because the data FloorPlan is known in advance, the composite order is also known in advance; and each time two adjacent imagelets are completed, a single imagelet can be composited. For n data blocks, then, at most n/2 imagelets must be accumulated at any one time. Rendering and compositing thus occurs on-the-fly as data blocks stream through the system. Direct volume rendering in this fashion can produce imagery of large datasets, which would otherwise not fit in core.

Navigation. With large data, interaction requires some sub-setting of the data. One approach is to construct a level-of-detail tree based on a statistical comparison of the volume data. This investigation, by Bailey at SDSC, gathers adjacent voxels of the volume that have similar values within specified bounds (e.g., standard deviation) into a single region, or supervoxel [3]. This process continues recursively, building up larger supervoxels for volume regions with high continuity, while keeping supervoxels small for regions of rapid change. Supervoxels can be arranged within a binary cut tree. At each tree node, the children are supervoxels, which in turn are subdivided into further supervoxels, and so on. While a traditional quadtree partitions space based on the spatial layout of data, this binary cut tree subdivides space based on variations in the data values themselves. Ongoing work is integrating this approach into the structure of the visualization pipeline, constructed on-the-fly as data blocks stream in from ADR and KeLP, to the volume renderer. Identified features can be tagged (cf. 4.1n), with meta-data being propagated back down the pipeline to archival storage.

4 Motivating Applications

SDSC's development of scalable visualization tools is closely coupled to elements of NPACI's Science Applications Thrusts. These were chosen to span multiple distance scales–astronomical, planetary, human–thereby ensuring robust algorithms and software.

4.1 Modeling Galactic Objects

Visualization of complex astronomical data, for the newly rebuilt Hayden Planetarium at the American Museum of Natural History in New York [12], has provided a robust test of our visualization pipeline concepts. A 3D model of the Orion nebula [27], integrated with Hubble telescope imagery [21], provided the data for the visualization. The surfaces of the polygonal model were perturbed to create realistic cloudy structures and then converted to a 3D voxel dataset. Each voxel is described now by {R, G, B, opacity, emissivity}. The latter feature was essential for modeling the glowing gas clouds found in nebulae. In addition, it enables the tagging of features, at the voxel level, in complex volume datasets. This exercise demonstrated that surface features could be integrated into volume visualizations, as required in the brain-mapping application below. An Orion fly-through sequence was rendered on the SDSC teraflops computer using MPIRE, generating about 27,000 images. A two-minute animated composite image, of 7.4 million pixels, will be displayed on the seven-projector Trimension system that drives the Hayden virtual planetarium dome (see Color Plate 10 on page 290).

4.2 Oceanographic Simulations

CFD simulations on a planetary scale [1] are used to study the earth's oceans, currently at 2° resolution in latitude and longitude. These require multiple 40-hour simulation runs on the SDSC Cray-T90. Each run yields 2 GB of data, which are amenable to visualization and analysis using current tools. The model is being ported to parallel platforms, including the SDSC teraflop computer, and testing of 1° resolution oceanographic simulations is under way. In the coming year, single simulation runs are expected to yield 200 GB datasets.

A consortium of scientists (including NPACI's Stammer and Wunsch) at MIT, JPL, the Scripps Institution of Oceanography (UCSD) and the Max-Planck Institut für Meteorologie of Hamburg are undertaking a five-year National Ocean Partnership (NOPP) Program. This aims to bring ocean state estimation from its current experimental status to a practical and quasi-operational tool for studying large-scale ocean dynamics. This program will lead to observational strategies for examining the ocean's role in climate variability. The central technical goal is to provide a complete global-scale ocean state estimation over at least the fifteen-year period, 1985–2000, at 0.25° resolution, with a complete error description and regional refinements. Their model will combine all available and anticipated large-scale observational datasets for in-situ and sea surface states, including the TOPEX/POSEIDON satellite observations [24]. The dynamic general circulation model will be used to estimate the evolving 3D physical state of the full oceanic circulation and to provide a testing ground for the visualization methods discussed herein (e.g., see Color Plate 11 on page 290).

4.3 Characterization of Small-Scale Turbulence

Direct Numerical Simulation (DNS) contributes to the study of fluid turbulence by solving the exact 3D, time-dependent Navier-Stokes equations. This requires complete resolution of all relevant space and time scales, generating an impetus for larger simulations with larger output data. As the oceanic simulations move to finer resolution, visualization of turbulent features will be important there also.

DNS studies have provided evidence of distinct, small-scale structures. Regions of intense ω^2 (vorticity = curl-velocity) concentrate into tube-like or filamentary structures [20], suggesting a view of small-scale turbulence as intermittent coherent structures embedded in a random or less-ordered background field. In fact, a precise definition of what constitutes a structure is a matter of controversy and presents a visualization challenge. While efforts have been made to formalize the definition of a vortex structure [14], typically identification of vortex structures in simulation results occurs through visualization of isoscalar surfaces of ω^2 that exceed an arbitrary threshold value (cf. panel (a) of Color Plate 9 on page 289), although no formal basis for this exists. It is possible that these intense ω^2 events contribute to the overall energy dissipation in a non-local manner, either in space or time. The sample volume visualization, shown in panel (b) of Color Plate 9 on page 289, reveals more of the continuous spatial distribution of vorticity than can be seen with the usual isosurfaces (Color Plate 9 on page 289). Correlating this with other quantities (e.g., energy dissipation and strain rate) characterize the flow field might illuminate the role of "distant" energy dissipation in the turbulence mechanism.

4.4 Mapping the Human Brain

Brain imaging and mapping are generating large datasets that are pushing graphics supercomputers to their limits [25]. The Laboratory of Neuro Imaging (LONI) at UCLA is comprehensively characterizing the human brain. One of the problems is the difference between datasets acquired for the same brain. For instance, brains can be studied by complementary but different techniques: Magnetic Resonance Imaging (MRI), which can generate three different datasets, depending on the mode chosen (T1, T2, or Proton Density); Positron Emission Tomography (PET); Computed Tomography of X-rays (CT); and high-resolution cryosectioning. Each taps different aspects of brain structure or function. The inherent resolutions and orientation of each technique differ, so analyzing or compositing these datasets is not straightforward [23]. Success will aid researchers in understanding brain function and will aid clinicians in diagnosing and treating problems. The SDSC volume visualization tools are being designed to meet such challenges.

5 Conclusion

Elements of the visualization pipeline have been demonstrated already. Nomura's parallelization of her CFD code shows how KeLP works in an application domain. New simulations, with $(1024)^3$ cells, will yield 40 GB of data per time step, clearly requiring the sort of visualization methodologies introduced here. The Hayden project demonstrated that expression trees can efficiently composite multiple large volumes for rendering by MPIRE and facilitate multiple level-of-detail comparisons of volume data. Interoperability of ADR and MPIRE has been demonstrated (at SC'99) for the astrophysical data. The scalable visualization methodology introduced here provides one route to understanding the new generation of scientific data.

Acknowledgments

Undertakings of this magnitude are inevitably the work of many hands. We gratefully acknowledge the collaborations and contributions of R. Moore, J. Genetti, D. Nadeau, M. Bailey and G. Johnson at SDSC; our NPACI partners, J. Saltz, S. Baden, D. Stammer, C. Wunsch, K. Nomura and A. Toga, and their research groups; and D. Davidson.

References

1. The ATOC Group. The general circulation and heat budget of the North Pacific Ocean from satellite altimetry, acoustic tomography, and a general circulation model. *Science*, 281:1327–1332, 1998.
2. S. B. Baden and S. J. Fink. Communication overlap in multi-tier parallel algorithms. In *SC '98 Proceedings*, Orlando, FL, 1998.
3. M. J. Bailey. Feature detection in large data sets by volume binary cutting. To be published in 2000.
4. C. Bajaj, V. Pascucci, and D. Schikore. *Data Visualization Techniques*, chapter Accelerated IsoContouring of Scalar Fields. John Wiley & Sons, 1998. `<http://www.ticam.utexas.edu/CCV/>`.
5. C. Baru, R. Moore, A. Rajasekar, and M. Wan. The SDSC storage resource broker. In *CASCON'98 Conference Proceedings*, Toronto, Canada, 1998. `<http://www.npaci.edu/DICE/>`.
6. R. Bennett, K. Bryant, A. Sussman, R. Das, and J. Saltz. Jovian: A framework for optimizing parallel I/O. In *1994 Scalable Parallel Libraries Conference Proceedings*, pages 10–20. IEEE Computer Society Press, 1994.
7. R. Ferreira, T. Kurc, M. Beynon, C. Chang, A. Sussman, and J. Saltz. Object-relational queries into multi-dimensional databases with the active data repository. To appear in the *Int. J. Supercomputer Applications and High-Performance Computing* in 1999.
8. S. J. Fink, S. R. Kohn, and S. B. Baden. Efficient run-time support for irregular block-structured applications. *Journal of Parallel and Distributed Computing*, 50(1–2):61–82, 1998.
9. J. C. Freytag, D. Maier, and G. Vossen, editors. *Query Processing for Advanced Database Systems*. Morgan Kaufmann Publishers, San Mateo, CA, 1994.
10. J. Genetti and G. Johnson. *MPIRE: Massively Parallel Interactive Rendering Environment*, 1999. `<http://mpire.sdsc.edu/>`.
11. J. Gray, A. Bosworth, A. Layman, and H. Pirahesh. Data cube: A relational aggregation operator generalizing group-by, cross-tab, and sub-totals. In *1996 International Conference on Data Engineering Proceedings*, pages 152–159. IEEE Computer Society Press, 1996.
12. Hayden Planetarium. On the Web, 1999. `<http://www.amnh.org/>`.
13. J. Howe, T. Grimmett, S. B. Baden, and K. K. Nomura. Modernization of legacy application software. In *Applied Parallel Computing: Large Scale Scientific and Industrial Problems*, volume 1541 of *Lecture Notes in Computer Science*. PARA'98, Springer-Verlag, 1998.
14. J. Jeong and F. Hussain. On the identification of a vortex. *J. Fluid Mech.*, 285:69–94, 1995.
15. G. Johnson and J. Genetti. Volume rendering of large datasets on the Cray T3D. In *Cray Users Group Proceedings*, pages 155–159, 1996.
16. S. R. Kohn, J. H. Weare, E. M. G. Ong, and S. B. Baden. Software abstractions and computational issues in parallel structured adaptive mesh methods for electronic structure calculations. In *Workshop on Structured Adaptive Mesh Refinement Grid Methods*. Springer-Verlag, 1997.

17. T. Kurc, C. Chang, R. Ferreira, A. Sussman, and J. Saltz. Querying very large multi-dimensional datasets in ADR. Technical Report in SC'99 Proceedings CS-TR-4022 & UMIACS-TR-99-29, University of Maryland, Dept. Computer Science, Portland, OR, 1999.
18. B. Moon and J. H. Saltz. Scalability analysis of declustering methods for multidimensional range queries. *IEEE Transactions on Knowledge and Data Engineering*, 10(2):310–327, March–April 1998.
19. D. R. Nadeau. Space-time expressions for visualization of large-scale data. To be published in 2000.
20. K. K. Nomura and G. K. Post. The structure and dynamics of vorticity and rate of strain in incompressible homogeneous turbulence. *J. Fluid Mech.*, 377:65–97, 1998.
21. C. R. O'Dell and S. K. Wong. Hubble space telescope mapping of the Orion Nebula. *Astrophys. J.*, 111:846–855, 1996.
22. J. Rantakokko. Comparison of partitioning strategies for PDE solvers on multiblock grids. In *Applied Parallel Computing: Large Scale Scientific and Industrial Problems*, volume 1541 of *Lecture Notes in Computer Science.* PARA'98, Springer-Verlag, 1998.
23. E. M. Santori and A. W. Toga. Superpositioning of 3D neuroanatomic datasets. *J. Neurosci. Methods*, pages 187–196, 1994.
24. D. Stammer. Global characteristics of ocean variability from regional TOPEX/POSEIDON altimeter measurements. *J. Phys. Oceanogr.*, 27:1743–1769, 1997.
25. A. W. Toga and J. C. Maziotta. *Brain Mapping: The Methods.* Academic Press, 1996.
26. J. D. Ullman. *Principles of Database and Knowledge-Base Systems*, volume II, pages 633–733. Computer Science Press, 1989.
27. Z. Wen and C. R. O'Dell. A 3-D geometric model of the Orion Nebula. *Astrophys. J.*, 438:784–793, 1995.

SIM-VR: Interactive Crash Simulation

Clemens-August Thole[3], Otto Kolp[3], Hans Georg Galbas[3], Stefanos Vlachoutsis[2], Heinrich Werner[1], Jürgen Wind[4], Jan Clinckemaillie[2], and Martin Göbel[4]

[1] BMW AG; Knorrstraße 147; D-80788 München; Germany
[2] ESI SA; Rue Saarinen 20, Silic 270; F-94578 Rungis Cedex; France
[3] GMD—Institute for Algorithms and Scientific Computing (SCAI); Schloß Birlinghoven; D-53754 Sankt Augustin; Germany
[4] GMD—Institute for Media Communication (IMK); Schloß Birlinghoven; D-53754 Sankt Augustin; Germany

Abstract. Investigating car crash simulation in a 3D virtual environment, changing the structure of the car, initiating a new simulation, and immediately obtaining new results for VR analysis–such an environment would drastically enhance the productivity in the design and construction of car structures. A consortium of BMW, ESI, TAN, and the institutes SCAI and IMK of GMD investigated the current situation as part of the Esprit project TTN@GMD in the context of the SIM-VR activity. A prototype developed in this project demonstrates that such an environment can be implemented. Simulation results produced by a 128-node SGI Origin 2000 are sent over the Internet and are fed directly into a VR environment (GMD's Responsive Workbench, RWB^{TM}), which supports the analysis of the results. Simulation parameters can be modified interactively. Acceptable response times of the crash simulation could only be achieved by improving the scalability even for medium-sized models.

1 Background

Crash simulations today are the most demanding simulations in the automotive industry. Typical models can have up to 400,000 elements and can demand about four days simulation time on a 4-node SGI Origin 2000. Simulations are usually done by experts in numerical crash analysis on a concrete design made by the construction department of a car manufacturer. Interactive simulation would allow the computing engineer and the construction engineer to find the best solution in a joint meeting.

For the SIM-VR project an industrial test case was selected, which reflects the typical situation for a technical designer: to select the optimal material qualities and sheet thickness for the contradicting requirements from the areas of noise, vibration and harshness, crashworthiness and lightweight design. The model has 60,658 shell elements, takes more than 80,000 time steps, and the simulation time is about thirteen hours on four processors of a 24-node Silicon Graphics Power Challenge XL R10000 (see Fig. 1). As an important parameter to be changed interactively, the thickness of the steel sheets in any part of the model was chosen.

Fig. 1. SIM-VR Testcase.

2 The SIM-VR Prototype

A simulation environment was set up, which uses parallel PAM-CRASH (mpp-version) for simulation on the 128-node SGI Origin 2000 (195 MHz) at the University of Bergen in Norway and GMD's Responsive Workbench for visualization at GMD (a 12-node SGI Onyx2 with 4 vector pipes). The data transfer between the two locations was based on CORBA over the Internet (see Fig. 2).

Simulation parameters like material thickness can be selected and modified using the virtual environment interface at the Responsive Workbench to directly feed new parameters into the remote simulation. Simulation results are postprocessed on-the-fly and visualized in the VR environment. The most time-consuming part of the demonstration was the execution of the numerical simulation itself.

2.1 mpp PAM-CRASH

PAM-CRASH is an industrial simulation code for vehicle crashworthiness. It is a member of the PAM-SYSTEM software family developed by the ESI group and distributed by Pam System International SA. PAM-CRASH is used by many automotive companies for crash simulation. The first message-passing version of PAM-CRASH was developed by ESI during the ESPRIT EUROPORT and CAMAS projects[1], and since that time it has been substantially enhanced. It is currently available in the V1996 MPP release, which shows superior performance compared to the SMP version even on shared-memory machines.

PAM-CRASH is parallelized using grid partitioning; the data-partitioning tool is DDT of the University of Southampton. Each processor handles the computation related to its subdomain.

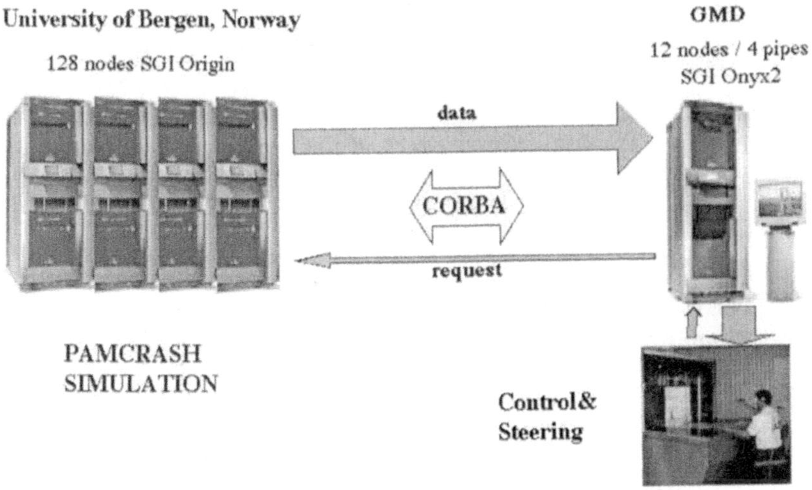

Fig. 2. Setup of the SIM-VR Prototype.

Achieving good speed-ups for crash-simulation is a big challenge in general, for the following reasons.

- Contact search requires some form of global communication.
- Contact search and treatment changes dynamically during the simulation and therefore causes load-balancing problems.

Performance benchmarks were carried out on SGI Origin machines with up to 126 nodes using an enhanced version of mpp-PAM-CRASH. The results in Fig. 3 show that a speed-up of 50 could be achieved, due to several enhancements:

- improved contact search algorithms,
- improved optimization of the SGI compiler (reducing the relative execution time of a nearly sequential part of the code),
- improved SGI message-passing libraries, and
- improved load balancing strategies.

The advanced contact search algorithms are detailed in [2]. With respect to the load balancing, overpartitioning strategies have been used as detailed in [3]. In this case, overpartitioning implies that for a 32-node execution, the computational grid is decomposed into 128 parts. Each node gets four parts assigned, each of which has different load characteristics. By combining subgrids from the rear and the front of the car, an improved load balancing for the calculation of forces and velocities, as well as for the contact search and treatment could be achieved. For a 32-node run, the overall execution time could be reduced from 4.208 seconds to 3.469 seconds.

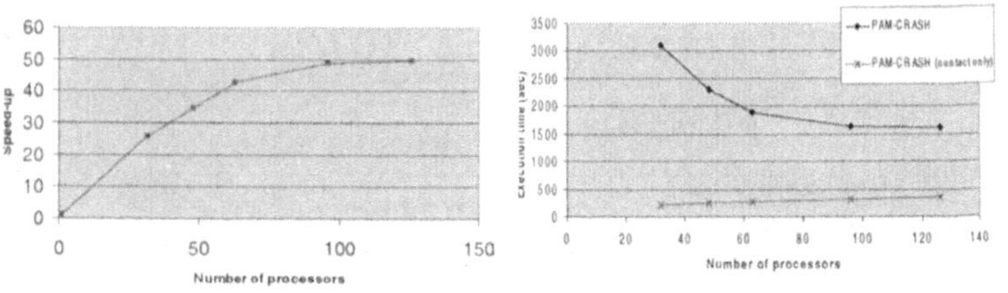

Fig. 3. Performance results for the BMW-CRASH50 testcase.

2.2 Visualization

GMD-IMK's dual-sided Responsive Workbench (RWB) [4] is specially designed for the VR visualization of engineering applications. It is commercially available from T.A.N., Düsseldorf (under the name HoloBench). GMD-IMK's AVANGO (formerly known as AVOCADO) visualization software [5] supports transparent use of different VR environments, such as walls, visionariums, CAVEs, and Workbenches. The RWB and similar VR environments are able to visualize models of this size without any need for data reduction.

A subset of the different time steps of the simulation results (about 40 steps) for a specific run are shown in an online animation on the RWB. The results of these simulation steps are immediately added to the animation sequences, as soon as they become available, and therefore the "movie" gets longer according to the progress of the simulation. Similar to a video player, it is possible to stop the visualization sequence and step forward and backwards for detailed analysis. As the RWB applies head-tracked stereo, the users may view inside the car by moving their head, or they can position the 3D visualization arbitrarily by using the novel cubic mouse input device.

Several simulation runs can be handled at the same time and a user can switch between the different sessions without any delay. The elements of the computational grid form a hierarchy of parts. Such a part represents, for example, the door of the car. The user can select such groups and move them away from the whole body for separate analysis or a better inspection of the interior of the car. A separate button on the workbench starts the PAM-VIEW analysis tool of ESI. This makes the standard analysis information on the current simulation available on a separate monitor close to the workbench.

2.3 Communication

A specific feature of the new environment is the interactive specification and start of new simulation runs. Groups of elements can be selected and material parameters for these groups can be changed. This allows, for example, a user to change the thickness of a specific part of the object. Running jobs can be terminated out of the VR-environment, and new jobs with the selected parameters can be started. The input deck for PAM-CRASH is updated automatically according to the interactive parameter selection.

The data transfer and the steering commands are exchanged between the compute server (at the University of Bergen) and the visualization server (at the GMD, St. Augustin) using the Internet. The actual implementation with the remote machine is performed via CORBA. A CORBA server at the University of Bergen receives start and stop commands. This server starts and stops the scripts, which actually initiate the parallel simulation. A continuous stream of simulation results is sent back also via the CORBA layer. A sustained data rate of 400–600 KBs between the two machines could be achieved on a regular basis. The size of the results for a specific time step of the simulation is about 2 MB. Measurements showed that the sequence of data transfer via Internet, data preparation for the visualization system, and the rendering can be performed in less than 10 seconds. Because 40–50 time steps are selected for visualization, the duration of the simulation itself is still the most time-consuming part.

3 User Evaluation

The environment was tested by engineers of BMW using the 128-node SGI at the University of Bergen. Three applications were executed in parallel using 63, 30, and 30 nodes for simulation. A single simulation took about 40 minutes. The check of simulation results, the discussion between construction engineer and simulation engineer, and the start of a new test took at least 10 minutes per run. In many cases, it was not necessary to actually run the simulation up to its end. This means that real interactive working was possible.

Current modeling techniques have been using meshes of more than 200,000 elements, and 400,000-element models are becoming more and more common. Because 100,000 elements currently seem to be a maximum for rapid interactive design iterations in this environment, it is particularly suited for the substructuring technique, in which the part of the model being analyzed is extracted from the full model, and continuity is ensured by the application of full simulation results along the boundaries of the subpart. This feature opens the usage of interactive simulation in practice. For a practical use, the environment should allow also changes in the topology of the car design, as is currently being developed within the nationally funded AUTOBENCH project.

4 Conclusion

The SIM-VR demonstrator has shown that interactive simulation can be implemented and used in industrial practice. This opens new opportunities for high-performance computing. It also opens new opportunities in the process of design of new cars. Construction engineers could test (with support of and in discussion with computing engineers) different alternatives of a design interactively, which results in a substantial reduction of development time.

Acknowledgments

The SIM-VR project was funded by the European Commission under the Esprit program as part of the TTN network. The reported work was performed jointly by the SIM-VR project partners (BMW, ESI SA, GMD-SCAI, and GMD-IMK). The partners thank SGI, in particular, Christian Tanasescu and Herve Chevanne for support and many hours of computing time, as well as Petter Bjørstad and his team at the University of Bergen for the use of their SGI platform in exclusive mode.

References

1. J. Clinckemaillie, B. Elsner, G. Lonsdale, S. Meliciani, S. Vlachoutsis, F. de Bruyne, and H. Holzner. Performance issues of the parallel PAM-CRASH code. *The International Journal of Supercomputing Applications and High-Performance Computing*, 11(1):3–11, Spring 1997.
2. B. Elsner, H. G. Galbas, B. Görg, O. Kolp, and G. Lonsdale. *Advances in Computational Structures Technology*, chapter A Parallel, Multi-Level Contact Search Algorithm in Crashworthiness Simulation. Civil-Comp Press, Edinburgh, 1996.
3. H. G. Galbas and O. Kolp. Dynamic load balancing in crashworthiness simulation. In *Vector and Parallel Processing-VECPAR'98*, volume 1573 of *Lecture Notes in Computer Science*. Springer-Verlag, 1999.
4. W. Krüger, C. A. Bohn, B. Fröhlich, H. Schüth, W. Strauss, and G. Wesche. The responsive workbench. *IEEE Computer*, July 1995.
5. H. Tramberend. AVOCADO: A distributed virtual reality framework. In *IEEE Virtual Reality Proceedings*, pages 14–21. IEEE, March 1999.

Visualization on the Grid of Virus-Host Interaction

R. Holland Cheng

Department of Biosciences Novum; Karolinska Institute; SE–141 57 Huddinge; Sweden; <rhc@csb.ki.se>

Abstract. The visualization of macromolecular structures on voxels provides the basis for insights in dynamic events in assembly. Certain macromolecules arise as well organized frameworks. Symmetry or crystalline arrays arise through a strict geometry of subunits. Single asymmetric motifs can be superimposed on themselves taking such geometric constraints into account. Biological molecules are often composed of subunits arranged in a symmetrical manner. Such conformations provide an opportunity for signal enhancement originating from the repetition of molecules in a larger configuration. Such repetitions require not only specific interactions among protein motifs, but also a built-in flexibility to accommodate conformational changes. Thus, it is advantageous to analyze critical molecular interactions with detailed protein folding information rooted in the volume density as a whole. Two-dimensional image projections can be collected in electron microscopes at temperatures of liquid nitrogen or helium. Using low-dose electron beams, viral assemblies of large lattices have been studied in vitreous ice to localize function-specific domains identified through difference imaging. Robust algorithms have been implemented based on various grid transformations to allow effective averaging of massive image data sets.

1 Introduction

Recent improvements of electron microscopes allow specimens to be cooled to the temperature of liquid helium, which eliminates thermal drift. The instrument stability during data acquisition has significantly improved the quality and resolution of the image data compared to the operation at liquid nitrogen temperatures. Compared to operating at the temperature of liquid-nitrogen, the liquid-helium temperature makes it possible to apply a 2–3 times higher electron dose to obtain better image quality revealed through centric rings in the Fourier spectrum (Fujiyoshi, Omura, and Wu, unpublished data in Fig. 1A; [8]). But, the electron dose is still somewhat limited in order to protect the specimen from radiation damage. Electron microscopes have also been improved in other ways. For instance, the use of field emission guns with their improved beam coherency has had a great impact on biological imaging. Techniques also continue to develop to achieve better stability in the specimen preparation stage.

One major disadvantage of low-dose imaging is that the unstained biological specimen gives rise to very weak electron scattering. As a result,

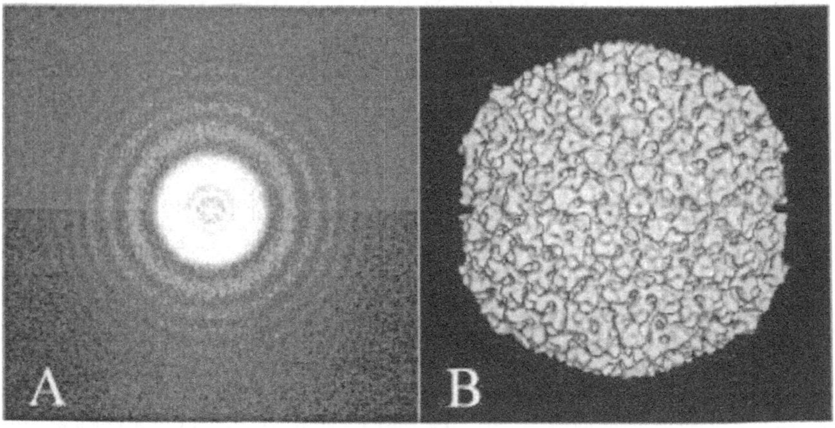

Fig. 1. A typical Fourier spectrum and the corresponding 3D reconstruction of phytoreovirus generated from data collected at a 300 kV FEG microscope.

the signal-to-noise ratio of such images is usually poor. Image enhancement through processing of very large data sets is required to achieve a useful signal-to-noise ratio. Only through the use of advanced algorithms that manages massive datasets from low-dose electron micrographs can a resolution beyond 10 Å (Fig. 1B) be achieved for large structures, such as the phytoreovirus. The use of such algorithms is also essential in understanding statistically significant dynamic events in molecular interactions at different functional states. In fact, low-dose electron microscopes and advanced image processing have significantly improved structural studies of biological macromolecules, because relatively small numbers of molecules are required compared to crystallographic studies.

2 The Projection Theorem and Image Reconstruction

Image reconstruction of transmission-electron micrographs is based on the projection theorem, where the projection image represents the scattering density of the specimen onto a plane normal to the direction of the electron beam. To determine a periodic detailed 3D structure from a set of such 2D images, Fourier transforms are made of images sampled at points of a regular Cartesian grid. Each such 2D Fourier transform is equivalent to the diffraction pattern of a central section of the 3D Fourier transform. By combining projections at different angles, the 3D Fourier space can be filled with central sections of 2D Fourier transforms, and the 3D information of the projected object restored (see Fig. 2).

Already recognized by Caspar and Klug in 1962 [2], the structural motifs can often self-assemble within a highly-ordered framework. Such symmetry information among the subunits can thus be used to reduce the images needed to fill up the 3D Fourier space. Alternatively, the symmetries can be used for averaging, similarly to what is done in crystallographic processing. The averaging is quite effective in enhancing the signal-to-noise ratio of individual motifs. However, in non-crystalline image processing, it is possible to further improve the quality of the reconstructed 3D structure by comparing the reprojections from the averaged 3D volume to the corresponding datasets.

Knowledge of 3D spatial relationships between the various parts of the specimen can only be obtained from one or more of these projected views of the object. Figure 3 shows typical steps in electron microscopy and associated digital image processing. First, a highly-optimized and homogeneous specimen is prepared before the image data—which consists of a set of 2D projections—is collected in an electron microscope operated at liquid nitrogen or helium temperature. The different projections may be collected either from a number of particles in different, but identifiable, orientations or from a single particle by tilting the goniometer stage in the microscope. With a densitometer, the micrograph of interest is digitized and converted to an array of optical density values by a computer-controlled scanner. Subsequent detection of specimen preservation during the imaging is performed by a computer in order to select images for further analysis. Fourier transforms of background-floated images are often used to determine spatial frequency during the process. Then, an iterative process of orientation determination starts to align the images, and to construct the 3D density distribution of the specimen.

3 A Robust Algorithm

Detection of rotational and translational symmetry elements through interpolation in the Fourier space can be used to uniquely determine the viewing axis. A. Klug and colleagues, at Cambridge MRC Lab, first developed Fourier image processing to analyze macromolecular assemblies. The projection theorem of Fourier transforms relates the transforms of 2D-projection sections to planar sections through the origin of the 3D Fourier space. In addition, and most meaningfully, the section direction is the same as the projection direction [9]. With sufficient image data, the quality and statistical significance of 3D reconstructions can be assessed by the self-consistency among the images to be averaged (e.g., [1]). Averaged information at points of a regular lattice provides sufficiently accurate approximations to aid the estimation of molecular mass, chemical stoichiometries, and thus the size and shape of subunits.

A robust algorithm is obviously needed in order to take advantage of the built-in repetition of features, for instance, rotational symmetry (Fig. 4). The task is to fill up the Fourier space by incorporating data from several orientations. Five parameters have to be determined for each particle in image

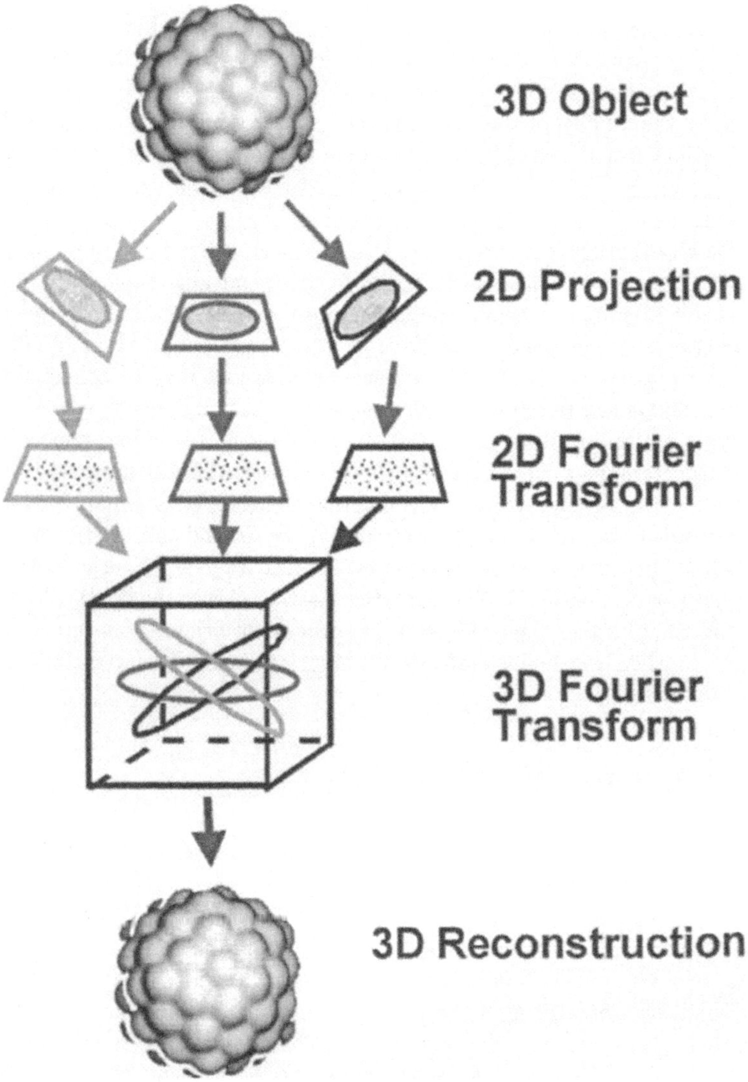

Fig. 2. A schematic illustration of the projection theorem. The image projection of a 3D object represents the density summation through the viewing axis. The Fourier transform of such a 2D image is equivalent to the crystallographic diffraction, but with added phase information. The Fourier transform of a 2D projection forms a *central* section of the 3D Fourier transform. Combining the 2D Fourier transforms of projections with different viewing angles can be made to fill up the 3D Fourier space, and thus enable 3D reconstructions.

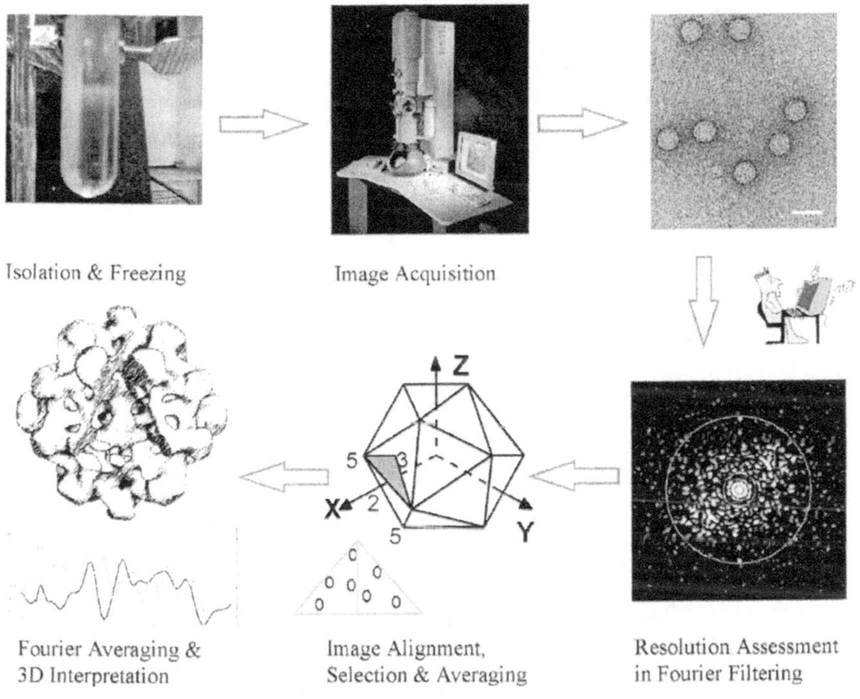

Fig. 3. Schematic representation of microscopy image gathering and reconstruction.

alignment and orientation. These are the angular parameters and the center parameters describing the spatial orientation of the image projection axis and the placement of the virus particle within the image array, respectively. Transforms of the raw image and the reference model are matched by a correlation coefficient (CC) in order to assign orientation parameters within the asymmetric unit. Such a model-based phasing approach is widely used in the structural determination of macromolecules from crystallographic and microscopic data (e.g., [12,1]).

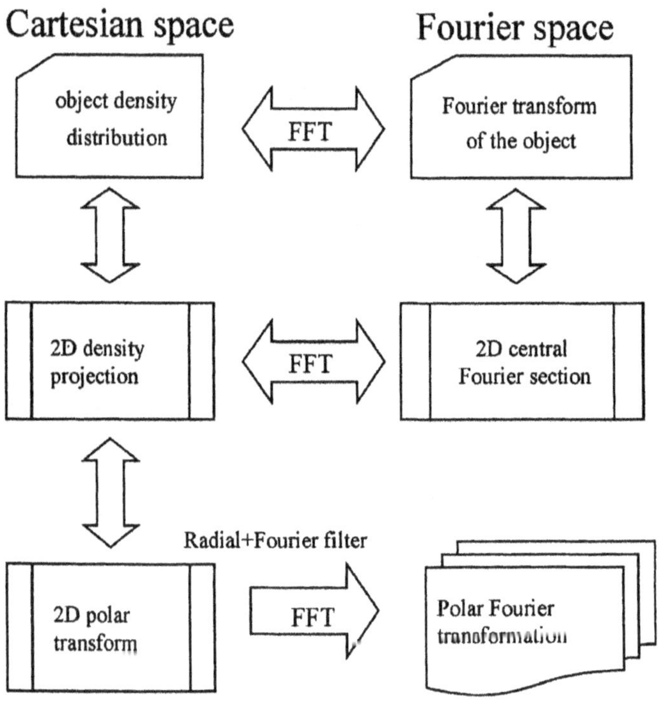

Fig. 4. A schematic drawing of the projection theorem and polar Fourier transforms. These 2D transformations use a radial filter as well as a Fourier band-pass filter in order to enhance the signal-to-noise ratio for phase alignment.

4 Phase Refinement and Validation

Once the orientations with the maximum CC are found, a figure of merit can be established through additional ranking among contour blocks with similar CC level. These measures, together, takes into consideration the orientation distribution and the correlation coefficient. An additional weighting function is constructed based on a probability assessment to facilitate the final assignment of projection directions for individual data images (Fig. 5; Wu and Bergman, unpublished data). Here, phase information based on image orientation can now be refined at smaller intervals by improving the agreement between the raw image and the re-projected model along the assigned image orientation. Intermediate reconstructions are built to substitute the initial and the subsequent models during the refinement. In addition, to be free from model bias, an averaged phase difference is calculated among the assigned orientations of images. By minimizing this phase deviation, orientation parameters of individual images are further fine tuned based on the consistency among the images to be included for reconstruction. This assesses not only the quality of individual data images, but also the consistency of the phase, between the low and the high spatial frequency, when it is extended progressively at higher resolution. Nevertheless, to reduce the noise level, the selection of effective band-pass filters, both in Fourier and Real space, is quite critical in performing a successful phase refinement. The best set of image data is combined to produce a 3D Fourier transform, and then the 3D electron density distribution through an inverse Fourier-Bessel method [7]. The reliability index [5] can be checked by the agreement among partial datasets to determine the maximal resolution.

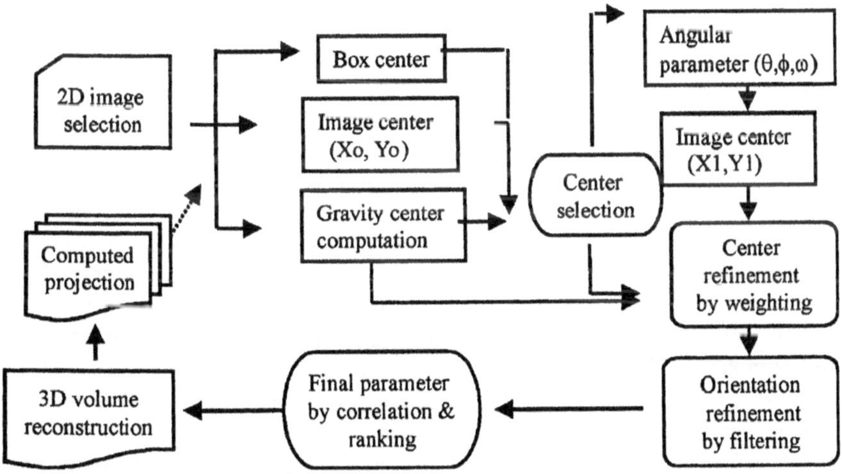

Fig. 5. Grid sampling and interpolation are implemented to search for the angular and the center parameters of the image projection axes by correlation. Weighting of the phase origin facilitates the refinement by sensitizing the angular alignment with model reconstructions.

5 Symmetry-Imposed Density Filling

In structure determination, building density models based on the symmetry-defined grid points proves to be useful for the orientation alignment of virus-like particles (Fig. 6). The particles of the human polyoma JC virus, produced in insect cells, has 72 pentameric capsomers arranged on a $T = 7d$ lattice. The morphological units of the JC virus capsid are arranged on an icosahedral surface (Fig. 6B; Kato and Xing, unpublished data). On top of a depth-cued volume, the 72 spheres are perfectly revealed according to the right-handed, $T = 7$ lattice of the Caspar-Klug theory [2]. Such 3D structures can then be reconstructed with imposed symmetry by individual single images to confirm the orientation before merging them into the set of multiple images. Subsequent models obtained during cycles of phase refinement were needed to converge to a final and correct solution. Improvement of the correlation coefficients is usually monitored as a function of spatial frequency as well radial spacing during the refinement.

6 Interpolation of Atom Scattering on the Grids

Three-dimensional structural models can be built to observe the fidelity between the microscopic image reconstruction and the atomic model-derived density of the viral capsid [3]. For finer details of residue coordinates, sequence-based information is aligned to predict the 3D structure of the subunits, because the JC capsid shares about 70 percent identity with the known atomic structure of SV40 [11]. From aligning the residues, models can be built up based on the identity or similarity of amino acids, genetic coding, and chemical property. Information of $C\alpha - C\alpha$ distance, secondary-structure conformation, as well as residue accessibility aids the success of the model construction (see Fig. 6E and Fig. 6F). To place the effective scattering amplitudes, all the non hydrogen atoms in the capsid protein are taken to produce all the equivalent atomic coordinates and temperature factors from an asymmetric unit with non-crystallographic symmetry operators [6,4].

Structure factors can be generated from the capsid protein of the JC virus out to a spatial frequency of 1/10 $Å^{-1}$. The scattering amplitudes are then derived from these calculated structure factors by Fourier synthesis to sample the scattering amplitudes at various grid sizes. An exponential damping function is usually introduced to the amplitudes of the structure factors near the resolution boundary. Similar to a temperature factor, this function eliminates the effect of Fourier ripples in the calculated density globally at lower spatial frequency for a less-constrained thermal motion. The density map is subsequently calculated by Fourier synthesis of the structure factors into cubic voxels under a defined dimension. At this stage, an empirical function to correct the microscope data can be derived to improve the correlation between the microscopic and predicted atomic data within the protein shell. A

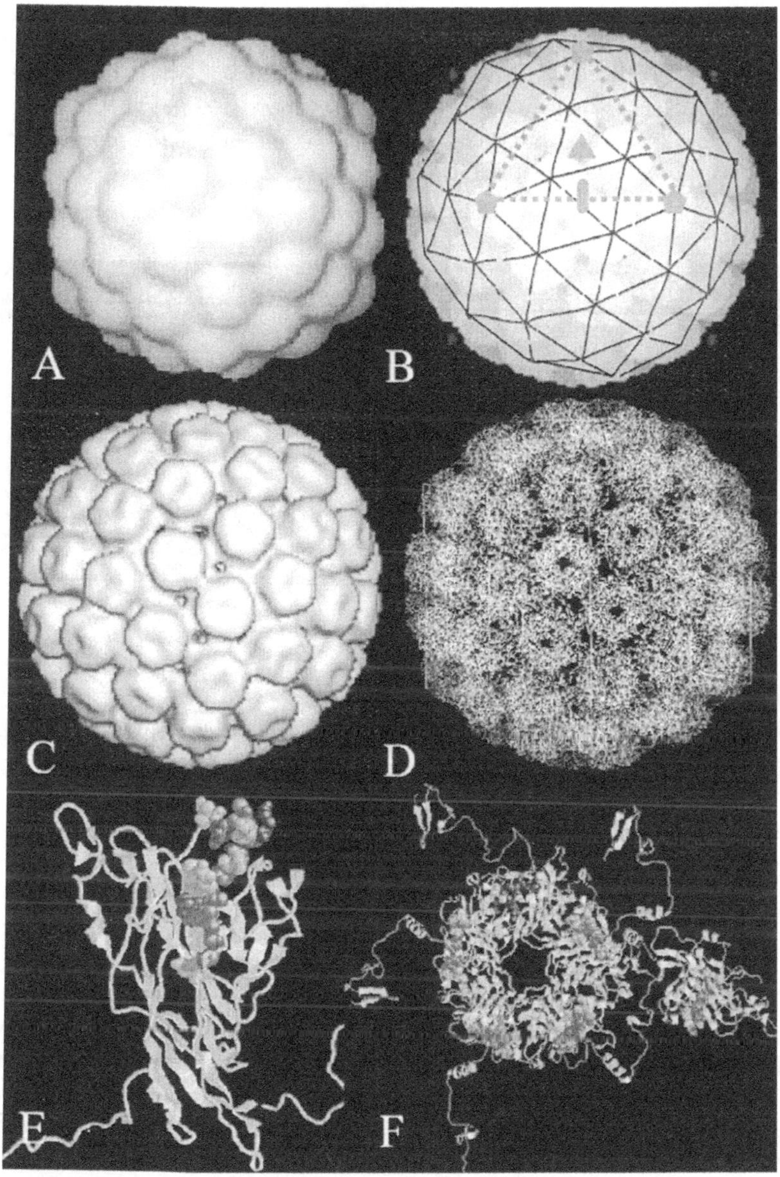

Fig. 6. The polyoma JC virus has 72 pentameric capsomers arranged on a $T = 7d$ lattice (A, B). The capsid is composed of 72 morphological units shown in both an EM map (C) and a model map (D). The essential antigenic epitope is exposed at the top region on the predicted structure, suggesting its function in the virus infection (E, F).

correction function can be simulated by a least-square fit of the Fourier amplitudes computed from the structures [3,6]. In Color Plate 12 on page 291, sequential radial shells of the density map are displayed from 210 Å to 172 Å at 3.5 Å intervals viewed from the twofold axis. Based on the defined spherical shells, density points can be summed within the radial bins in order to enable the analysis of protein-protein interaction among the morphological units based on the predicted model.

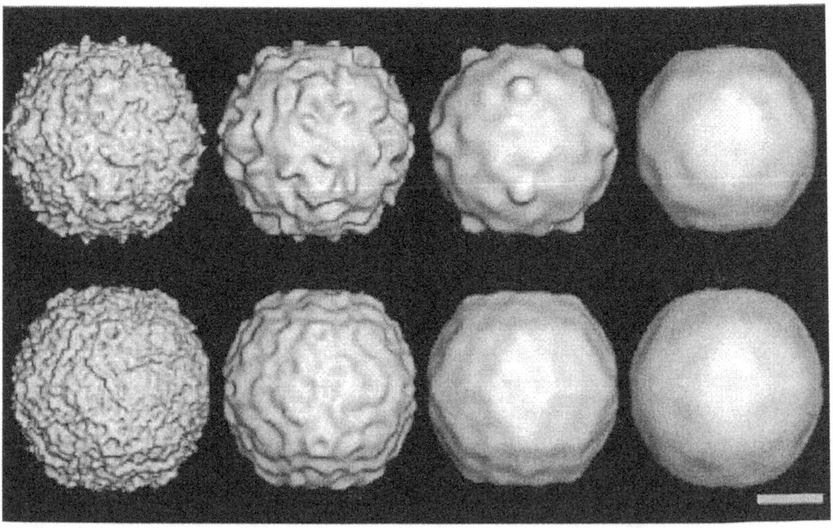

Fig. 7. A potential map on variously sampled grids, based on the atomic model of human poliovirus and the insect virus, displayed as shaded, solid-surface grayscale representations at resolutions progressing from 10, 25, 40, and 55 Å. The receptor-binding sites of poliovirus, the depressed "canyon" regions, become invisible at the lowest resolution. Scale bar = 10 nm.

In a similar manner, the scattering amplitudes of two structurally related picornaviruses, human poliovirus and an insect virus (cricket paralysis virus [13], are interpolated to illustrate features at resolutions of 10, 25, 40, and 55 Å. Both capsids are ≈ 300 Å in diameter and contain 12 protruded five-fold vertices (Color Plate 13 on page 292, top panel). About 20 Å deep, the receptor binding residues in poliovirus have been shown to locate in the depressed region of the five-fold canyon (arrowhead marked) clearly seen at the resolution of 25 Å. In the insect virus, without knowing where the receptor would bind, the canyon is entirely absent and the homologous residues are completely buried by the surface loops from the capsid (Color Plate 13 on page 292, bottom panel). However, such differences in fine details cannot be distinguished at the panels of resolutions below 40 Å.

7 Difference Imaging: Localization of Bound Molecules

Receptor binding is the first step in a virus infection. This is followed by the virus disruption and the RNA uncoating, which is a process that requires a lot of conformational changes of the capsid protein. Image reconstruction of receptor-bound poliovirus reveals the possibility to observe not only the binding site at surface protrusions (Color Plate 13 on page 292, [10,15,14]), but also the outcome after the depressive region of the capsid is bound by the molecule. Thus, being able to localize the elements unique for the binding is essential in analysis of this type. Primarily, the interpolation of the electron scattering on an accordant grid sampling provides an excellent model for difference imaging and for the identification of essential contacts between molecules. The difference map reveals the density of the bound receptor by subtracting the viral capsid from the receptor-bound virus complex map. Thus, the difference map guides the fitting of the receptor.

The density volume of image reconstruction can be used to confine the model structure of the receptor onto the viral capsid. Energy minimization and rigid body motion is usually employed after a visual check of the fit, using an interactive GUI (graphical user interface). The receptor domain fits well into the electron density of the N-terminal region of the receptor in the receptor-bound viruses. The density volume sets a much higher accuracy of modeling under the symmetry-confined grids in the analysis of the contact residues (Color Plate 13 on page 292). In the case of the receptor-bound virus, a nanometer-resolution reconstruction provides the framework for constraining, to within a few Ångströms, the docking of the atomic model of the receptor to the atomic model of the virus (Fig. 8).

From the 3D reconstruction, the receptor binding in poliovirus is clearly distinguished from the receptor binding in rhinovirus. The long axis of the receptor is along the five-fold axis to the three-fold axis in poliovirus, while the receptor is more perpendicular to the surface of rhinovirus. As a result, the binding site in poliovirus is more accessible and will be less protected from immune surveillance, which facilitates virus uncoating, than the binding site in rhinovirus. The structure of the virus-receptor complex lays the foundation for receptor-mediated capsid uncoating at the entry. Although poliovirus and rhinovirus share strong sequence identity and commonly antiparallel β-barrel configuration, evolution forced them to select different but related cellular surface molecules as their receptors. The differences, revealed in the receptor-binding orientations and footprints, seem to explain why poliovirus is less protected from immune surveillance, but more readily prone to receptor-mediated virus uncoating and entry at the cell surface.

8 Conclusions

Image reconstruction of microscope data allows one to observe biological macromolecules in their unstained state. When we interpret these images,

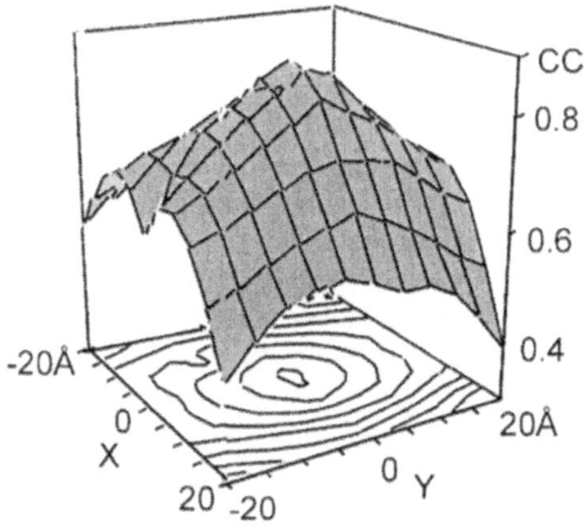

Fig. 8. The quality of fit between crystal structures and microscopy density maps. Molecular models for poliovirus and receptor fragments were placed into the reconstruction grids using automatic fitting procedures, after manual docking. Model reconstructions were implemented as the rigid body in reciprocal space for refinement. The quality of fit, measured with correlation coefficients (CC) computed to 26 Å resolution, was estimated with different positioning of the docking model; it was detectable with an accuracy of less than 5 Å along the density grids of the X-Y plane. The final docking of the receptor models is consistent with the predicted positions with mutagenesis experiments.

the combination of the 3D density of microscopy and atomic models of crystallography allows a higher resolution of information to be revealed. Receptor-bound structures of human rhinovirus and poliovirus have been comparatively studied to uncover the receptor-mediated recognition and entry in virus infection. Similarly, structural determination of antibody-bound virus complexes can be studied at acidic pH in order to understand how the same virus takes a reversed path and fuses with the host membrane for its next replication. With elaborated image processing and statistics-based interpolation, images taken at the temperature of liquid nitrogen or liquid helium will allow us to study a snapshot of infection events before and after the particle disruption. An environment of advanced rendering techniques would also be essential in conferencing and exploring the sophisticated volumetric data sets effectively.

Acknowledgments

I thank all the members of the Structural Virology and Cell Biology group at the Karolinska Institute for sharing their data and stimulating ideas. The study is sponsored by the Natural Science Research Council (NFR-11691), the Medical Research Council (MFR-12175), and the Structural Biology Network in Sweden.

References

1. T. S. Baker and R. H. Cheng. A model-based approach for determining orientations of biological macromolecules imaged by cryoelectron microscopy. *J. Struct. Biol.*, 116:120–30, 1996.
2. D. L. D. Caspar and A. Klug. Physical principles in the construction of regular viruses. *Cold Spring Harbor Symp. Quant. Biol.*, 27:1–24, 1962.
3. R. H. Cheng. Correlation of cryo-electron microscopic and X-ray data and compensation of the contrast transfer function. *Proc. Elec. Microsc. Soc. Am.*, 50:996–997, 1992.
4. R. H. Cheng, R. J. Kuhn, N. H. Olson, M. G. Rossmann, H. K. Choi, T. J. Smith, and T. S. Baker. Nucleocapsid and glycoprotein organization in an enveloped virus. *Cell*, 80:621–30, 1995.
5. R. H. Cheng, N. Olson, and T. Baker. Cauliflower mosaic virus: a 420 subunit ($t = 7$), multilayer structure. *Virology*, 186:655–68, 1992.
6. R. H. Cheng, V. Reddy, N. Olson, A. Fisher, T. Baker, and J. Johnson. Functional implications of quasi-equivalence in a $t = 3$ icosahedral animal virus established by cryo-electron microscopy and x-ray crystallography. *Structure*, 2:271–82, 1994.
7. R. A. Crowther. Procedures for three-dimensional reconstruction of spherical viruses by Fourier synthesis from electron micrographs. *Philos. Trans. R. Soc. Lond. Ser. B*, 261:221–230, 1971.
8. Y. Fujiyoshi, T. Mizusaki, K. Morikawa, H. Yamagishi, Y. Aoki, H. Kihara, and Y. Harada. Development of a superfluid helium stage for high-resolution electron microscopy. *Ultramicroscopy*, 38:241–251, 1991.
9. S. D. Fuller, S. J. Butcher, R. H. Cheng, and T. S. Baker. Three-dimensional reconstruction of icosahedral particles: The uncommon line. *J. Struct Biol*, 116:48–55, 1996.
10. J. M. Hogle, M. Chow, and D. J. Filman. Three-dimensional structure of poliovirus at 2.9 Å resolution. *Science*, 229:1358–1365, 1985.
11. R. C. Liddington, Y. Yan, J. Moulai, R. Sahli, T. L. Benjamin, and S. C. Harrison. Structure of simian virus 40 at 3.8 Å resolution. *Nature*, 354:278–84, 1991.
12. M. G. Rossmann. *The Molecular Replacement Method.* Gordon & Breach, New York, 1972.
13. J. Tate, L. Liljas, P. Scotti, P. Christian, T. Lin, and J. E. Johnson. The crystal structure of cricket paralysis virus: The first view of a new virus family. *Nat. Struct. Biol.*, 6:765–74, 1999.
14. L. Xing, K. Tjarnlund, B. Lindqvist, G. Kaplan, R. H. Cheng, and J. M. Casasnovas. Distinct cellular receptor binding modes in poliovirus and rhinoviruses. Submitted, 1999.
15. R. Zhao, D. C. Pevear, M. J. Kremer, V. L. Giranda, J. A. Kofron, R. J. Kuhn, and M. G. Rossmann. Human rhinovirus 3 at 3.0 Å resolution. *Structure*, 4:1205–1220, 1996.

GISMOS: Graphics and Interactive Steering of MOlecular Simulations

Calle Lejdfors[1], Malek O. Khan[1], Anders Ynnerman[2], and Bo Jönsson[1]

[1] Theoretical Chemistry, Chemical Center; Lund University; P.O.Box 124; SE–221 00 Lund; Sweden
[2] National Supercomputer Center; Linköping University; SE–581 83 Linköping; Sweden

Abstract. A modular software package for computational steering and visualization of molecular simulations is presented. The package, GISMOS, builds on instrumentation of existing Fortran programs to accomplish data extraction for visualization as well as run time input of control parameters for steering. A set of template files for instrumentation is provided to minimize the effort required by programmers to attach GISMOS to existing codes. Standard libraries and procedures for user interfaces, graphics production, and inter-process communication are used to ensure flexibility and portability of the package. The interactive steering environment contains a GUI for static and dynamic properties of the simulation as well as tools for 2D and 3D visualization. A Monte Carlo simulation code has been used as a case study. The implementation clearly shows the potential value of GISMOS in research and education. In addition, GISMOS has proven to be a powerful tool for debugging purposes.

1 Introduction

In the wake of the "biological revolution," molecular simulations are becoming increasingly attractive as useful tools. While numerical methods have reached a certain degree of maturity, tools for intuitively grasping these simulations have so far been sparse. GISMOS is an effort to develop a toolkit that will help experts as well as novices to explore the field of molecular simulations more effectively.

As the name implies, GISMOS provides a means for both visualization and control of molecular simulations, all on-the-fly. The modular toolkit is built on the standard for 3D graphics, OpenGL, and the popular open source user interface toolkit gtk+. This will ensure maintainability and portability. Currently GISMOS exists for SGI Irix and Linux platforms; on the latter, an open-source implementation of OpenGL, called MESA, is used. There should be no problems porting GISMOS to other Unix-based systems and even to the Windows platform, owing to the portability of OpenGL(MESA) and gtk+.

The concept of a scientist interactively working with an ongoing computation is known as computational steering. Though widely acknowledged as a useful tool [8,11,13,10], it still has not found widespread use in the field of molecular simulations. The reason for this is the gap between application scientists and computer specialists. It has simply been too cumbersome for a person working in the field of molecular simulations to implement the necessary software. Toolkits for computational steering have been created—for example, HERMES [3], CUMULVS [2], and CSE [13]—but these have generally been targeted at a broader audience and are thus more complex in their nature. The visualization is provided by standard programs, such as AVS[1] and Data Explorer[2], which have steep learning curves. By combining VMD [4], a visualization package for molecular simulations, and Sigma [3], a molecular dynamics program, an environment for visual steering of molecular dynamics, SMD [6], has been created in order to enhance protein design. SMD can also be used in an immersive virtual 3D environment [12]. While the above packages have approached the problems from a somewhat computer-science–centric view, and are either very general or specifically bound to certain simulation codes, our goal is to create a toolkit that will provide visualization and steering at a very small effort for any molecular simulation programmer.

One of the governing ideas is that GISMOS should be used together with existing, well-verified programs written in Fortran, with only minute changes to the numerical code. This can be achieved by making small modifications in the source to provide access points for the parameters and results. This process is called instrumentation and typically takes the form of subroutine calls inserted in the code [11]. These calls can transmit data to and from GISMOS. The communication scheme has been designed so that GISMOS can be used to start and stop a simulation as well as take control of a running simulation.

GISMOS provides both a GUI (graphical user interface) to, and visual output from, the simulation. The GUI lets users enter initial values for the simulation as well as change these parameters at run-time. The output from the simulation is visualized with a 3D tool that displays the molecules in real time as well as a 2D tool that provide plots of interesting variables and accumulated averages of, for example, distribution functions. These two parts of GISMOS allow the user to directly monitor a simulation as well as the effects of changing simulation parameters. This kind of computational steering of molecular simulations is surprisingly helpful for debugging simulation programs, for identifying the interesting parameter space, and for gaining general insight into molecular systems.

1 <http://www.avs.com>

2 <http://www.research.ibm.com/dx/>

3 <http://femto.med.unc.edu/SIGMA/index.html>

2 Architecture

2.1 Architectural Overview

A complete application, that is, a simulation coupled with GISMOS, consists of three parts: simulation, visualization, and steering. These parts are connected via different communication mechanisms. Fig. 1 shows a block diagram of the overall data flow. In the following sections, these different parts will be described in more detail.

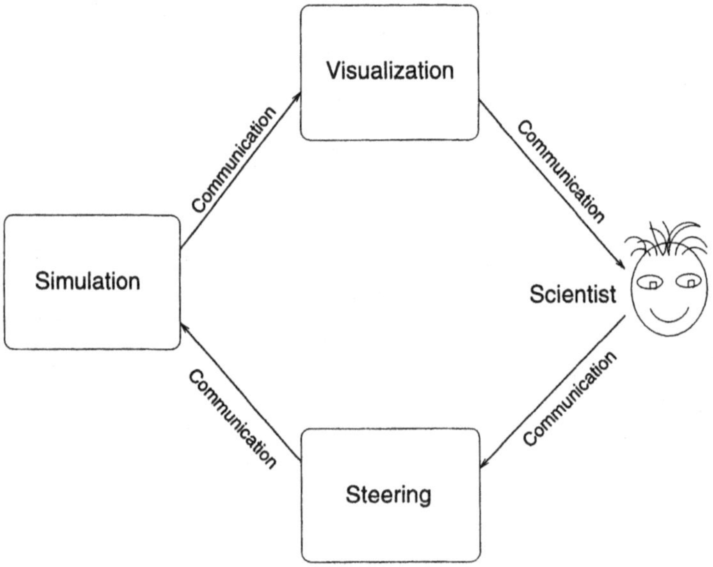

Fig. 1. Schematic picture of an application using GISMOS. Each arrow represents some kind of communication between the different parts of the application.

2.2 Implementation Overview

GISMOS is built around the concept of modules. These modules are thought of as independent, self-contained programs executing synchronously inside GISMOS's address space. In that way, they are somewhat similar, even though more complex, to the object abstraction used in modern programming languages. The GISMOS core is responsible for managing these modules as well as providing them with an interface broad enough to be useful and restricted enough to prevent access to core operations, something of which modules should make no assumptions. Thus, the following are the only parts of the core that are visible from the modules.

- The global data repository, containing the string to data mappings and some related convenience functions. All data variables are thus accessible via a symbolic name.
- The modules handler, dealing with module loading, unloading etc.
- A number of signals responsible for two-way message passing.

The separation of functionality into different modules is critical for code maintenance, debugging, and extensibility.

2.3 Numerical Simulations

All kinds of molecular simulations [1] can be used together with GISMOS. At an appropriate point in the program, GISMOS subroutines must be called in order to send and/or receive information needed for visualization and changing parameters in the simulations. The different data types and their calls are given in Table 1. These calls can be used from both Fortran and C/C++ programs. For the application programmer, it is thus enough to just add a few `import` and `export` calls. Care must be taken to ensure proper handling of the steered parameters.

Table 1. GISMOS library calls. These calls are used from within the simulation program.

Fortran type	GISMOS library call
int	gismosexportint
long	gismosexportlong
real	gismosexportreal
double	gismosexportdouble
real[ ][ ]	gismosexportrealvector
double[ ][ ]	gismosexportdoublevector
real, real, real[a]	gismosexporttrirealvector
double, double, double	gismosexporttridoublevector

[a] Cartesian coordinates stored in three different vectors.

2.4 Communication

In the initial phase of development, the simulations communicated with GISMOS via files. This approach gives an acceptable speed and has proven both robust and transparent for inter-program information transfer. Currently, GISMOS can also use shared memory areas for communication. This approach gives a substantial speed increase without causing any extra complexities for the user.

Shared memory is a method for sharing physical memory between unrelated processes on the same machine that is available in most, if not all, modern operating systems.[4] This eliminates the need to copy information to an intermediate format, like a file, and instead provides almost instantaneous information transfer. It also has the distinct advantage of having easier error handling than, for instance, pipes or sockets. The only concern is keeping exported data within the segment, and such a scheme is easy to implement. Care must be taken, of course, to prevent simultaneous reads and writes to this segment; this kind of locking is trivially implemented using either a mutex semaphore or a read/write pair of semaphores (in the case of multiple programs being interested in the data residing in the segment).[5]

2.5 Visualization of Particle Simulations

GISMOS supports a number of different visualization modules as well as the ability to easily create and integrate new modules for different purposes. Currently, a 3D module is used to display "snapshots" of the ongoing simulation. This module is built around the concept of shapes as being able to accept data in one form and render geometric shapes in accordance to these data. Each of these different shapes defines its own way of interpreting and graphically representing data. This, together with the shape-drawing code's location in a shared library that can be updated independently of the main 3D visualization code, makes upgrading old shapes and adding new ones easy.

All visualization code is written in OpenGL using a thin C++ wrapper called Ivf++ [7] implementing the trivial scene-graph.[6] The reason for not using a more advanced strategy is mostly the lack of need.

Routines for exporting the generated image in different formats are available. These formats include PNG and JPEG for simple screen shots, VRML for Web-based publications, and POV-Ray for high-quality output (`<http://www.povray.org/>`).

The 2D module provides plotting capabilities and is mostly used to display important variables, derived and/or mean values from the simulation, and their evolution over time as well as their response to parameter steering. It includes a wide variety of different options, including spline curve-fitting and raw PostScript export routines for easy printing.

---

[4] There are ways to use shared memory in a distributed environment; however, this is not considered here.

[5] A semaphore is an integer with an initial value of 1 that can be accessed atomically using the two operations P and V. The P operation decrements the count. If the result is non-negative, the call returns successfully; otherwise, it blocks until the count has become non-negative. The V operation increments the count and can thus release a blocked P operation. A mutex, *mut*ual *ex*clusion, semaphore is a semaphore that can be held by at most one process at any time.

[6] A directed acyclic graph where each node contains rendering information, matrix transformations etc. Enables logical grouping of the different parts of a scene.

2.6 Steering of Particle Simulations

The steering capabilities of GISMOS are designed for flexibility and ease of use, and not so much for speed. They use a simple file-based system, where feedback is provided through the normal communications channels. The reasons for using this overly simple method are twofold. First, a more efficient approach using, for instance, shared memory would lead to tight coupling between the simulation and the visualization code. Second, the use of a more complex method would require far greater incisions in the simulation code, not to mention having to learn yet another Application Programming Interface (API).

The steering system is built around the concept of template files. These files contain information on how to generate the needed communications file. These template files are then processed by GISMOS, and the correctly formatted communications file is generated. The format is similar to that of GNU autoconf's `.in` files.

After the template files have been written, users of the program do not need in-depth knowledge of the actual code. Users will be able to manipulate both the initial state of the simulation and the running state of the system, without hampering so-called super-users.

In practice we see two different modes of using GISMOS. The first is as a scientific tool, wherein the application programmer will have to create template files. Doing this will allow control of exactly those parts of the simulation that should be visualized and steered, gaining both a simulation code debugger and a tool for understanding. The second mode is as an educational tool, where the student will not need to tamper with the technical aspects of the simulation code and can use pre-configured systems to explore the system behavior.

The visual feedback provided make grasping a complex system much easier and the student thus gains an intuitive knowledge of how different parameters interact to create effects often visible and used daily in the macroscopic world.

3 An Example of Using GISMOS

A Monte Carlo (MC) program for simulating singly charged polymers together with counterions and salt was modified to work with GISMOS [5,9]. This program has been used to investigate the electrostatic properties of DNA. Due to the long-range nature of electrostatic interactions, these simulations were very time consuming (in the order of weeks), and the need for monitoring ongoing simulations is evident. The GUI and the output that GISMOS provides for the MC program is shown in Fig. 2.

The following are the different components, as identified by their title bar.

GISMOS v0.01 The main window where the user choose which modules to use.

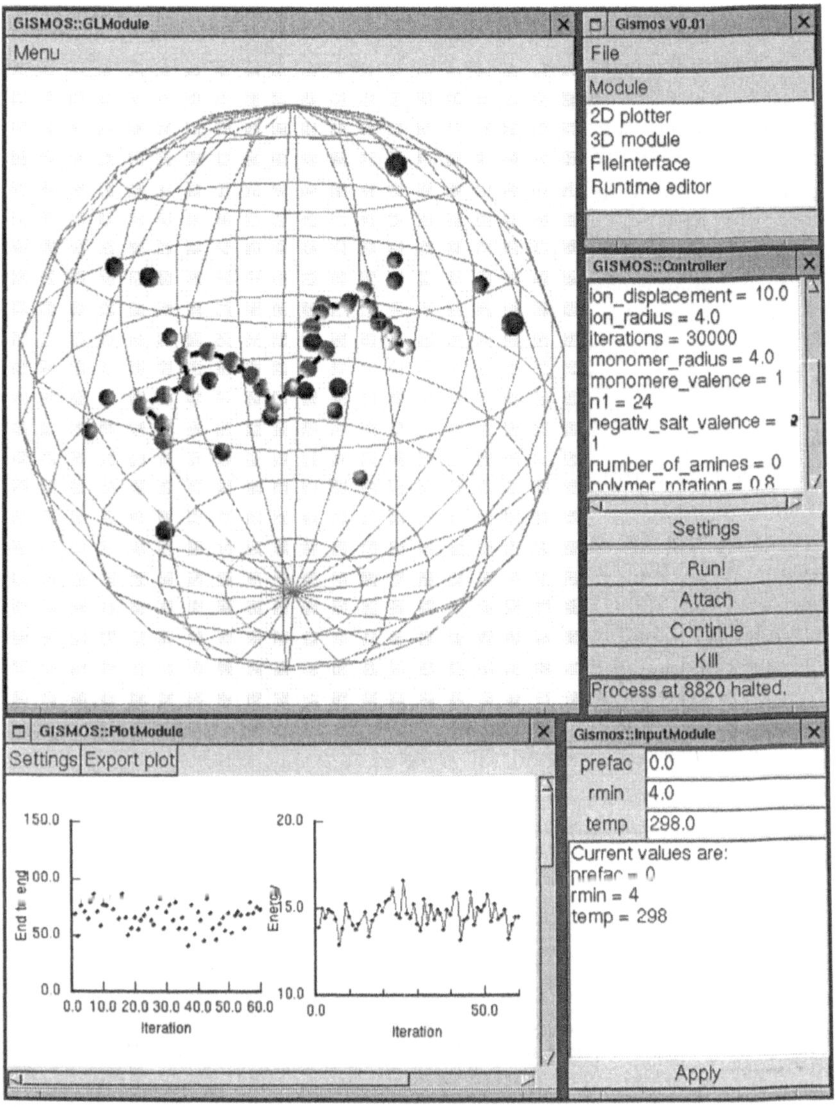

Fig. 2. Screen dump from a simulation of a DNA model. The charged phosphate groups and the counterions are shown.

GISMOS::Controller A module that shows the current setting of the simulation. From here it is also possible to start, stop, and attach to a simulation.

GISMOS::InputModule This is the place for the steered parameters.

GISMOS::GLModule The 3D visualization of the particles simulated is shown here.

GISMOS::PlotModule Used for presentation of certain parameters from the ongoing simulation.

To fully integrate the MC program with GISMOS, three steps had to be taken. Two template files had to be created, one for the initial input values and one for the steering parameters. Last, the code had to be modified in order to send and receive the relevant information. These steps will be covered below.

The template file for the input parameters are shown in Fig. 3. The parameters that are preceded by $ are shown in the GISMOS-window GISMOS::Controller (see Fig. 2), and they can be changed at the beginning of the simulation. The default value is given after the parameter name. The parameters with only a number cannot be changed by GISMOS.

```
$n1:24,$temperature:298.0,$uncorrected_bond_length:4.0,
$amine_bond_length:2.7,$iterations:30000,$ink:1,0,
$polymer_rotation:0.8,$amine_displacement:10.0,
$amine_rotation:2.0,$ion_displacement:10.0,$cell_size:60.0,
$salt_concentration:0.00,999,1,$ion_radius:4.0,
$monomer_radius:4.0,60,$monomere_valence:1,
$counterion_valence:1,$negativ_salt_valence:1,
$pos_salt_valence:1,$epsilon:78.4,400,$number_of_amines:0,
$amine_valence:1,999,50,$square_well_potential:0.0
```

Fig. 3. A template file for the input parameters used in the DNA simulations.

A template file for the steering parameters are shown in Fig. 4. These parameters appear in the GISMOS window GISMOS::InputModule (see Fig. 2), and it is possible to change them throughout the simulation. As mentioned above, the communication involved in the steering process is handled via files. GISMOS updates a file according to the description in the template file, which then can be read by the simulation program (see Fig. 5).

The small modifications that have to be made in the simulation program are outlined in Fig. 5. All of the export calls have a tag and a value. In GISMOS, this tag is associated with a certain way of handling the value that is exported. This is of course defined by the user and can be changed during the simulation. The import of the steering parameters is a straightforward `read` of the file created by GISMOS.

```
#temp:298.0        #rmin:4.0        #prefac:0.0
```

Fig. 4. A template file for the parameters that are supposed to be steered in the DNA simulations.

```
*     VISUALIZATION PART
*     export of the parameters that are to be visualized
*     The shared memory version is used
      call gismosexportbegin(ind) !ind is a counter

*     export of parameters needed for the 2D plots
      call gismosexportdouble('temperature', dtemp)
      call gismosexportdouble('endend\_distance', Ree)
      call gismosexportint('Energy', xww1)
      call gismosexportdouble('box\_size', Rc)

*     export of the particle coordinates needed for the 3D plots
      call gismosexporttridoublevector('coord\_poly', xpoly, ypoly,
     &     zpoly, n1)
      call gismosexporttridoublevector('coord\_ion', xion, yion,
     &     zion, nion)
      call gismosexportend

*     STEERING PART
*     updating of the steering parameters from the file steer.inp
      read(FILE='steer.inp')dtemp,rmin,prefac
```

Fig. 5. Pseudo-code representing the calls that have to be added to the simulation program in order to interact with GISMOS.

To illustrate the flexibility of GISMOS, snapshots from three different simulations programs have been included (Color Plate 14 on page 293).

4 Conclusion and Future Directions

The rapid development of powerful computers is making it possible to computationally study systems of increasing complexity. Along with the complexity of the studied systems comes an increasing amount of computational data. Visualization provides researchers with a powerful tool to analyze this data rapidly, using the visual capacity of the human brain. With the GISMOS package, we have shown how the insights gained through 2D and 3D visualizations can be used to steer parameters of molecular simulations interactively.

One of the most attractive features of GISMOS is the minimal effort required to attach the package to existing simulation codes. The modular nature and use of standard procedures and libraries for graphics generation and communications make GISMOS easily extendible, portable, and robust. Early results from the case studies clearly show that GISMOS is a valuable research tool and that students can reach a steeper learning curve by solving problems using GISMOS. It is therefore concluded that GISMOS provides a general alternative to more complex real-time visualization solutions and provides a platform for handling today's demands, as well as tomorrow's.

The next logical development step is to supply GISMOS with a network layer, that is, the ability to visualize and steer simulations located on remote machines in a client-server scheme. This will allow GISMOS to make use of high-end computational resources as well as high-end graphics rendering hardware, which may reside on different systems. The creation of such a layer is non-trivial. To maintain real-time characteristics across network boundaries, care and effort has to be put into design as well as implementation. It would naturally use the same library calls as described above (Fig. 1). The actual implementation, local or networked, would thus be hidden from the user. This logically extends to introducing a scheme for automatic dynamic loading of different communication interfaces from within the simulation program at request from GISMOS.

The program is available from `<http://www.teokem.lu.se/gismos/>`. All questions, comments, and requests are welcome.

References

1. M. P. Allen and D. J. Tildesley. *Computer Simulation of Liquids.* Oxford University Press, Oxford, England, 1989. Paperback.
2. G. A. Geist II, J. A. Kohl, and P. M. Papadopoulos. Cumulvs: Providing fault tolerance, visualization, and steering of parallel applications. *Int. J. Supercomputer Applications and High Performance Comp.*, 11, 1997.

3. G. Hanyzewski and M. Mull. Hermes: A tool kit for developing distributed modelling applications. *Nanotechnology*, 7, 1996.
4. W. F. Humphrey, A. Dalke, and K. Schulten. VMD: Visual molecular dynamics. *J. Mol. Graphics*, 14, 1996.
5. M. O. Khan and B. Jönsson. Electrostatic correlations fold DNA. *Biopolymers*, 49, 1999.
6. J. Leech, J. F. Prins, and J. Hermans. SMD: Visual steering of molecular dynamics for protein design. *IEEE Comp. Sci. Eng.*, 3, 1996.
7. Lund University, Division of Structural Mechanics, Lund, Sweden. *Interactive Visualization Framework Ivf++: User's Guide*. Under development.
8. B. H. McCormick, T. A. Defanti, and M. D. Brown. Visualization in scientific computing. *Computer Graphics*, 22, 1987.
9. S. Mel'nikov, M. O. Khan, B. Lindman, and B. Jönsson. Phase behaviour of single DNA in mixed solvents. *J. Am. Chem. Soc.*, 121, 1999.
10. J. D. Mulder, J. J. van Wijk, and R. van Liere. A survey of computational steering environments. *FGCS*, 15, 1999.
11. S. G. Parker, C. R. Johnson, and D. Beazley. Computational steering software systems and strategies. *IEEE Comp. Sci. Eng.*, 4, 1997.
12. J. F. Prins, J. Hermans, G. Mann, L. S. Nyland, and M. Simons. A virtual environment for steered molecular dynamics. *FGCS*, 15, 1999.
13. R. van Liere, J. D. Mulder, and J. J. van Wijk. Computational steering. *FGCS*, 12, 1997.

Monte Carlo Simulation of Solutions of Like-Charged Colloidal Particles

Per Linse and Vladimir Lobaskin

Physical Chemistry 1; Center for Chemistry and Chemical Engineering; Lund University; P.O. Box 124; SE–221 00 Lund; Sweden

Abstract. Model systems of charged spherical macroions possessing 60 elementary charges with either monovalent, divalent, or trivalent counterions interacting solely through hard-core and Coulomb forces have been investigated by Monte Carlo simulations. Ewald summation of the long-range Coulomb interactions and cluster moves to improve the macroion mobility were found to be two indispensable techniques for these studies. Although the direct macroion-macroion interaction is purely repulsive, we find that for sufficiently strong electrostatic coupling, there exists an effective attraction that separates the solution into two fluid phases. The attraction is short-ranged and arises from correlations among counterions localized near different macroions. This mechanism for attraction differs from that attraction inferred to operate in solutions of charged latex particles, which is presently challenging the classical DLVO theory.

1 Introduction

The physico-chemical properties of like-charged colloids of the size 1–1000 nm, in a wide variety of biological and technologically important systems, are dominated by electrostatic forces. Although the electrostatic interactions in such systems have been extensively studied over the past 50 years, they remain a central problem in colloidal science [3,2]. A large body of experimental evidence confirms the intuitive expectation that purely repulsive forces operate between like-charged colloidal particles. (As usual, the forces between the colloidal particles discussed here are statistically averaged over the positions of the small ions.) However, in recent years it has been experimentally inferred that attractive forces may exist between like-charged objects in the colloidal domain, for example, from the reduced swelling of lyotropic liquid lamellar phases when monovalent counterions are replaced by divalent ones [9,8] as well as from the observed phase separation [20,7] and the deduced interaction potentials [11,6] in concentrated suspensions of charged latex particles. At present these and related phenomena are being intensively studied by various experimental techniques.

Monte Carlo (MC) and molecular dynamics (MD) simulations are powerful tools for obtaining accurate solutions of model systems of different degrees of sophistication [1]. Within the framework of the primitive model, the most accurate description of the structure of colloidal solutions is based on the so-called two-component model. In this approach, both the charged colloidal particles (later referred to as macroions) and the small counterions are explicitly represented by charged species, whereas the solvent is treated as a dielectric medium. Accurate computer simulations of the two-component model are of great importance for assessing the accuracy of (i) more approximate statistical mechanical methods, such as liquid state theories, and (ii) simpler models, such as the one-component model. In the latter case, only the macroions enter the model explicitly now interacting with a potential preaveraged over the locations of the small ions. A comprehensive review of this area was recently made by Vlachy [21].

Until recently, however, simulations of like-charged colloids employing the two-component model have been impeded by two technical obstacles. These are (i) the long-range nature of the Coulomb force making a normal potential cutoff not applicable, and (ii) the accumulation of counterions close to the macroion—which is a physically correct behavior, but one that leads to very low mobility of the macroions in a Metropolis MC or MD simulation [14]. The aim of the present contribution is to describe the use of the two methods that circumvent these two obstacles and to present novel results of these systems.

2 Model System

Consider a model system representing aqueous solutions of ionic surfactant micelles in the absence of added salt. (A micelle is composed of self-assembled surfactant molecules, and it is often spherical.) The model contains two types of spherical charged particles: (i) macroions of diameter $\sigma_{\mathrm{MM}} = 40$ Å and charge $Z_{\mathrm{M}} = -60$, representing the micelles, and (ii) small ions of diameter $\sigma_{\mathrm{II}} = 4$ Å and charge $Z_{\mathrm{I}} = +1, +2$, or $+3$, representing the counterions of the macroions. The solvent enters the model only by its relative dielectric permittivity ε_{r}. The interaction between the particles is pairwise additive, and for pair ij, where i and j denote either macroion or counterion, it is given

$$U_{ij}(r) = \begin{cases} \infty, & r < \sigma_{ij} \\ \frac{Z_i Z_j e^2}{4\pi\varepsilon_0\varepsilon_{\mathrm{r}}}\frac{1}{r}, & r \geq \sigma_{ij} \end{cases}, \tag{1}$$

where e is the elementary charge, ε_0 the permittivity of vacuum, r the center-to-center separation between the particles, and $\sigma_{ij} = (\sigma_{ii} + \sigma_{jj})/2$.

We have considered macroion number densities in the range $\rho_{\mathrm{M}} = 2.5 \times 10^{-7}$–$20 \times 10^{-7}$ Å^{-3}, corresponding to macroion volume fractions of 0.84–6.7 percent. For simplicity, below we will denote the macroion density in units of $\rho_{\mathrm{M}}^0 = 2.5 \times 10^{-7}$ Å^{-3}. The temperature $T = 298$ K and $\varepsilon_{\mathrm{r}} = 78.4$ have been used throughout. The main results were obtained from systems containing $N_{\mathrm{M}} = 80$ macroions and a sufficient number of counterions $N_{\mathrm{I}} = \mid Z_{\mathrm{M}}/Z_{\mathrm{I}} \mid N_{\mathrm{M}}$ to ensure overall electroneutrality.

3 Simulation Details

The model was solved by MC simulations using the canonical ensemble. All particles were enclosed in a cubic box of length L, and periodic boundary conditions were applied. The use of the Ewald summation and a cluster-move algorithm are two important techniques improving such simulations, and below they are described in more detail.

3.1 Ewald Summation

The long-range Coulomb interaction was handled by the Ewald summation technique with conducting boundary conditions [1,4,17]. In this approach, the system is made infinite by a replication of the system in all three directions (see Fig. 1). By superimposing each charged particle with a Gaussian charge distribution of the width $\alpha^{-1/2}$ and opposite charge, the total Coulomb interaction can be transformed to

$$\sum_{i<j} \frac{q_i q_j}{r_{ij}} = \sum_{i<j} \frac{q_i q_j}{r_{ij}} \operatorname{erfc}(\alpha r_{ij}) + \frac{2\pi}{L^3} \sum_{\mathbf{k} \neq 0} k^{-2} \exp(-k^2/4\alpha^2) \left[\sum_j q_j \exp(-i\mathbf{k} \cdot \mathbf{r}_j) \right]^2 - \frac{\alpha}{\sqrt{\pi}} \sum_j q_j^2 \; , \tag{2}$$

where the first term on the right-hand side denotes the Coulomb interaction among the charged particles (now screened), the second term denotes the interaction among the compensating Gaussian charge distributions expressed in the reciprocal space, and the third term denotes a correction for the self-energy of the Gaussian charge distributions. In (2), $q_i = Z_i e$, $\mathbf{r}_{ij} = \mathbf{r}_i - \mathbf{r}_j$, and $\mathbf{r}_i$ denotes the location of particle i.

The Ewald summation is controlled by two parameters: s, which is related to the relative accuracy of a single evaluation of the potential energy ε as $\varepsilon \approx \exp(-s^2)/s^2$; and the parameter α, which also partitions the computational load between the real and reciprocal spaces. For given s and α, the cutoff distance of the pair interactions in the real space becomes [17]

$$R_{\text{cut}} = \frac{s}{\alpha} \tag{3}$$

and the cutoff in the reciprocal space

$$k_{\text{cut}} = 2s\alpha \; . \tag{4}$$

Here, R_{cut} becomes smaller than half of the box length $L/2$, and the summation in the real space becomes restricted to the nearest image of the particles.

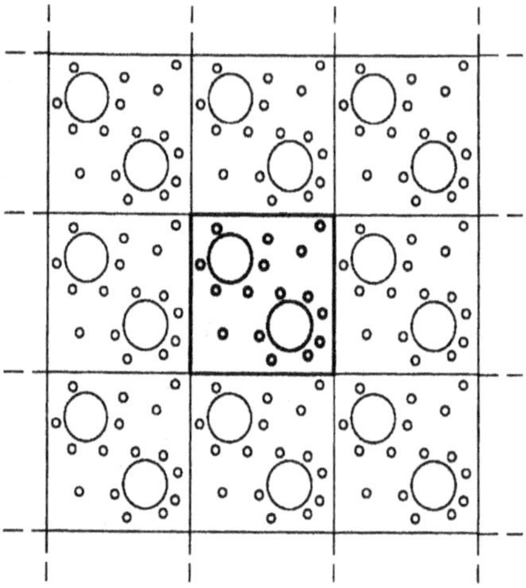

Fig. 1. Schematic illustration showing the primary (central) simulation box containing charged macroions and the charged counterions as well as the infinite replication of the primary box.

On the contrary, $k_{\text{cut}} > 2\pi/L$, and the number of replicas along one direction used in the summation in the reciprocal space becomes

$$n_{\text{cut}} = 1 + \left[\frac{k_{\text{cut}}}{2\pi/L}\right] = 1 + \left[\frac{s^2 L}{\pi R_{\text{cut}}}\right] , \tag{5}$$

where $[\ldots]$ here denotes a truncation to an integer number and where (3) and (4) have been used. Moreover, we have reduced the number of replicas by using a spherical cutoff of $\mathbf{n}$ according to $\mid \mathbf{n} \mid = n_{\text{cut}}$, where $\mathbf{n} = (n_x, n_y, n_z)$ is a box translation vector, and $n_x, n_y,$ and n_z are integers. In the present work, $s = 3.0$ has been used throughout, giving $\varepsilon \sim 10^{-5}$. A larger s would have improved the accuracy of the energy evaluation but at the expense of longer computation time.

In the real space, a neighbor-list technique was used to find pairs of interacting particles. The frequency of generating new neighbor lists was controlled automatically by employing the displacements of the two particles with the largest net displacements since the previous generation of the neighbor list. When generating the neighbor list, one typically includes particles within $R_{\text{NL}} = R_{\text{cut}} + \Delta_{\text{NL}}$. The use of $R_{\text{NL}} > R_{\text{cut}}$ is motivated by the movement of the particles. Because the simulation time depends on R_{NL}, Δ_{NL} has to be optimized; a small value implies that the neighbor list has to be generated often, whereas a large value leads to an unnecessarily large computational load in the real space. The pairwise energy was evaluated from a look-up table, and a quadratic interpolation scheme was used. In the reciprocal space, there is only a single loop over the particles, and no neighbor list is involved.

Thus, regarding the energy evaluation, the computational efficiency depends on α (or the related R_{cut}, which will be used below) and Δ_{NL}.

3.2 Cluster-Move Technique

The standard Metropolis MC procedure [1,19] involves attempts to make moves of the particles in a random fashion. At each move, translational displacements in the x-, y-, and z-directions are made with step lengths uniformly sampled from $[-\Delta_{\text{trans}}/2, \Delta_{\text{trans}}/2]$. The trial move is accepted with the probability

$$P_{\text{acc}} = \min\{1, \exp[-(U_{\text{new}} - U_{\text{old}})/kT]\} \ , \tag{6}$$

where U_{new} and U_{old} are the potential energies of the trial and old configurations, respectively. The convergence rate of the simulation depends on the translational displacement parameter Δ_{trans}. The particle to be moved is normally selected randomly, and one attempted move counted per particle is referred to as one MC pass.

Due to the strong accumulation of the counterions near the macroions (see Fig. 5), the standard MC method leads to a very restricted mobility of the macroions and hence a poor statistics of the sampled properties [14]. When a hard-core overlap is encountered (i.e., $r < \sigma_{ij}$), U_{new} becomes formally infinite (1) and P_{acc} becomes zero. The efficiency of the simulation can be increased by applying an MC cluster-move (MCCL) algorithm. Such ideas have previously been applied to MC simulations of strongly coupled systems (see, e.g., [5]).

In our implementation [18], we define a cluster by selecting the central macroion and particles within the distance R_{cl} from the macroion center, the latter with a probability $p \leq 1$, when a macroion is subjected to a trial move. The trial move of the whole cluster is accepted with the probability

$$P_{\text{acc}} = \min\{1, (1-p)^{n_{\text{new}} - n_{\text{old}}} \exp[-(U_{\text{new}} - U_{\text{old}})/kT]\} \ , \tag{7}$$

where n_{old} and n_{new} are the numbers of particles (excluding the central macroion) inside the cluster volume before and after the trial move, respectively, and U_{old} and U_{new} have the same meaning as in (6). The acceptance rule with the additional factor prior to the Boltzmann weight [cf. (6) and (7)] preserves the detailed balance condition. The cluster-move procedure is further illustrated in Fig. 2. In our MCCL algorithm, there are no trial moves involving a macroion only, and the trial moves of individual counterions are unaffected.

The MCCL algorithm can be optimized with respect to the macroion mean-square displacement reflecting the evolution of the system in the configurational space. The algorithm is controlled by three tunable parameters: the particle inclusion probability p, the cluster radius R_{cl}, and the translational displacement parameter of the cluster Δ_{cl}.

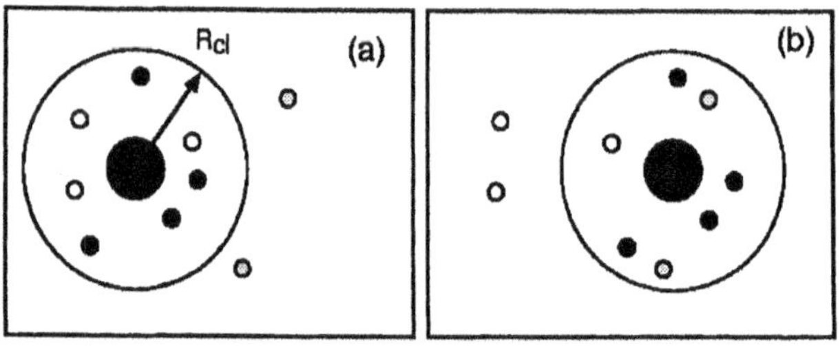

Fig. 2. Schematic representation of a cluster move showing the location of a macroion (big filled circle) and counterions (small circles) (a) before and (b) after the cluster move. Ions moved together with the macroion are denoted by filled circles, those within R_{cl} before the move but not moved by open circles, and new ions within R_{cl} after the move by shaded circles. In this example, we have $n_{new} = 7, n_{old} = 7$, and $n_{new} - n_{old} = 0$.

Finally, macroions with nearby counterions (dressed macroions) may at some conditions form aggregates. The mobility of a dressed macroion in an aggregate is low, now due to hard-core overlap with the nearby dressed macroions. By making a simultaneous trial move of all the dressed macroions forming an aggregate (in the same spirit as in the MCCL algorithm), the macroion mobility may again increase by several orders of magnitude. In this generalized MCCL algorithm, we have trial moves of aggregates of dressed macroions, individual dressed macroions, and individual counterions.

4 System-Size Effects

The system-size effects have been investigated by comparing properties of systems of different sizes at constant number density. We have considered the reduced energy and the macroion-macroion radial distribution function (rdf) of the 60:1 system. The rdf provides information on the spatial correlations in a system.

The reduced energies of the 60:1, 60:2, and 60:3 systems at the macroion number density $\rho_M = 2\rho_M^0$ with $N_M = 20$, 40, and 80 are shown in Table 1. For monovalent and divalent counterions, no significant system-size effects appear for $N_M \geq 40$. This is in accordance with a previous study of a 20:1 asymmetric electrolyte, where we found no significant system-size effect with $N_M \geq 40$ when applying the Ewald summation with conducting boundary conditions [17]. Figure 3 displays the macroion-macroion rdf's with monovalent counterions obtained at $\rho_M = 2\rho_M^0$ with $N_M = 20$, 40, and 80 macroions in the primary box. Also here we have a very good agreement between simulations with 40 and 80 macroions. The rdf obtained with 20 macroions is still very respectable, but it overestimates the structure somewhat. The macroion-counterion and counterion-counterion rdf's display no significant system-size dependence for $N_M \geq 20$ (data not shown). Their faster system-size convergence is in accordance with their shorter correlation lengths.

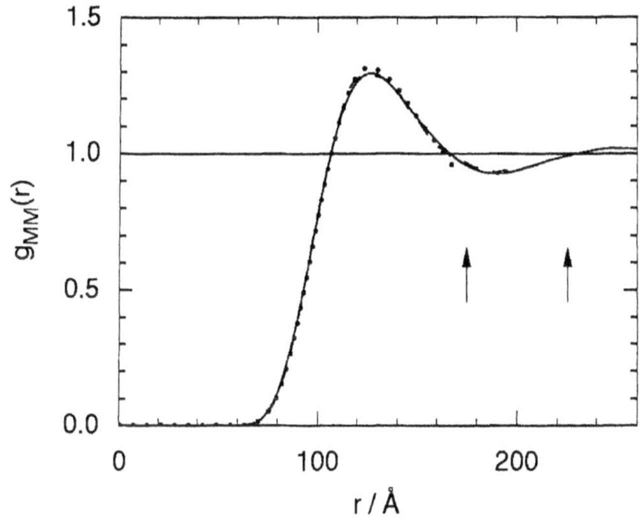

Fig. 3. Macroion-macroion radial distribution functions for monovalent counterions at macroion number density $\rho_M = 2\rho_M^0$ with $N_M = 80$ (solid curve), 40 (dashed curve), and 20 (dotted curve) macroions in the primary box. The arrows indicate the ends of the calculated distribution functions for $N_M = 20$ and 40, respectively. The horizontal line at $g = 1$ corresponds to a system of uncorrelated macroions. Data taken from [16].

Regarding the case with trivalent counterions, Table 1 shows a weak lowering of the energy with increasing system size. This is related to the reduced surface-to-volume ratio of the aggregated state appearing for the 60:3 system (cf. Fig. 6c). Additional aspects on the system-size effect are also given in Sect. 6.

Thus, despite the very strong direct electrostatic interaction between the macroions ($> 100kT$ at half of the box size, k being the Boltzmann constant), convergent energies and structures are obtained for a reasonably small number of macroions. The Ewald summation technique is hence a very attractive approach to deal with the long-range interactions in these systems. Other boundary conditions, such as the minimum image convention (which is useful for simple electrolyte systems) fails completely for highly asymmetric electrolytes.

Table 1. Reduced energy U/NkT of different systems and system sizes at $\rho_M = 2\rho_M^0$.[a]

System	60:1	60:2	60:3
$N_M = 20$	-8.658(1)	-20.327(1)	-33.78(2)
$N_M = 40$	-8.661(1)	-20.335(2)	-33.88(1)
$N_M = 80$	-8.659(1)	-20.335(2)	-33.93(1)

[a] U denotes the total potential energy, N the total number of particles, k the Boltzmann constant, and T the absolute temperature. The uncertainty given is based on a subdivision of the run into 10 sub-batches. $\rho_M^0 = 2.5 \times 10^{-7}$ Å^{-3}.

5 Simulation Efficiency

MC simulations produce averages that are subjected to statistical uncertainties. The precision of an average is improved by increasing the simulation length. Generally, the precision obtained for a given computer time can be used as a measure of the efficiency. Here, however, the efficiency will be measured by considering the macroion mean-square displacement $\langle \Delta r_M^2 \rangle$. We have previously shown that a large $\langle \Delta r_M^2 \rangle$ is well correlated with a high efficiency of the evaluation of the internal energy and structural properties [13]. Moreover, the use of $\langle \Delta r_M^2 \rangle$ facilitates the detection of possible underestimation of statistical uncertainties arising from suppressed system mobility in the configurational space due to low acceptance ratio of the trial moves. Besides the system as such, the efficiency depends on the algorithm used as well as its implementation.

Because the MC algorithm has four parameters to be optimized ($R_{cut}, \Delta_{NL}, \Delta_M$, and Δ_I) and the MCCL algorithm has six ($R_{cut}, \Delta_{NL}, p, R_{cl}, \Delta_{cl}$, and Δ_I), a full optimization is not practical. Instead, values of some of the parameters have been assigned based on physical arguments or optimized in a restricted number of simulations. First, we have used $\Delta_I = \Delta_M (\equiv \Delta_{trans})$ for the MC and $\Delta_I = \Delta_{cl} (\equiv \Delta_{trans})$ for the MCCL algorithms. The use of $\Delta_I \gg \Delta_M$ and $\Delta_I \gg \Delta_{cl}$, respectively, would require a large Δ_{NL} with an accompanying increased computational load in the real space due to larger counterion mobility without improving the slow macroion mobility. Second, initial simulations showed that the maximal $\langle \Delta r_M^2 \rangle$ was achieved for a particle inclusion probability near one. For simplicity, we have selected $p = 1$ throughout, implying that all moves leading to an inclusion of a new particle into the cluster are rejected. On decreasing p, the simulation efficiency deteriorates, owing to hard-core overlaps between the macroion and the particles located inside the cluster volume but not moving with the cluster. For $p = 1$, the content of clusters is renewed during the counterion moves.

In Table 2, we have compiled data for the 60:1 and 60:2 systems using the standard MC and the MCCL algorithm as described above. For the 60:1 MC case, the remaining parameters R_{cut}, Δ_{NL}, and Δ_{trans} were fully optimized [13]. In the other three cases, the same values of R_{cut} and Δ_{NL} as in the 60:1 case were used, because their optimal values were found to be only weakly dependent on the counterion valency and the simulation algorithm. Hence, for the 60:2 MC case, only Δ_{trans} was optimized. The optimal Δ_{trans} was close to that for the 60:1 MC case, and for simplicity $\Delta_{\mathrm{trans}} = 1$ Å was selected. Finally, for the MCCL simulations, the parameters R_{cl} and Δ_{trans} were fully optimized for the 60:1 system, and the same values were used for the 60:2 system, although a fourfold increase in $\langle \Delta r_M^2 \rangle$ was achieved by increasing Δ_{trans} from 15 to 50 Å.

Table 2. Simulation parameters and performance data of simulations using different combinations of systems and simulation algorithms at $\rho_{\mathrm{M}} = 8\rho_{\mathrm{M}}^0$ with $N_{\mathrm{M}} = 40$[a].

System/method	60:1 MC[b]	60:2 MC	60:1 MCCL	60:2 MCCL
R_{cut} / Å	90[c]	90[d]	90[d]	90[d]
Δ_{NL} / Å	10[c]	10[d]	10[d]	10[d]
Δ_{trans} / Å	1[c]	1[c]	15[c]	15[e]
R_{cl} / Å	-	-	30[c]	30[e]
n_{pass}	10×10^4	5×10^4	2×10^4	5×10^4
Fraction accepted macroion moves	0.11	0.06	0.12	0.20
Average number of particles in cluster	-	-	48.5	28.4
$\langle \Delta r_M^2 \rangle$ / (MC pass Å^2)	0.005	0.003	1.9	6.7
t_{CPU} / h[f]	101	18	32	24

[a] $\rho_{\mathrm{M}}^0 = 2.5 \times 10^{-7}$ Å^{-3}.
[b] From [13].
[c] Optimized by maximizing $\langle \Delta r_M^2 \rangle$.
[d] Taken the same as for the 60:1 MC case, but the optimal values are only weakly dependent on the counterion valency and the simulation algorithm.
[e] Taken the same as for the 60:1 MCCL case.
[f] The integrated Monte Carlo/molecular dynamics/Brownian dynamics simulation package MOLSIM [12] for molecular systems was employed. The reported CPU times refer to the use of an IBM RS/6000 595 workstation.

Simulation data are given in Table 2, and the macroion mean-square displacements as a function of the elapsed simulation time are displayed in Fig. 4. Regarding the 60:1 system, we find that the MCCL algorithm increases the macroion mobility 400-fold ($\langle \Delta r_M^2 \rangle = 1.9$ Å^2 as compared to 0.005 Å^2 for the standard MC algorithm). The reason is that the simultaneous move of a macroion and its nearby counterions into the space outside the cluster can be performed with a much large displacement step and larger acceptance probability than a move of a single macroion. In the MCCL algorithm, the trial move is made into a region of relatively low counterion density as compared to the high local counterion density in the standard MC algorithm. Even taking into account the 50 percent increase of the computational load in performing the cluster moves, the efficiency of the MCCL algorithm is still two orders of magnitude higher than standard MC (see Fig. 4).

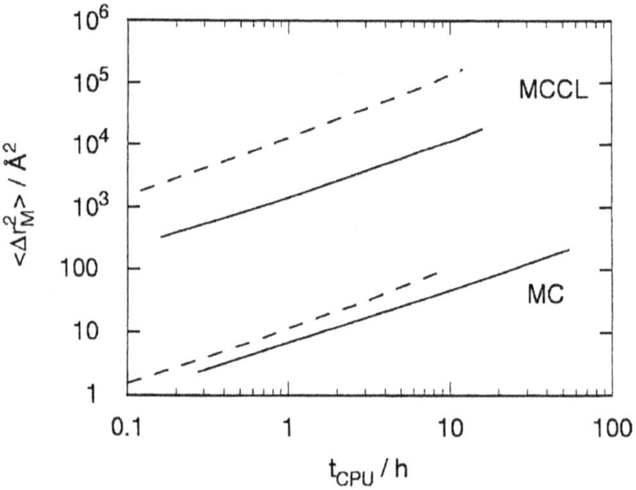

Fig. 4. Macroion mean-square displacements vs. the CPU time for the 60:1 (solid curves) and 60:2 (dashed curves) at indicated algorithm. Simulation parameters are given in Table 2. The data for the 60:1 MC simulation are taken from [13].

Moreover, this improvement accelerates for divalent counterions, due to reduced mobility in the standard MC simulation ($\langle \Delta r_M^2 \rangle = 0.005$ to 0.003 Å^2) and an increased mobility in the MCCL simulation ($\langle \Delta r_M^2 \rangle = 1.9$ to 6.7 Å^2). The former effect is a consequence of the higher local number density of counterions near the macroions, and the latter effect is a result of the lower counterion density beyond the cluster radius $R_{\rm cl}$ for the 60:2 system. Finally, despite the lower mobility found in the 60:2 MC case, the efficiencies of these simulations are comparable to those in the 60:1 MC case due to the smaller number of particles in the former system, which makes the computational load smaller (see Table 2 and Fig. 4).

To conclude, the MCCL algorithm strongly improves the macroion mobility at a moderate increase of the computational effort and makes simulation of highly asymmetric electrolytes feasible.

6 Properties of the Model System

The properties of the 60:1, 60:2, and 60:3 systems have been investigated by extensive MCCL simulations using $N_{\mathrm{M}} = 80$ at different densities [15,16]. The results at $\rho_{\mathrm{M}} = 2\rho_{\mathrm{M}}^0$ will be given here. Figure 5 shows two-dimensional projections of typical system configurations from the end of the simulations. With monovalent counterions (Fig. 5a), the macroions are well separated (in three dimensions), and a considerable fraction of the counterions are accumulated near the surfaces of the macroions, as previously alluded to, although many counterions are distributed throughout the intervening solution. In the case of divalent counterions (Fig. 5b), we observe a tendency of macroion pairing and a stronger counterion accumulation near the macroions. Finally, with trivalent counterions (Fig. 5c), the macroions form a large aggregate (occasionally with one or two macroions separated) and nearly all counterions are associated with the macroions. For monovalent and divalent counterions, the equilibrium is achieved after 10,000 passes, starting from a random distribution; whereas, for trivalent counterions, about 500,000 passes are required. Simulations were also performed where the macroions initially formed a dense primitive cubic lattice structure. In the case of monovalent and divalent counterions, the macroions are completely dissociated after 5,000 passes; in the case of trivalent counterions, however, the aggregate remains but is essentially melted after 200,000 passes.

The promotion of the macroion aggregation with increasing counterion valency has further been documented by investigating the macroion-macroion structure factor $S_{\mathrm{MM}}(k)$ at different system sizes at the macroion number density $\rho_{\mathrm{M}} = 2\rho_{\mathrm{M}}^0$. Figure 6 shows that $S_{\mathrm{MM}}(k)$ for monovalent and divalent counterions is independent of the system size. In contrast to this observation, for trivalent counterions equilibrium $S_{\mathrm{MM}}(k \to 0)$ grows fairly linearly with the number of macroions, indicating a large-scale instability of the system. In fact, the structure factors are similar to the form factor of a homogeneous sphere consistent with the configuration shown in Fig. 5c. We also note that the aggregates are fluid-like, because higher-order maxima are very weak. Hence, our results obtained with trivalent counterions using different start configurations and different system sizes unambiguously demonstrate a large-scale aggregation of the macroions. For a macroscopic system, the equilibrium state would be two coexisting fluid phases of different number density of the asymmetric electrolyte.

To conclude, the presented results unambiguously demonstrate that a solution of like-charged colloidal particles plus counterions, interacting solely through hard-core repulsive forces and long-range Coulomb forces (the so-called primitive model) displays conditions where two phases of different electrolyte density are in equilibrium. Because a system of uncharged hard spheres does not exhibit such a coexistence of two fluids of different density, we infer that an attractive force of electrostatic origin is by necessity acting among the macroions, which moreover is sufficiently strong to induce a phase

(a)

(b)

(c)

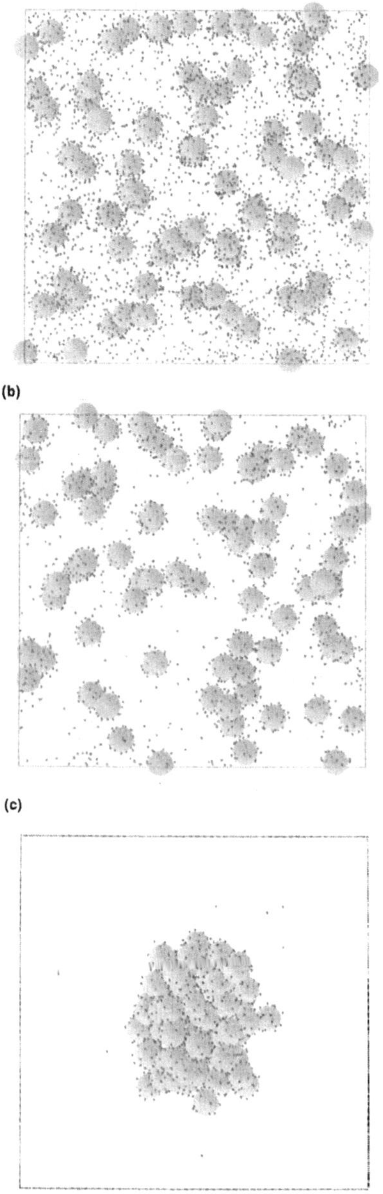

Fig. 5. Typical configurations of the asymmetric electrolyte at macroion number density $\rho_M = 2\rho_M^0$. The simulated system contains 80 macroions, each carrying 60 elementary charges (large spheres), and a corresponding number of (a) monovalent, (b) divalent, or (c) trivalent counterions (dots). The side of the box is 430 Å, and particles are drawn to scale. Periodic boundary conditions are applied in all three directions to avoid surface artifacts. With permission from the American Physical Society [15].

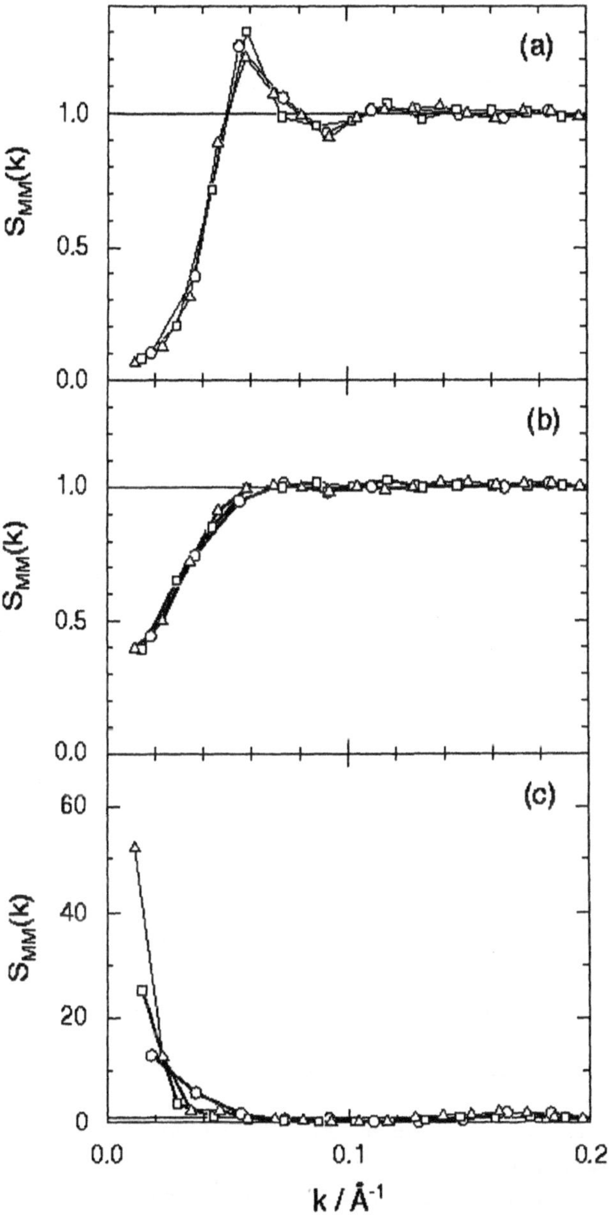

Fig. 6. Macroion-macroion structure factors $S_{MM}(k)$ for (a)monovalent (b)divalent or (c)trivalent counterions at macroion number density $\rho_M = 2\rho_M^0$ with $N_M = 20$ (spheres), 40 (squares), and 80 (triangles) macroions. The horizontal line at $S_{MM} = 1$ corresponds to systems of uncorrelated macroions. Data taken from [15].

separation. Thus, at low electrostatic coupling (monovalent counterions), a single-phase appears, which is characterized by repulsive forces among the macroions. At high electrostatic coupling (trivalent counterions), two coexisting phases are present, owing to effectively attractive forces among the macroions. For the intermediate system (divalent counterions), we have still a one-phase system, but it is close to the two-phase boundary. The attraction arises from spatial correlations of counterions associated with different macroions [10], and the effect of these correlations becomes more important as the counterion valency is increased. With increasing counterion charge, the electrostatic double layer becomes thinner, and this reduces the range of the double layer repulsion. At a sufficiently high counterion charge, the attractive correlation force dominates at short separations.

Finally, our results are consistent with experimental data for solutions of charged surfactant micelles—a coexistence of two fluid micellar solutions with either monovalent or divalent counterions has never been observed. Moreover, we would like to stress that our findings should not be directly compared with solutions of charged latex particles, owing to the size and charge differences. Nevertheless, from further analysis [16], we concluded that the recently reported attraction in aqueous solutions of latex systems [20,7] cannot be rationalized by the mechanisms controlling the aggregation operating here, and in these systems other explanations of the attraction have to be considered.

7 Summary

We have performed MC computer simulations of a model electrolyte with charge asymmetry 60:1, 60:2, and 60:3, the first representing an aqueous micellar solution, and the latter two used to investigate the effect of increased electrostatic coupling. In the case of trivalent counterions, the like-charge macroions form essentially a single aggregate, and the attractive force originates from correlation among counterions associated with different macroions. Moreover, we have employed the Ewald summation for handling the long-range Coulomb interactions and a cluster-move algorithm to improve the mobility of the macroions, and we have demonstrated the necessity of these two approaches. For example, systems with only 40 macroions are sufficient to get system-size independent results, and the cluster-move technique improves the simulation efficiency with more than two orders of magnitude. The use of these two approaches opens up the possibility of performing MC simulations of asymmetric electrolytes with macroions carrying several hundreds of elementary charges, and we can then address the structure of solutions of new classes of colloidal particles.

Acknowledgment

This work was supported by the Swedish National Research Council (NFR) and the Swedish Research Council for Engineering Science (TFR); and with computing resources from the Swedish Council for Planning and Coordination of Research (FRN), and the Center for Parallel Computers (PDC) at the Royal Institute of Technology.

References

1. M. P. Allen and D. J. Tildelsley. *Computer Simulation of Liquids.* Clarendon Press, Oxford, England, 1987.
2. A. K. Arora and B. V. R. Tata, editors. *Ordering and phase transitions in charged colloids.* VCH Publisher, Inc., New York, 1996.
3. D. F. Evans and H. Wennerström. *The Colloidal Domain Where Physics, Chemistry, Biology, and Technology Meet.* VCH Publishers, Inc., New York, 1994.
4. P. Ewald. Die Berechnung optischer und elektrostatischer Gitterpotentiale. *Ann. Phys.*, 64:253–287, 1921.
5. H. L. Gordon and J. P. Valleau. A "partially clothed" pivot algorithm for model polyelectrolyte solutions. *Mol. Simul.*, 14:361–379, 1995.
6. D. G. Grier. A surprisingly attractive couple. *Nature*, 393:621–623, 1998.
7. K. Ito, H. Yoshida, and H. Ise. Void structure in colloidal crystals. *Science*, 263:66–68, 1994.
8. C. Kang and A. Khan. Self-assembly in systems of didodecyldimethylammonium surfactants: binary and ternary phase equilibria and phase structures with sulphate, hydroxide, acetate, and chloride counterions. *J. Colloid Interface Sci.*, 156:218–228, 1993.
9. A. Khan, K. Fontell, and B. Lindman. Liquid crystaillinity in systems of magnesium and calcium surfactants. *J. Colloid Interface Sci.*, 101:193–200, 1984.
10. R. Kjellander and S. Marcelja. Inhomogeneous coulomb fluids with image interactions between planar surfaces III: Distribution functions. *J. Chem. Phys.*, 88:7138–7146, 1988.
11. A. E. Larsen and D. G. Grier. Like-charge attractions in metastable colloidal crystallites. *Nature*, 385:230–233, 1997.
12. P. Linse. *MOLSIM.* Lund University, Sweden, 1999.
13. P. Linse. On the convergence of simulation of asymmetric electrolytes with charge asymmetry 60:1. *J. Chem. Phys.*, 110:3493–3501, 1999.
14. P. Linse and B. Jönsson. A Monte Carlo study of the electrostatic interaction between highly charged aggregates: A test of the cell model applied to micellar systems. *J. Chem. Phys.*, 78:3167–3176, 1983.
15. P. Linse and V. Lobaskin. Electrostatic attraction and phase separation in solutions of like-charged colloidal particles. *Phys. Rev. Lett.*, 83:4208–4211, 1999.
16. P. Linse and V. Lobaskin. Electrostatic attraction and phase separation in solutions of like-charged colloidal particles. In press in *J. Chem. Phys.*, 2000.
17. V. Lobaskin and P. Linse. Accurate simulation of highly asymmetric electrolytes with charge asymmetry 20:1 and 20:2. *J. Chem. Phys.*, 109:3530–3541, 1998.
18. V. Lobaskin and P. Linse. Simulation of asymmetric electrolytes with charge asymmetry 60:1 using hard-sphere and soft-sphere models. *J. Chem. Phys.*, 111:4300–4309, 1999.

19. N. Metropolis, A. W. Rosenbluth, M. N. Rosenbluth, A. H. Teller, and E. Teller. Equation of state calculation by fast computing machines. *J. Chem. Phys.*, 21:1087–1092, 1953.
20. B. V. R. Tata, M. Rajalakshmi, and A. K. Arora. Vapor-liquid condensation in charged colloidal suspensions. *Phys. Rev. Lett.*, 69:3778–3781, 1992.
21. V. Vlachy. Ionic effects beyond Poisson-Boltzmann theory. *Annu. Rev. Chem.*, 50:145–165, 1999.

Towards Large Eddy Simulation of Complex Flows

Niklas Alin, Magnus Berglund, Christer Fureby, and Eric Lillberg

Warheads and Propulsion Department; Weapons and Protection Division; Defence Research Establishment (FOA); SE–172 90 Stockholm; Sweden

Abstract. Since the Navier-Stokes Equations cannot generally be directly solved for complex flows, one is left with two alternatives: a Reynolds Averaged Simulation, which is the conventional approach, or a Large Eddy Simulation. We have focused on the general applicability of the Very Large Eddy Simulation approach for three cases: flow around a prolate spheroid at incidence, flow over a fully appended submarine, and supersonic flow over a cavity in a curved surface. We have made comparisons between the predictions and experimental data for some statistical quantities.

1 Introduction

Fluid dynamics covers a variety of cases, including incompressible and compressible flows, as well as flows involving such things as chemical reactions, electro- and magnetohydrodynamics, phase changes, and solid-fluid interactions. In many real-life problems, complications arise from the complex geometries of man-made constructions that interact with the fluid. At the high *Re* of practical interest, Direct Numerical Simulations (DNS) cannot be used to resolve all scales of motion, and modeling becomes unavoidable.

Depending on the degree of sophistication and accuracy required, we separate between Reynolds Averaged Simulation (RAS) and Large Eddy Simulation (LES). In RAS the Navier-Stokes Equations (NSE) are averaged over time, homogeneous directions, or across an ensemble of equivalent flows, [32], whilst in LES the NSE are averaged over regions of space—small enough to ensure that the cut-off wavelength is within the inertial subrange, [21]. Subgrid models are required to ensure the accurate evolution of the 3D time-dependent details of the large-grid–scale motions responsible for the primary transport.

An additional issue is the adequate modeling of the boundary conditions (i.e., supergrid modeling [17]) to emulate, for example, inlet turbulence or transmissive conditions. For many practical applications, neither RAS nor LES is appropriate—RAS because of its averaging definitions, limited validity and accuracy; and LES because of its high resolution requirements. Although LES alleviates the computational demands as compared to DNS, the high *Re*/low resolution regime is still inaccessible for LES. During the last few years this problem has attracted interest, and hybrid models such as Very

Large Eddy Simulation (VLES) [33], Detached Eddy Simulation (DES), [30], and Coherent Scale Capturing (CSC) [6] have been proposed.

In the present study, a similar approach is used to calculate the evolution of the large-scale energetic eddies, with the effects of the small-scale eddies represented by a model. Filtering the NSE, explicitly or implicitly, by means of the discretization results in equations for the large-scale eddies. The effects of the subgrid eddies are modeled, either explicitly by a subgrid model or implicitly by numerical methods with built-in subgrid models [2]. The explicit model proposed is based on hierarchical simplification of the differential stress equation model of Deardorff [4], leading to an algebraic model, whilst the implicit modeling is based on the Flux-Corrected Transport (FCT) approach. The concept of (V)LES is supported by Woodruff et al. [42], documenting the effects of subgrid models for resolutions between λ_K and $2\lambda_I$.

2 The (V)LES Simulation Model

The fluid dynamic model used is the NSE [22], describing conservation of mass and balance of momentum and energy of a linear viscous fluid with Fourier heat conduction. In the explicit approach, the NSE are low-pass filtered using the decomposition $f = \overline{f} + f'$ or $f = \tilde{f} + f''$, depending on whether the flow is compressible or not, where $\overline{f} = G * f$, $\tilde{f} = \overline{\rho f}/\overline{\rho}$, and $G = G(\Delta)$. If G is assumed distributive, associative, and commutative with differentiation,

$$\begin{cases} \partial_t \overline{\rho} + \operatorname{div}(\overline{\rho}\tilde{\mathbf{v}}) = 0 \\ \partial_t(\overline{\rho}\tilde{\mathbf{v}}) + \operatorname{div}(\overline{\rho}\tilde{\mathbf{v}} \otimes \tilde{\mathbf{v}}) = -\operatorname{grad}\overline{p} + \operatorname{div}(\overline{\mathbf{S}} - \mathbf{B}) + \overline{\rho}\tilde{\mathbf{f}} \\ \partial_t(\overline{\rho}\tilde{e}) + \operatorname{div}(\overline{\rho}\tilde{\mathbf{v}}\tilde{e}) = \operatorname{div}(\overline{\mathbf{h}} - \mathbf{b}) - \overline{p}\operatorname{div}\tilde{\mathbf{v}} + \overline{\mathbf{S}} \cdot \tilde{\mathbf{D}} + \pi + \varepsilon + \overline{\rho}\tilde{\sigma} \end{cases}, \tag{1}$$

where ρ is the density, $\mathbf{v}$ the velocity, p the pressure, $\mathbf{S} = \eta \operatorname{tr}\mathbf{D}\mathbf{I} + 2\mu\mathbf{D}_D$, where $\mu = \rho\nu$ is the kinematic viscosity, η the bulk viscosity, $\mathbf{D}_D = \mathbf{D} - \frac{1}{2}(\operatorname{tr}\mathbf{D})\mathbf{I}$, $\mathbf{D} = \frac{1}{2}(\mathbf{L} - \mathbf{L}^T)$, and $\mathbf{L} = \operatorname{grad}\mathbf{v}$, $\mathbf{B} = \overline{\rho}(\widetilde{\mathbf{v} \otimes \mathbf{v}} - \tilde{\mathbf{v}} \otimes \tilde{\mathbf{v}})$ is the subgrid stress tensor, $\mathbf{f}$ the specific body force, e the specific internal energy, $\mathbf{h} = \kappa \operatorname{grad} e$ with κ the thermal conductivity, $\mathbf{b} = \overline{\rho}(\widetilde{e\mathbf{v}} - \tilde{e}\tilde{\mathbf{v}})$ the subgrid flux vector, $\pi = \overline{p \operatorname{div}\mathbf{v}} - \overline{p}\operatorname{div}\tilde{\mathbf{v}}$ the subgrid pressure dilatation, $\varepsilon = \overline{\mathbf{S} \cdot \mathbf{D}} - \overline{\mathbf{S}} \cdot \tilde{\mathbf{D}}$ the subgrid dissipation, and σ is the non-mechanical net power. The LES equations have been extensively studied, including investigations of the discretization errors [14], the commutation error [11,15], the use of frame indifference [11,31], and realizability [11,28].

The most popular closure models are based on gradient hypotheses so that $\mathbf{B} = \frac{2}{3}(\overline{\rho}k - \mu_k \operatorname{div}\tilde{\mathbf{v}})\mathbf{I} - 2\mu_k\tilde{\mathbf{D}}_D$ and $\mathbf{b} = -2\kappa_k \operatorname{grad}\tilde{e}$, where k, μ_k and κ_k are the subgrid kinetic energy, eddy-viscosity, and diffusivity, respectively, which must be modeled separately [21]. Such models perform well on coarse isotropic grids, but better models are required for coarse anisotropic grids. Such improvement may be based on the transport equations for $\mathbf{B}$ and $\mathbf{b}$. Although such models have been proposed [4] and used

[13], they put sufficient demands on computer resources to make it desirable to consider the use of simpler models. Adopting the ideas of Rodi [25] reduces the partial differential equation for $\mathbf{B}$ to a set of algebraic equations, $\mathbf{B} = k(\mathbf{P} + \Pi - \mathbf{E})/(p - \varepsilon)$, where $p = \frac{1}{2}\operatorname{tr}\mathbf{P}$ and $\varepsilon = \frac{1}{2}\operatorname{tr}\mathbf{E}$, with $\mathbf{P}$ representing production, Π pressure-strain, and $\mathbf{E}$ dissipation of $\mathbf{B}$. Given sub-models for Π and $\mathbf{E}$, an implicit algebraic model for $\mathbf{B}$ is obtained. Using the differential stress model of Deardorff [4], we have an Algebraic Stress Model (ASM) with $\mathbf{B} = \frac{2}{3}\overline{\rho}k\mathbf{I} + \frac{2}{5}\mu_k\tilde{\mathbf{D}}_D - \beta\tau(\tilde{\mathbf{L}}\mathbf{B}^T + \mathbf{B}\tilde{\mathbf{L}}^T)_D$, where $\beta = 1/(p/\varepsilon - 1 + c_1)$ and $\tau = k/\varepsilon$. Assuming equilibrium, it can be shown that $k = c_I\Delta^2\|\tilde{\mathbf{D}}\|^2$, $\mu_k = c_D\Delta^2\|\tilde{\mathbf{D}}\|$, and $\beta\tau = c_3/\|\tilde{\mathbf{D}}\|$, where $c_I = 0.13$, $c_D = 0.09$, and $c_3 = 0.66$. If $c_3 = 0$ or if $\tau \to 0$, the well-known Smagorinsky model (SMG) [29] or the One-Equation Eddy Viscosity Model (OEEVM) [27] are recovered. A similar analysis can be made for $\mathbf{b}$, but usually the gradient hypothesis is used with $\kappa_k = \kappa_k/Pr_T$, where Pr_T is the Prandtl number. Models for π are not extensively elaborated on [5], whilst considerable attention has been paid to ε [19].

Equations (1) are in this study discretized using the finite volume method, dividing the domain D into cells Ω_P so that $\cup_P\Omega_P = D\cup\partial D$ and $\cap_P\Omega_P = \emptyset$. Letting $f_P = \frac{1}{\delta V_P}\int_{\Omega_P} f\,dV$ represent the approximation to the cell-average of f over the Pth cell, Gauss theorem is used to derive the semi-discretized LES equations. Following [24], $\operatorname{grad}\overline{p}$ is not discretized at this stage, and the discretized equations are integrated in time using a multi-step method [7],

$$\begin{cases} \sum_{i=1}^{m}\left(\alpha_i\overline{\rho}_P^{n+i} + \frac{\beta_i\Delta t}{dV_P}\sum_f(\overline{\rho}\tilde{\mathbf{v}}\cdot d\mathbf{A})_f^{n+i}\right) = 0 \\ \sum_{i=1}^{m}\Big(\alpha_i(\overline{\rho}\tilde{\mathbf{v}})_P^{n+i} + \frac{\beta_i\Delta t}{dV_P}\sum_f\left((\overline{\rho}\tilde{\mathbf{v}}\cdot d\mathbf{A})\tilde{\mathbf{v}} - (\overline{\mathbf{S}} - \mathbf{B})d\mathbf{A}\right)_f^{n+i} \\ \qquad + \beta_i(\overline{\rho}\tilde{\mathbf{f}})_P^{n+i}\Delta t\Big) = -\beta_i(\operatorname{grad}\overline{p})_P^{n+i}\Delta t \\ \sum_{i=1}^{m}\Big(\alpha_i(\overline{\rho}\tilde{e})_P^{n+i} + \frac{\beta_i\Delta t}{dV_P}\sum_f\left((\overline{\rho}\tilde{\mathbf{v}}\cdot d\mathbf{A})\tilde{e} - (\overline{\mathbf{h}} - \mathbf{b})\cdot d\mathbf{A}\right)_f^{n+i} \\ \qquad - \beta_i(\overline{p}\operatorname{div}\tilde{\mathbf{v}} + \overline{\mathbf{S}}\cdot\tilde{\mathbf{D}} + \pi + \varepsilon + \overline{\rho}\tilde{\sigma})_P\Delta t\Big) = 0 \end{cases}, \tag{2}$$

where α_i and β_i are parameters and Δt the time step. Let $\tilde{\mathbf{f}}_f^C(\tilde{\mathbf{v}})$ and $\tilde{f}_f^C(\tilde{e},\tilde{\mathbf{v}})$ represent approximations to the convective fluxes $\mathbf{f}_f^C(\tilde{\mathbf{v}}) = \left((\tilde{\mathbf{v}}\cdot d\mathbf{A})\tilde{\mathbf{v}}\right)_f$ and $f_f^C(\tilde{e}) = \left((\tilde{\mathbf{v}}\cdot d\mathbf{A})\tilde{e}\right)_f$, and $\tilde{\mathbf{f}}_f^D(\tilde{\mathbf{v}})$ and $\tilde{f}_f^D(\tilde{e},\tilde{\mathbf{v}})$ approximations to the combined viscous and subgrid fluxes. The functional reconstruction of $\tilde{\mathbf{f}}_f^C(\tilde{\mathbf{v}})$, $\tilde{f}_f^C(\tilde{e},\tilde{\mathbf{v}})$, $\tilde{\mathbf{f}}_f^D(\tilde{\mathbf{v}})$, and $\tilde{f}_f^D(\tilde{e},\tilde{\mathbf{v}})$ is performed with linear interpolation across cell-faces f of Ω_P and central difference approximations to the inner gradients. To decouple the pressure-velocity system a Poisson equation for $\overline{p}$ is derived from (2_1)–(2_2), and the resulting system is solved sequentially with iteration over the explicit coupling terms using an implicit method.

A recent alternative to LES is Monotone Integrated LES (MILES) [2,9], in which the NSE are solved using high-resolution monotone methods with embedded non-linear filters providing implicit subgrid models, thus making explicit models obsolete. In MILES, the finite volume discretization effectively filters the NSE over cells Ω_P using an anisotropic kernel. The functional reconstruction of the convective terms $\mathbf{f}_f^C(\mathbf{v})$ and $f_f^C(e,\mathbf{v})$ is performed with a flux-limiting method combining a high-order flux-function with a low-order dispersion-free flux-function using a nonlinear flux-limiter Γ. The FCT concept [1], which attempts to incorporate as much as possible of the anti-diffusion term, without violating causality, positivity, or monotonicity, is used. The high-order flux-function is based on linear interpolation, the low-order flux-function on first-order upwinding, and the flux-limiter on the normalized kinetic energy and a parameter. The functional reconstruction of $\tilde{\mathbf{f}}_f^D(\tilde{\mathbf{v}})$ and $\tilde{f}_f^D(\tilde{e},\tilde{\mathbf{v}})$, the handling of the pressure-velocity system, and the solvers are the same as for the (V)LES model.

A comparison between the (V)LES and MILES models is best obtained from the modified equations, involving Taylor series expansions of the discretized equations. For (V)LES,

$$\begin{cases} \partial_t\bar{\rho} + \operatorname{div}(\bar{\rho}\tilde{\mathbf{v}}) = \tau_t(\bar{\rho}) + \tau_c(\bar{\rho}) + \tau_d(\bar{\rho}) + O(\Delta t^3, |\mathbf{d}|^3) \\ \partial_t(\bar{\rho}\tilde{\mathbf{v}}) + \operatorname{div}(\bar{\rho}\tilde{\mathbf{v}}\otimes\tilde{\mathbf{v}}) = -\operatorname{grad}\bar{p} + \operatorname{div}(\overline{\mathbf{S}} - \mathbf{B}) + \bar{\rho}\tilde{\mathbf{f}} \\ \qquad\qquad +\tau_t(\tilde{\mathbf{v}}) + \tau_c(\tilde{\mathbf{v}}) + \tau_d(\tilde{\mathbf{v}}) + O(\Delta t^3, |\mathbf{d}|^3) \\ \partial_t(\bar{\rho}\tilde{e}) + \operatorname{div}(\bar{\rho}\tilde{\mathbf{v}}\tilde{e}) = \operatorname{div}(\overline{\mathbf{h}} - \mathbf{b}) - \bar{p}\operatorname{div}\tilde{\mathbf{v}} + \overline{\mathbf{S}}\cdot\tilde{\mathbf{D}} + \bar{\rho}(\pi + \varepsilon + \tilde{\sigma}) \\ \qquad\qquad +\tau_t(\tilde{e}) + \tau_c(\tilde{e}) + \tau_d(\tilde{e}) + O(\Delta t^3, |\mathbf{d}|^3) \end{cases}, \tag{3}$$

where $\tau_t(\cdot)$, $\tau_c(\cdot)$, and $\tau_d(\cdot)$ are the leading-order temporal convective and diffusive truncation errors. Similarly for the MILES model,

$$\begin{cases} \partial_t\rho + \operatorname{div}(\rho\mathbf{v}) = \tau_t(\rho) + \tau_c(\rho) + \tau_d(\rho) + O(\Delta t^3, |\mathbf{d}|^3) \\ \partial_t(\rho\mathbf{v}) + \operatorname{div}(\rho\mathbf{v}\otimes\mathbf{v}) = -\operatorname{grad}p + \operatorname{div}\mathbf{S} + \rho\mathbf{f} + \tau_t(\mathbf{v}) - \operatorname{div}\mathbf{B} \\ \qquad\qquad +\tau_c(\mathbf{v}) + \tau_d(\mathbf{v}) + O(\Delta t^3, |\mathbf{d}|^3) \\ \partial_t(\rho e) + \operatorname{div}(\rho\mathbf{v}e) = \operatorname{div}\mathbf{h} - p\operatorname{div}\mathbf{v} + \mathbf{S}\cdot\mathbf{D} + \rho\sigma + \tau_t(e) - \operatorname{div}\mathbf{b} \\ \qquad\qquad +\tau_c(e) + \tau_d(e) + O(\Delta t^3, |\mathbf{d}|^3) \end{cases}, \tag{4}$$

where $\mathbf{B} = \mathbf{C}\mathbf{L}^T + \mathbf{L}\mathbf{C}^T + \beta^2\mathbf{L}\mathbf{d}\otimes\mathbf{L}\mathbf{d}$ and $\mathbf{b} = \rho\mathbf{C}\operatorname{grad}e$, where $\mathbf{C} = \beta(\mathbf{v}\otimes\mathbf{d})$ and $\beta = (1-\Gamma)\operatorname{sgn}\mathbf{v}$, is the leading-order truncation error of the convective terms, which acts as the implicit subgrid model.

3 Code Description and Validation Studies

A significant validation effort has accompanied the development of the present simulation capacity. The code FOAM (Field Operation And Manipulation) is designed as a C++ class library for Computational Continuum Mechanics with which it is easy to develop codes to investigate modeling and simulation of fluid or solid mechanics problems. As well as simplifying implementation of contemporary models, this makes checking the modeling more straightforward. This aspect is enhanced by the inclusion of features such as automatic dimension-checking. As well as this, the object-oriented paradigm methodology has enabled the dissociation of different levels of the code, thus minimizing unwanted interaction and permitting research into the numerics to be largely divorced from that of modeling. In addition, the fluid dynamic simulation models used are carefully tested and validated against simple flows. The overall design of the code and some illustrative simulation cases are presented in [40]. Furthermore, the code has been used to investigate several flow problems, for example, [12,8,10,39]. The code is parallelized using domain decomposition to partition and map the data space onto a set of processors. Static load balancing is done prior to the domain decomposition stage using information on the available computer system and the grid. FOAM employs PVM for message passing in view of its availability, portability, and functionality. The parallel code typically has communications overheads of less than 5 percent and scalability studies indicate that state-of-the-art hardware can achieve parallel efficiencies of 99 percent on up to 256 processors for 10^6 grid points.

4 Flow around a Prolate Spheroid at Incidence

The first case considered is the incompressible flow around a prolate spheroid, with a 6:1 major-minor semi-axis ratio, mounted in a cylindrical wind-tunnel. Despite the simplicity of the geometrical configuration, the flow field shows complex behavior with, at least for high angles of attack, several separations. This type of flow has been studied both experimentally (e.g., [3,16,41]) and numerically (e.g., [23,36]), for various angles of attack and body-length *Re*-numbers. We have in the present investigation focused on a 20° angle of attack and a body-length *Re*-number of $4.2 \cdot 10^6$.

The computational domain consists of the interior of the wind-tunnel, with the same hydraulic radius as used in the experimental studies [3,41], except for the space occupied by the prolate spheroid. For this configuration we have used block-structured O-grids, composed of three blocks, with approximately $1.5 \cdot 10^5$ and $4.5 \cdot 10^5$ cells. In an attempt to resolve as much as possible of the boundary layer, the grid is clustered towards the prolate spheroid and stretched in other regions, keeping moderate cell aspect ratios. The outer boundary is divided in three parts—an inlet boundary, a wind-tunnel wall boundary, and an outlet boundary—whilst the inner boundary consists of the surface of the prolate spheroid. At the inlet boundary we set $\text{grad}\,\overline{p} \cdot \mathbf{e}_1 = 0$, where $\mathbf{e}_k$ denotes the unit vector in the x_k-direction, and $\overline{\mathbf{v}} = u_0\mathbf{e}_1 + \overline{\mathbf{u}}'$, where $\overline{\mathbf{u}}'$ denotes a random perturbation in the form of Gaussian noise with variance $0.01u_0$ in the $\mathbf{e}_1$-direction and $0.005u_0$ in the $\mathbf{e}_2$- and $\mathbf{e}_3$-directions to emulate the upstream turbulence. To reduce the computational cost, we used slip conditions on the wind-tunnel walls and, thus, no effort was made to resolve the flow-field in the proximity of the wind-tunnel. At the outlet, we assumed $\overline{p} = const.$ and $(\text{grad}\,\overline{\mathbf{v}})\mathbf{e}_1 = \mathbf{0}$, and for the inner boundary we used no-slip conditions. The simulations were initiated with quiescent conditions, and the unsteady flow then evolved naturally. For the coarse grid, simulations with SMG, OEEVM, and MILES were performed; while for the fine grid, only OEEVM and MILES were used.

This type of flow is both difficult to model and poorly understood, because it exhibits all the fundamental transition and open separation phenomena of 3D flows. Due to the high *Re* we cannot resolve the flow in the proximity of the prolate spheroid; and we must rely on the subgrid model to give a proper relaxation of the velocity, and hence a well predicted velocity gradient, towards the body. This demand is put forward because the accurate prediction of the subgrid kinetic energy, the subgrid viscosity, and the surface wall shear stress (all depending on $\overline{\mathbf{v}}$ and $\overline{\mathbf{L}}$) are of utmost importance in predicting the viscous drag and the hydro-acoustic characteristics of the body. Color Plate 15 on page 294 shows a perspective view of the prolate spheroid with the hull colored by the static pressure and the flow pattern represented by stream ribbons and the streamwise vorticity $\overline{\omega}_1 = (\operatorname{curl}\overline{\mathbf{v}})_1$ in three equidistant cross sections. As can be seen, the most prominent phenomenon in this flow is the strong vortices on the leeward side of the spheroid, generated by separation on the forebody. Further, these primary vortices are accompanied by several smaller vortices, and the vortex interactions give rise to a highly skewed 3D shear layer. Clearly visible is the downstream development of the primary vortices. In the fore–cross section, the vortex is rather flat and located close to the body. In the mid–cross section, the vortex is still located close to the body, but it is larger in both the radial and circumferential directions, and the vortex core has become more circular. In the aft–cross section the vortex is even larger and its core is even more circular. Moreover, the vortex has clearly detached from the body at this stage. Also visible in this cross section is the secondary vortex located underneath the primary one.

Fig. 1 shows a comparison of experimental data from [3] and [41] with computed circumferential pressure coefficient distributions for the two grids and the three subgrid models (a) at $X/L = 0.600$ and (b) at $X/L = 0.772$, where X denotes the major axial distance from the nose and L denotes the length of the body. It is clear from these results that the computations can be considered to be grid independent on the windward side and almost grid independent on the leeward side. On the windward side there are only minor differences between the subgrid models used while the differences on the leeward side are more pronounced. In this study we make no attempt to explain the origin of these differences, but merely notice their existence.

It is also clear that we have good quantitative agreement with experimental data for the windward side at $X/L = 0.772$. On the leeward side we have good qualitative and reasonable quantitative agreement with experimental data at both $X/L = 0.600$ and $X/L = 0.772$. Also note that C_p is somewhat overestimated at both locations and also suffers from a phase shift at $X/L = 0.600$. Possible reasons for these discrepancies can be found by comparing both parts of Color Plate 16 on page 294 with experimental data from [3] and [41], respectively. Let us denote the radial distance from the body normalized with the radius by r/R. For the location of the primary vortex center at $X/L = 0.600$, we find $(r/R, \phi) \approx (0.10, 148°)$ and $(r/R, \phi) \approx (0.16, 157°)$ for prediction and experiments, respectively, while the corresponding values

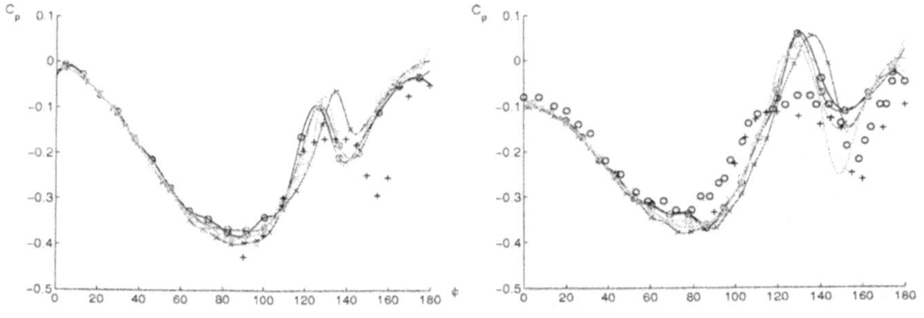

(a) (b)

Fig. 1. Comparison of measured and predicted circumferential static pressure coefficient distributions (a) at $X/L = 0.600$ and (b) at $X/L = 0.772$. Experimental data from [3] are denoted with "+" and from [41] with "∘." For the computed results, continuous lines denote SMG, OEEVM, and MILES, while "o" denotes the coarse grid and "×" the fine grid.

for $X/L = 0.772$ are $(r/R, \phi) \approx (0.26, 152°)$ and $(r/R, \phi) \approx (0.31, 155°)$, respectively. It is natural to assume that the phase shift is caused by the error in the predicted circumferential location of the primary vortex center. Likewise, the error in the predicted radial location of the primary vortex center is believed to be responsible for the overestimation of C_p.

5 Flow over a Fully Appended Submarine Hull

As an example of a complex geometry high-Re number flow, we consider the flow around a fully appended submarine [18,20]. The hull, sail, and stern appendices are created from analytical descriptions, and the 1:24 scale model is mounted in a wind tunnel with an almost circular cross section (see panel (a) of Color Plate 17 on page 295). For the hull an O-grid topology is used, and for the sail and stern appendages C-O grid topologies are adopted, resulting in block-structured grids of 10 blocks with about $1.0 \cdot 10^6$ cells. Care is taken to ensure that the first cell spacing and the aspect ratios are suitable for capturing the viscous boundary layer over the geometry. The simulations start from rest, and the unsteady flow evolves naturally. The boundary conditions for the submarine simulations are similar to those of the prolate spheroid, with the only exception being that the inlet turbulence is isotropic with a variance of $0.01u_0$.

For the purpose of this study we have focused on the SMG, ASM, and MILES models. Panel (b) of Color Plate 17 on page 295 shows a perspective view of the flow at 0° pitch and 0° drift, with the hull colored by the static pressure coefficient and the flow represented by stream ribbons and the streamwise vorticity $\overline{\omega}_1$. It is evident that the flow is unsteady with distinct boundary layer development along the hull, and that the presence of the sail introduces additional complexities to the flow along the midbody section of the hull, manifested by the sail-tip vortices and the necklace-shaped edge horseshoe vortex. The main feature of the flow around the forebody is a transition region in which the flow over the bow becomes turbulent. The flow around the sail and along the midbody section of the hull is dominated by a leading-edge horseshoe vortex, originating at the stagnation point at the junction between the sail and the hull. The sail-tip vortex pair is stable and maintains its coherence far downstream where it eventually breaks. The leading-edge horseshoe vortex persists at the junction between the sail and the hull until it starts to interact with the vortex street formed at the trailing edge of the sail, where it becomes unstable and detaches from the hull.

Further towards the stern this vortex-pair impinges on the hull, generating a secondary longitudinal vortex on either side of the primary vortex. As the leading-edge horseshoe vortex is advected further aft, it partly loses its coherence and flattens. During this process this vortex pair interacts with other vortical structures produced mainly from the curved hull boundary layer, which thickens with downstream distance from the sail. The flow in the region just ahead of the propulsor plane is complex, with multiple streamwise vortices from the hull interacting with the appendage-generated vortices whilst redistributing axial-radial Reynolds stresses.

Figure 2a shows the measured and predicted pressure coefficient $C_p = 2(p-p_0)/(\rho\langle v_1\rangle^2)$ along the upper meridian line for different subgrid models. The overall agreement is good for all three subgrid models; however, some differences can be detected at the leading and trailing edges of the sail and at the leading edge of the top dead center stern appendage. Correct prediction of the static pressure coefficient is important to predict occurrence and location of cavitation, which increases resistance and the acoustic signature. Figure 2b shows a comparison of the measured and predicted circumferentially averaged streamwise velocity profile, that is, $\langle\langle v_1\rangle_t/u_0\rangle_\varphi$, at $x_1/L = 0.978$. It appears that the (V)LES models successfully predict the stern boundary layer profile and the recovery to the freestream flow, with only minor differences between the models. The lack of influence of the fairwater on the circumferentially averaged boundary layer profiles should also be observed, given the additional flow complexity it causes.

Figures 3a, 3b, and 3c show contours of the predicted mean axial velocity, vorticity, and axial rms-velocity fluctuations at the propulsor plane $x_1/L = 0.978$. Each appendage results in one longitudinal vortex-pair from its tip and one from its juncture to the hull. The centers of the eight vortex-pairs shed from the four stern appendages are estimated to be at $r/R \approx 0.24$ and $\varphi \approx 18°$ and at $r/R \approx 0.36$ and $\varphi \approx 22°$ from the center plane of each appendage. These results seems to be in good agreement with the experimental data of Huang et al. [20]. The sail-tip vortices and the leading-edge horseshoe vortex appear to be weak at this location. The velocity contours, however, show typical evidence of the sail. Pairs of counter-rotating streamwise vortices, generated by each appendage, bring high momentum fluid behind the

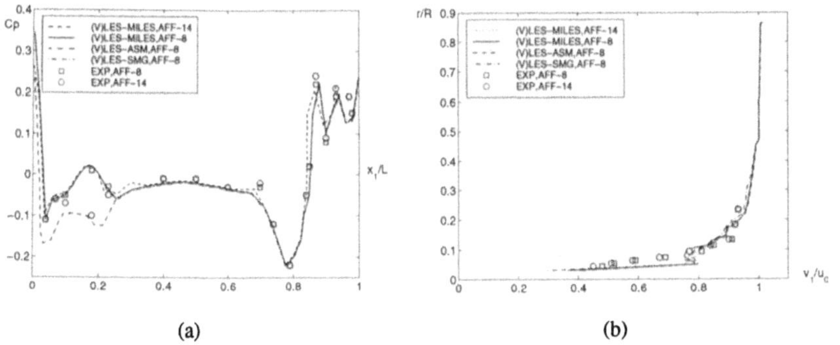

Fig. 2. Comparison of (a) the measured and predicted static pressure coefficient C_p along the upper meridian line for the AFF-8 case, and (b) the circumferentially averaged velocity profiles at the propulsor plane $x_1/L = 0.978$.

appendage down towards the hull. Hence, in the inner stern boundary layer the streamwise velocities behind the appendages are observed to be higher than those on either side of the appendage. The streamwise rms-velocity fluctuations show clear evidence of the leading-edge horseshoe vortex in terms of regions of high velocity fluctuations on the sail-side of the hull.

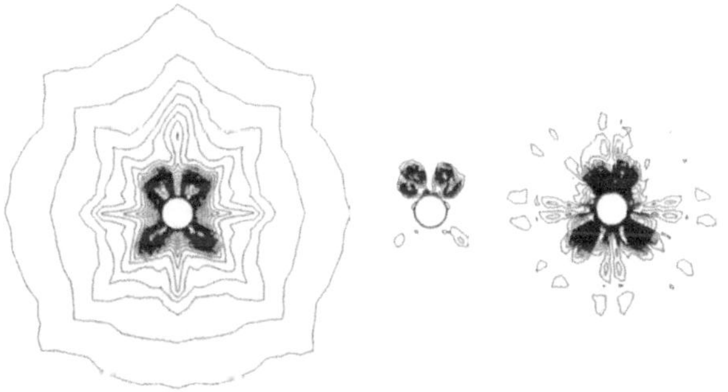

(a) (b) (c)

Fig. 3. Predicted mean flow in the propulsor plane $x_1/L = 0.978$ in terms of (a) streamwise velocity, (b) rms-velocity fluctuations, and (c) streamwise vorticity.

6 Supersonic Flow over a Cavity in a Curved Surface

The ability to accurately predict the fluctuating transonic or supersonic flow over rectangular cavities are not only of interest from a fluid dynamics or numerical point of view, but also for engineers in their effort to design air breathing engines and vehicles for transonic or supersonic speeds. The rapid pressure fluctuations inside the cavity give rise to acoustic phenomena and time-dependent structure loads due to the interactions of vortices, shear layers, and shock waves inside the cavity [26,37,38].

In the present study, (V)LES is performed for a flow over a rectangular cavity with a length-to-depth ratio of 2 embedded in a 3D arc-shaped body at *Ma* 1.52 and a *Re* of $60 \cdot 10^6$ based on the cavity length. Because the study of oscillatory turbulent flow fields in cavity flows involves capturing small-scale time-dependent structures in the supersonic flow, it puts high demands on the numerical scheme to be used. The problem requires a high-resolution scheme with at least second-order accuracy in smooth regions and with a well resolved, non-oscillatory behavior near discontinuities and sharp gradients. The numerical computations use the previously presented MILES approach on two different 3D grids, with the number of nodes varying from $3.0 \cdot 10^5$ to $7.0 \cdot 10^5$, as well as a 2D grid, with $6.8 \cdot 10^4$ for comparison.

Previous 2D investigations on cavity flows, (see [35]), have shown a characteristic flow pattern inside the cavity and a strong shear layer extending from the leading edge. The FFT analysis of the pressure field at the trailing edge of the cavity shows periodic phenomena due to standing acoustic waves and the transversal movement of the shear layer. The computations made in [35] show good agreement with experiments (see [26]) on small cavities at moderate *Re*. In the presented work the effort is extended to 3D high-*Re* supersonic flow over a rectangular cavity embedded in a non-flat surface. The simulations are made without the interaction of any wind-tunnel walls to observe the evolution of the undisturbed flow field downstream of the cavity. The MILES approach, using a Courant number of 0.2, produces energy spectra with a clear asymptotic $|\mathbf{k}|^{-5/3}$ behavior in the inertial subrange (see Fig. 4).

Of special interest in the cavity flow pattern is the periodic pressure fluctuation inside the cavity, because it can cause structural fatigue. An FFT analysis of the pressure fluctuations, sampled at the midpoint of the downstream edge of the cavity, for the different computations performed, shows similar results for the 3D simulations, regardless of the number of nodes; whereas the frequency content obtained in two dimensions is quite different. The fundamental peaks in the 3D cases are predicted at approximately the same frequencies, whilst the number of peaks as well as their distribution in the 2D case differs significantly, (see Fig. 5). The mechanisms responsible for the periodic pressure oscillations, as described in [37] and predicted by 2D Euler and RAS simulations in [35], are standing acoustic waves inside the cavity, giving peak frequencies at $f_n = na/2L$, and shedding of vortices in the shear layer. These mechanisms are reproduced by the 2D simulation within this study, but in the 3D cases, the characteristic flow pattern differs greatly. The transversal motion of the shear layer is still present but now accompanied by longitudinal edge vortices along the cavity sides as well as longitudinal,

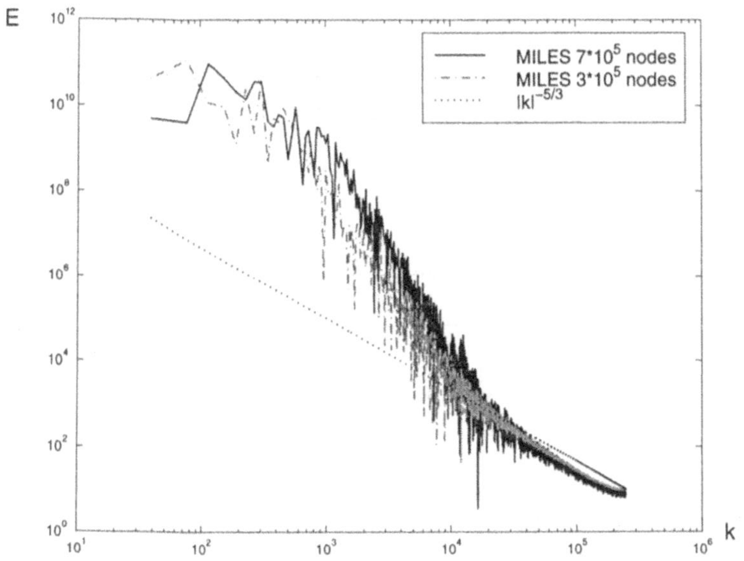

Fig. 4. Energy spectra for the computed 3D supersonic flow over the cavity using two different grids.

transversal, and vertical vortices downstream of the cavity (see panel (a) of Color Plate 18 on page 295). Also important is that the low-pressure regions inside the cavity are less stationary than in two dimensions, and the internal flow pattern is characterized by complex 3D vorticity comprising longitudinal, vertical, and transversal structures (see panel (b) of Color Plate 18 on page 295).

It could be argued from these results that the effects of 3D turbulence on the frequency content of the fluctuating pressure should not be disregarded. The rotational properties of 2D turbulence, as shown analytically in [34], are considerably different from those of "real" 3D turbulence where, in two dimensions, the rotational dependence of the momentum equation is absorbed by the fluctuating pressure.

7 Concluding Remarks

In many flow problems the *Re*-number is high and the geometry so complicated that traditional RAS models are insufficient and conventional DNS and LES models are too costly. Further, non-uniform grids are often required so that the spatial resolution and the quality of the grid vary widely over the computational domain. In this study, an emerging computational capability for physics-based simulation of flows using a (V)LES model has been demonstrated. The key feature of this model is that the low-pass filtered

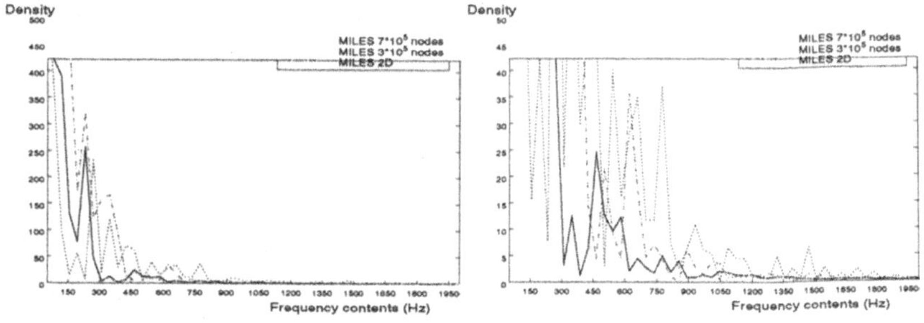

(a) (b)

Fig. 5. Pressure frequency contents for the 3D and 2D simulations, (a) showing density peaks and (b) showing a zoom-in on the smaller peaks.

NSE are solved together with an explicit or implicit subgrid model emulating the macroscopic effects of the energy cascade. With current computers, *Re*-numbers of about $10 \cdot 10^6$ are within reach and therefore so are problems of practical relevance. Following Woodruff et al. [42], we have mainly used simple subgrid models of generalized eddy-viscosity type. Pending a more detailed analysis of the requirements for the subgrid model, a generalized eddy-viscosity model—with the viscosity being a fourth-order tensor (thus facilitating simulation of anisotropic turbulence)—may be considered sufficient for capturing the first- and second-order statistics of the velocity. However, for more complex problems and for problems requiring additional quantities to be evaluated, more advanced subgrid models should probably be used. In this study we have experimented with both explicit and implicit models, giving no indication of the most appropriate approach. Three flow problems have been analyzed: flow around a prolate spheroid at incidence, flow over a fully appended submarine, and supersonic flow over a cavity in a curved surface. Comparisons between the predictions and experimental data for some statistical quantities show satisfactory agreement.

Acknowledgments

H. G. Weller is acknowledged for the initiation and continuous development of FOAM.

References

1. J. P. Boris, J. Gardner, A. Landsberg, G. Patniak, and E. S. Oran. LCPFCT: A monotone algorithm for solving continuity equations. NRL Memorandum Report 6410, Naval Research Laboratory, Washington, D.C., 1993.
2. J. P. Boris, F. F. Grinstein, E. S. Oran, and R. L. Kolbe. New insights into large eddy simulation. *Fluid Dyn. Res.*, 10(199), 1992.

3. C. J. Chesnakas and R. L. Simpson. Detailed investigation of the three-dimensional separation about a 6:1 prolate spheroid. *AIAA J.*, 35(990), 1997.
4. J. W. Deardorff. The use of subgrid transport equations in a three-dimensional model of atmospherical turbulence. *J. Fluids Engng. Trans ASME*, 95(429), 1973.
5. G. Erlebacher, M. Y. Hussaini, C. G. Speziale, and T. A. Zang. Toward the large-eddy simulation of compressible turbulent flows. *J. Fluid Mech.*, 238(155), 1992.
6. J. H. Ferziger. Simulation of complex turbulent flows: Recent advances and prospects in wing engineering. In S. Murakami, editor, *Cpm. Wind Eng.*, volume 1, 1993.
7. C. A. J. Fletcher. *Computational Methods in Fluid Dynamics.* Springer-Verlag, Berlin & New York, 1992.
8. C. Fureby. Large eddy simulation of rearward facing step flow. Accepted for publication in *AIAA J.*, November 1999.
9. C. Fureby and F. F. Grinstein. Monotonically integrated large eddy simulations. *AIAA J.*, 37(544), 1999.
10. C. Fureby, Y. Nilsson, and K. Andersson. Large eddy simulation of supersonic base flow. Paper 99-0426, AIAA, 1999.
11. C. Fureby and G. Tabor. Mathematical and physical constraints on large eddy simulations. *J. Theoretical Fluid Dyn.*, 9(85), 1997.
12. C. Fureby, G. Tabor, H. Weller, and A. D. Gosman. A comparative study of sub grid scale models in homogeneous isotropic turbulence. *Phys. Fluids*, 9(5):1416–1429, 1997.
13. C. Fureby, G. Tabor, H. G. Weller, and A. D. Gosman. On differential sub grid scale stress models in large eddy simulations. *Phys. Fluids*, 9(11):3578–3580, 1997.
14. S. Ghosal. An analysis of numerical errors in large-eddy simulations of turbulence. *J. Comp. Phys.*, 125(187), 1996.
15. S. Ghosal and P. Moin. The basic equations for the large eddy simulation of turbulent flows in complex geometry. *J. Comp. Phys.*, 118(24), 1995.
16. M. Goody, R. L. Simpson, M. Engel, C. Chesnakas, and W. Devenport. Mean velocity and pressure and velocity spectral measurements within a separated flow around a prolate spheroid at incidence. Paper 98–0630, AIAA, 1998.
17. F. F. Grinstein. Open boundary conditions in the simulation of subsonic turbulent shear flows. *J. Comp. Phys.*, 115(43), 1994.
18. N. C. Groves, T. T. Huang, and M. S. Chang. Geometric characteristics of DARPA SUBOFF models. Report DTRC/SHD-1298-01, DARPA, 1989.
19. C. Härtel, L. Kleiser, F. Unger, and R. Friedrich. Subgrid-scale energy transfer in the near wall region of turbulent flows. *Phys. Fluids*, 6(3130), 1994.
20. T. T. Huang, H.-L. Liu, N. C. Groves, T. J. Forlini, J. Blanton, and S. Gowing. Measurements of flows over an axisymmetric body with various appendages (DARPA SUBOFF experiments). In *19th Symposium on Naval Hydrodynamics Proceedings*, Seoul, South Korea, 1992.
21. M. Lesieur and O. Metais. New trends in large-eddy simulations of turbulence. *Annu. Rev. Fluid Mech.*, 28(45), 1996.
22. P.-L. Lions. *Mathematical Topics in Fluid Mechanics, Vol. 1: Incompressible Models.* Oxford Lecture Series in Mathematics and Its Applications. Oxford University Press, Oxford, England, 1996.

23. V. C. Patel and J. H. Baek. Boundary layers and separation on a spheroid at incidence. *AIAA J.*, 23(55), 1985.
24. C. M. Rhie and W. L. Chow. A numerical study of the turbulent flow past an isolated airfoil with trailing edge separation. Paper 82–0998, AIAA, 1982.
25. W. Rodi. A new algebraic relation for calculating the Reynolds stresses. *ZAMM*, 56(219), 1976.
26. J. E. Rossiter. Wind tunnel experiments on the flow over rectangular cavities at supersonic and transonic speeds. Reports and Memoranda 3438, Royal Aircraft Establishment, 1964.
27. U. Schumann. Subgrid scale model for finite difference simulation of turbulent flows in plane channels and annuli. *J. Comp. Phys.*, 18(376), 1975.
28. U. Schumann. Realizability of Reynolds stress turbulence models. *Phys. Fluids*, 20(721), 1977.
29. J. Smagorinsky. General circulation experiments with the primitive equations, i: The basic experiment. *Month. Wea. Rev.*, 91(99), 1963.
30. P. R. Spalart, W.-H. Jou, M. Strelets, and S. R. Allmaras. Comments on the feasibility of LES for wings, and on a hybrid RANS/LES approach. In *Advances in DNS/LES (1st AFOSR Int. Conf. on DNS/LES Proceedings)*, Ruston, LA, August 1997.
31. C. G. Speziale. Invariance of turbulent closure models. *Phys. Fluids*, 22(1033), 1979.
32. C. G. Speziale. Modeling of turbulent transport equations. In T. B. Gatski, M. Y. Hussaini, and J. L. Lumley, editors, *Simulation and Modeling of Turbulent Flows*. Oxford University Press, Oxford, England, 1996.
33. C. G. Speziale. Turbulence modelling for time dependent RANS and VLES: A review. *AIAA J.*, 36(173), 1999.
34. C.G. Speziale. Some interesting properties of two-dimensional turbulence. *Phys. Fluids*, 24(1425), 1981.
35. Y. Takakura, F. Higashino, T. Yoshizawa, and S. Ogawa. Numerical study on unsteady supersonic cavity flows. Paper 96-2092, AIAA, 1996.
36. C.-Y. Tsai and A. K. Whitney. Numerical study of three-dimensional flow separation for a 6:1 ellipsoid. Paper 99–0172, AIAA, 1999.
37. Y. Watanabe. *Characteristics of Pressure Oscillations in a Supersonic Flow around Cavities*. PhD thesis, Tokyo Noko University, March 1997.
38. Y. Watanabe, S. Kudo, and F. Higashino. Effects of cavity configurations on supersonic internal flow. In *20th ISSW Proceedings*, 1995.
39. H. G. Weller, G. Tabor, A. D. Gosman, and C. Fureby. Application of a flame-wrinkling LES combustion model to a turbulent shear layer formed at a rearward facing step. In *The 27th Int. Symp. on Comb.* The Combustion Institute, 1998.
40. H. G. Weller, G. Tabor, H. Jasak, and C. Fureby. A tensorial approach to CFD using object oriented techniques. *Comp. in Phys.*, 12(629), 1997.
41. T. G. Wetzel, R. L. Simpson, and C. J. Chesnakas. Measurement of three-dimensional crossflow separation. *AIAA J.*, 36(557), 1998.
42. S. L. Woodruff, J. M. Seiner, and M. Y. Hussaini. Grid-size dependence considerations for sub grid scale models for LES of Kolmogorov flows. Presented at the APS DFD Meeting, San Francisco, CA, USA, 1997.

Computation of Dendrites on Parallel Distributed Memory Architectures

Christer Andersson

Department of Numerical Analysis and Computing Science; Royal Institute of Technology; SE–100 44 Stockholm; Sweden; <christe@nada.kth.se>

Abstract. A code for simulating the solidification of a pure material from its undercooled melt based on a phase field approach has been written for parallel distributed memory architectures using MPI. The numerical scheme is based on finite differences and results in large sparse non-linear systems, which are then solved by a backtracking line search modification of Newton's method combined with GMRES. Experiments conducted on an IBM SP2 and networks of Sun Ultra 5 workstations show that the code scales well with the number of processors.

1 Introduction

We are interested in accurately simulating the solidification of a pure material from its undercooled melt, while taking into account the effects of surface tension and kinetic undercooling. This is perhaps the simplest physical process in which one can observe the growth of dendritic crystals, characterized by complicated tree-like patterns. Some properties of solids, such as electric conductivity and mechanical strength, are determined by the micro-structures produced upon solidification, and one would like to better understand the pattern formation in order to obtain desired properties in manufacture.

The classical formulation of the solidification problem results in heat equations defined in the solid and melt, coupled with boundary conditions at the solid-melt interface. Solving this problem has proved difficult, because it implies explicit tracking of a moving boundary. To avoid this difficulty, the implementation has been based on a phase field approach.

By introducing a smooth approximation of a thermodynamic order function taking on distinct values in solid and melt, the so-called phase field, a set of reaction-diffusion equations valid in the entire computational domain can be derived. The model used here is based on a derivation from basic thermodynamic principles [17], which ensures that the entropy of the system is monotonically increasing in time. In this formulation, the moving boundary is replaced by a diffuse interface of width ϵ, in which the phase field changes value from zero (solid) to one (melt). One can prove by rigorous asymptotic analysis that the sharp interface model is recovered as the interface thickness

$\epsilon \to 0$ [3]. The governing equations can be formulated as

$$\frac{\epsilon^2}{m}\frac{\partial \phi}{\partial t} = \phi(1-\phi)\big(\phi - \frac{1}{2} + 30\epsilon\alpha\Delta u\phi(1-\phi)\big) + \epsilon^2\nabla_a^2\phi \ , \tag{1}$$

$$\frac{\partial u}{\partial t} = -\frac{1}{\Delta}\frac{\partial}{\partial t}\Big(\phi^3(10 - 15\phi + 6\phi^2)\Big) + \nabla^2 u \ , \tag{2}$$

where $\phi(\mathbf{x},t)$ is the phase field and $u(\mathbf{x},t)$ is the dimensionless temperature scaled such that the melting temperature is 0 and the initial temperature is -1 on the boundaries. α, Δ and m are dimensionless positive physical constants. The Laplacian in (1) is non-linear in ϕ and has been modified to account for preferred growth directions in the material (see Sect. 2.1).

There is no need for explicit representation of the solid-melt interface, because its location is implicitly given by level sets of ϕ, for example $\{\mathbf{x}(t) : \phi(\mathbf{x}(t),t) = 1/2\}$, which greatly simplifies the handling of merging interfaces and computation in two or three space dimensions. In the sharp interface formulation, there is a discontinuity in the normal derivative of the temperature across the interface, which is replaced here by a rapid transition, due to the smoothness of the approximation. This makes it simpler to formulate numerical approximations.

However, in order to accurately resolve the interface, the grid resolution must be of order ϵ. This implies large computational grids, because ϵ should be small in order to be close to the sharp interface formulation. Also, the introduction of the phase field variable increases the dimensionality of the problem. In three dimensions, the solid-melt interface is given by a surface (2D); but to avoid explicit representation of this surface, one introduces a variable defined in the entire computational domain (3D). These are the major drawbacks of the phase field approach and some motivations for writing a parallel implementation.

Several papers have been published in which both qualitative [11] and quantitative [1,9,19] agreement between phase field models and sharp interface formulations have been investigated. Physical problems, such as solidification of binary alloys [18] and dendritic solidification under external flow [16], have also been addressed using a phase field approach. Other approaches include direct simulation of the sharp interface equations based on front tracking [15] or level set methods [4]. Some of the papers [1,9,15,16] describe methods based on adaptive meshing.

This implementation is part of an ongoing work aimed at finding computationally efficient numerical methods for simulation of dendritic growth. Because the phase field variable is not conserved, an accurate numerical method must be used in order to obtain the correct propagation speed of the solid-melt interface. The main purpose here is to investigate whether an implicit second-order scheme can be used to obtain reliable results within times comparable to semi-explicit first-order schemes based on a splitting approach.

2 Numerical Method

In this section, the numerical method used to solve (1) and (2) is introduced. We will first see how anisotropy can be included in (1) and complete the problem formulation by giving initial and boundary conditions. Via the method of lines, the partial differential equations can be replaced by a large non-linear system of equations, which must be solved every time step. This is done using a modification of Newton's method, which basically reduces the numerical method to solving sparse linear systems. For this purpose, the Generalized Minimal Residual Method (GMRES) [14] has been employed.

2.1 Mathematical Formulation

For many materials, including metals, the surface energy of the solid-melt interface will depend on orientation. To account for these effects in the phase field model, one needs to modify the parameter ϵ for the gradient terms, or equivalently [19], use

$$\nabla_a^2 = -\frac{\partial}{\partial x}\Big(\eta(\theta)\eta'(\theta)\frac{\partial}{\partial y}\Big) + \frac{\partial}{\partial y}\Big(\eta(\theta)\eta'(\theta)\frac{\partial}{\partial x}\Big) + \nabla\big(\eta^2(\theta)\nabla\big) \ , \qquad (3)$$

in (1), where η is the function used to model the anisotropy and θ is the angle between the normal to the interface and some reference direction,

$$\eta(\theta) = 1 + \gamma\cos(k\theta), \qquad \theta = \arctan\Big(\frac{\partial\phi/\partial y}{\partial\phi/\partial x}\Big) - \theta_0 \ . \qquad (4)$$

γ determines the magnitude of anisotropy, and k is the mode number. Because θ is determined by derivatives of ϕ, the anisotropic Laplacian (3) is non-linear.

The phase field equations are solved on rectangular domains with zero Neumann conditions prescribed for both ϕ and u at the boundaries ($(\nabla\phi)\cdot\hat{n} = (\nabla u)\cdot\hat{n} = 0$). The implementation can handle a wide variety of initial conditions, the most common choice being

$$\phi(x,y,0) = \frac{1}{2}\Big[\tanh\Big(\frac{\sqrt{x^2+y^2} - r_0(\tan^{-1}(y/x))}{2\sqrt{2}\epsilon}\Big) + 1\Big], \qquad (5)$$

$$u(x,y,0) = -1 \ , \qquad (6)$$

which corresponds to a solid whose shape is given (in polar coordinates) by the function r_0, and a uniform temperature distribution equal to the undercooling. The exact form of (5) is taken from the asymptotic approximation to the solution [3,17].

2.2 Discretization

Equations (1)–(4) have been discretized in space using second-order finite differences on uniform Cartesian grids of $N_x \times N_y$ points,

$$\mathbf{x}_{i,j} = \{x_{min} + i\Delta x \, , \, y_{min} + j\Delta y\} \quad , 0 \le i < N_x \, , \, 0 \le j < N_y \ . \qquad (7)$$

The value of a gridfunction f defined on (7) in the gridpoint $\mathbf{x}_{i,j}$ will be denoted $f_{i,j}$. It is desirable to minimize the width of the stencil for several reasons, one being that the amount of communication needed to compute the action of the difference operator decreases. For this purpose, approximations to certain gridfunctions will be needed between gridpoints, as denoted by non-integer indices in the formulas below. The standard five-point stencil can be used to approximate the ordinary Laplacian. For (3) we use,

$$\begin{aligned}\nabla_a^2\phi_{i,j} = &- D_{-x}\Big((\eta\eta')_{i+1/2,j}D_{0y}\frac{\phi_{i,j}+\phi_{i+1,j}}{2}\Big)\\ &+ D_{-y}\Big((\eta\eta')_{i,j+1/2}D_{0x}\frac{\phi_{i,j}+\phi_{i,j+1}}{2}\Big)\\ &+ D_{-x}\Big((\eta^2)_{i+1/2,j}D_{+x}\phi_{i,j}\Big) + D_{-y}\Big((\eta^2)_{i,j+1/2}D_{+y}\phi_{i,j}\Big) \ ,\end{aligned} \tag{8}$$

where the difference operators are defined as

$$D_{+x}f_{i,j} = (f_{i+1,j}-f_{i,j})/\Delta x \ , \qquad D_{+y}f_{i,j} = (f_{i,j+1}-f_{i,j})/\Delta y \ , \tag{9}$$

$$D_{-x}f_{i,j} = (f_{i,j}-f_{i-1,j})/\Delta x \ , \qquad D_{-y}f_{i,j} = (f_{i,j}-f_{i,j-1})/\Delta x \ , \tag{10}$$

$$D_{0x}f_{i,j} = (D_{+x}+D_{-x})f_{i,j}/2 \ , \qquad D_{0y}f_{i,j} = (D_{+y}+D_{-y})f_{i,j}/2 \ , \tag{11}$$

$\eta(\theta)$ is given by (4), and θ approximated by

$$\theta_{i+1/2,j} = \arctan\left(\frac{D_{0y}(\phi_{i,j}+\phi_{i+1,j})}{2D_{+x}\phi_{i,j}}\right) - \theta_0 \ , \tag{12}$$

$$\theta_{i,j+1/2} = \arctan\left(\frac{2D_{+y}\phi_{i,j}}{D_{0x}(\phi_{i,j}+\phi_{i,j+1})}\right) - \theta_0 \tag{13}$$

between the gridpoints. This expression requires all eight neighboring points, which should be compared to just four neighboring points for the standard five-point stencil.

Inserting the spatial approximations in (1)–(4), one obtains a large system of ordinary differential equations (ODE), where the solution consists of approximations to ϕ and u in all grid points. The ODE can be solved using standard methods found in the literature. Here, we have chosen to work with the backward difference formulas (BDF) [8]. For a general ODE $y'(t) = f(t, y)$, where $y(t_0), \dots, y(t_{n-1})$ are known, they take the form

$$\sum_{i=0}^{k} \alpha_i^{(k)} y(t_{n-i}) = \Delta t f(t_n, y(t_n)) \ , \tag{14}$$

where $1 \le k \le 6$ is the order of accuracy, and $\alpha_i^{(k)}$ are known constants. Thus one can obtain a high order of accuracy in the temporal discretization with very little work. The generalization to adaptive time stepping is also relatively straightforward [8]. The BDF methods have desirable stability properties. In the sense of theory of numerical methods for ODE, the first- and second-order methods are A-stable.

2.3 Non-Linear Solver

With the discretization of space and time discussed in Sect. 2.2, the solution can be advanced in time by solving a non-linear system of equations,

$$\sum_{i=0}^{k} \alpha_i^{(k)} \begin{bmatrix} \Phi_{n-i} \\ U_{n-i} + P(\Phi_{n-i})/\Delta \end{bmatrix} - \Delta t \begin{bmatrix} m\widetilde{\nabla}_a^2 \Phi_n + mF(\Phi_n, U_n)/\epsilon^2 \\ \widetilde{\nabla}^2 U_n \end{bmatrix} = 0 \; . \tag{15}$$

For brevity of notation, all non-linear terms in (1), except the anisotropic Laplacian, have been collected in F; the non-linear terms in (2) collected in P; and approximations of spatial derivatives are denoted by a tilde. The vectors Φ_n and U_n contain grid values of ϕ and u at time t_n to be discussed in more detail in Sect. 3.1 below.

Equation (15) is solved by Newton iteration. The well-known advantages are quadratic convergence in the vicinity of a root and the fact that the solution can be determined by solving "a few" linear systems with the Jacobian as the matrix.

Unfortunately, Newton's method will only converge if the initial guess is sufficiently close to the root. A backtracking line search technique [2,5] is used to increase the radius of convergence. The underlying idea is to adaptively choose a damping factor such that there is a sufficient decrease in the residual, and the directional derivative for the next iterate is not too small. This increases the radius of convergence while quadratic convergence is retained in the vicinity of the root. The extra work consists mostly of evaluating the non-linear function.

The major part of the computational work in Newton's method is needed for solving linear systems. For this application, the matrices will be sparse and banded (see Sect. 3.1). The storage requirements of a direct solver are prohibitive in this case, even if the matrix is permuted to decrease the bandwidth.

Krylov subspace projection methods [12] have several attractive properties for solving sparse linear systems. The storage requirements are only a few vectors, and they do not need the matrix itself but only its action on vectors[1]. In the context of Newton iteration, the action of the Jacobian can be approximated by a difference quotient, and the linear systems need not be solved exactly. In [2] it is shown how non-linear Krylov subspace correction

[1] Some Krylov methods also need the action of the matrix transpose and have not been considered here.

methods can be formulated, including a Newton-GMRES algorithm based on backtracking line search techniques.

Since the matrices considered here are non-symmetric, GMRES [12,14] has been chosen. Other possible choices include the Conjugate Gradient Squared Method (CGS) and the Biconjugate Gradient Stabilized Method (Bi-CGSTAB) [12]. GMRES requires more storage than both CGS and Bi-CGSTAB, but only half the number of matrix-vector products need to be computed. Restarting is used to limit the memory usage. The current implementation has Jacobi preconditioning only [12].

The numerical method employed here is second-order accurate and different from the one chosen by many other scientists (e.g., [9,19]). Instead of solving (15), Φ_n and U_n are updated separately according to

1. Given Φ_{n-1} and U_{n-1}, compute Φ_n from (1) using an explicit method (e.g., Forward Euler).
2. Solve (2) for U_n using an implicit method (e.g., Crank-Nicholson), approximating the time derivative of $P(\Phi)$ using Φ_n (as computed in 1) and Φ_{n-1}.

This approach is more efficient when it comes to floating point operations as well as storage requirements, because no non-linear equations have to be solved and because the linear systems have dimension $N_x N_y$ instead of $2N_x N_y$. There are, however, step size restrictions due to stability and due to the fact that the order of accuracy is only one.

3 Parallelization

The numerical method described in Sect. 2 has been implemented in C for parallel distributed memory architectures based on an SPMD (Single Program Multiple Data) model. One of the critical issues for good performance is to choose a data distribution that minimizes communication, as discussed in Sect. 3.1. The more technical details of the implementation, such as memory requirements and communication, are discussed in Sect. 3.2. The performance of the code is experimentally investigated in Sect. 4.1.

3.1 Data Distribution

Apart from evaluating (15) and the Jacobian of (15), a few key computational kernels can be identified from Newton's method and GMRES; namely, computing matrix-vector products, vector updates, norms, and scalar products. Because the last two are global operations and independent of data distribution, and because the vector update is strictly local, it makes sense to choose a data distribution such that (15) can be efficiently evaluated. The communication pattern of this operation is identical to that of computing the matrix-vector products.

Assuming that the solution is represented on the grid (7), and that the vectors $\Phi_n, U_n \in \mathbb{R}^{N_x N_y}$ contain grid values sorted in increasing order of jN_x+i ("natural ordering" [12]), the Jacobian of (15) will have the structure shown in Fig. 1. For a direct solver one would typically use another ordering of the elements to reduce the bandwidth. For GMRES this will not alter the storage requirements and the simple ordering has therefore been kept to simplify the implementation of the difference approximation.

The matrix in Fig. 1 can be seen as consisting of four block matrices showing dependencies of data defined on the grid. The structure of the upper left and lower right blocks are determined by the difference approximations of spatial derivatives. The remaining two blocks are diagonal and contain terms coming from the coupling between (1) and (2). Distributing the vectors Φ_n and U_n row-wise across the processes will give a row-cyclic distribution of the Jacobian (as indicated by the shaded area in Fig. 1) and will minimize communication.

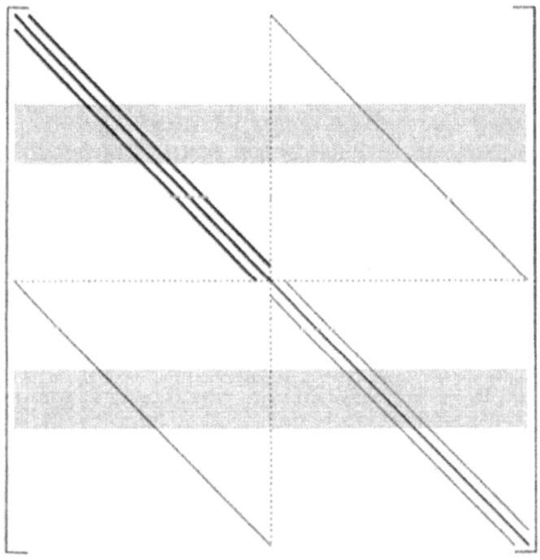

Fig. 1. Jacobian of (15). The shaded area shows the part of the matrix residing on the same process.

To evaluate (15), each process would have to send $2N_x + 1$ elements to its two neighbors. The communication can be overlapped by computation because only a relatively small fraction of the elements depend on external data (approximately $2p/N_y$ on p processes). This is accomplished by non-blocking MPI routines [6].

3.2 Implementation

The numerical method used here is more expensive per time step than the common choice discussed in Sect. 2.3, which means that the primary focus has been to write an efficient implementation to allow simulation within comparable time. To this end, existing numerical libraries have been used to a large extent. The only numerical routine written specifically for this application is the backtracking line search modification of Newton's method. All calls to basic linear algebra routines are made to ESSL (IBM SP2) or BLAS (Sun workstations). The GMRES implementation is taken from the PSPARSLIB library [13].

Inter-process communication has been implemented using MPI [6]. Non-blocking routines are used wherever possible to overlap computation and communication. Many of the operations needed for GMRES and the Newton iteration are inherently global and will introduce synchronization points. These are unavoidable in the Newton iteration, because all global reductions are used either to determine whether convergence has occurred or to update the approximate solution.

The GMRES implementation uses the classical Gram-Schmidt algorithm [12] to orthogonalize the newly computed vector against the previously computed vectors in each inner iteration. While this is advantageous for numerical stability concerns, it might result in poor parallel performance because two global reductions are used for each orthogonalization. It is possible to orthogonalize the entire Krylov basis at once; for example, using the m-step GMRES(m) algorithm [10], thus reducing the number of synchronization points. Such algorithms are not supported in the current release of PSPARSLIB.

Table 1 shows the distribution of the total run time over the different routines for one particular simulation. Notably, computing the analytic Jacobian of (15) accounts for almost half of the total time. Still, using the analytic expression instead of difference approximations significantly reduces the run time for most simulations at the expense of larger storage requirements. A more detailed look at the routines for computing (15) and its Jacobian show that almost 75% of the time is spent inside the trigonometric functions `sin`, `cos`, and `atan2` from the C math library. Thus there is a large part of the code that cannot be optimized further. Temporary storage of computed values is used to avoid excessive calls to the trigonometric functions.

Both GMRES and the Newton solver are based on reverse communication protocols. The internal work in these routines makes up the major part of the time used by ESSL. The 9% not accounted for by Table 1 are overhead, initialization, and unidentified routines. This fraction agrees well with the measured speedup (see Table 3).

The approximate memory usage of the program is shown in Table 2. The reciprocals of the matrix diagonal are stored for the Jacobi preconditioner to avoid divisions. Simpler versions of Newton iteration will require less stor-

Table 1. Distribution of total run time over communication and computational routines for a simulation on a 400×400 grid running on 4 IBM SP2 processors. Results were obtained by `gprof`. The time used for call counts was excluded before fractions were computed.

Routine	Fraction of run time
Analytic Jacobian of (15)	47.7 %
Evaluating (15)	23.8 %
ESSL	10.9 %
Matrix-vector products	5.7 %
Preconditioning	1.4 %
Communication	1.1 %

age, as well as using difference quotients to approximate the action of the Jacobian. One could also decrease the needed storage by choosing another Krylov method. Typically, the dimensions of the Krylov subspaces are small, say 5–20, and the gain in storage is not sufficient to justify the increase in computational complexity.

Table 2. Approximate storage requirements per process measured in `sizeof(double)`. Simulation on $N_x \times N_y$–grid using a BDF(k) method and Krylov subspaces of dimension m running on p processes.

Purpose of allocated storage	Size
Solution at different time levels	$2(k+1)\, N_x N_y/p$
Action of difference operator	$12\, N_x$
Newton solver	$10\, N_x N_y/p$
Krylov solver (PSPARSLIB [13])	$(10+4m)\, N_x N_y/p$
Jacobian of (15)	$16\, N_x N_y/p$
Preconditioner (Jacobi)	$2N_x N_y/p$
Total	$\approx (40+2k+4m)\, N_x N_y/p$

An effort has been made to write modular code, and the numerical methods (apart from the discretization) can be replaced with a relatively small amount of work. The generalization to similar physical problems, such as solidification of binary alloys [18], is also straightforward. Both numerical and physical parameters used for simulation are read from an input file for flexibility.

4 Computational Results

4.1 Performance

The scalability of the implementation has been investigated on two different architectures, the IBM SP2 *Strindberg* located at PDC, Stockholm, and networks of Sun Ultra 5 workstations. The (approximate) computational performance lies in the range of 80–150 MFlop/s on a single thin IBM SP2 node, depending on grid size and data. Peak performances of up to 230 MFlop/s have been obtained.

Tables 3 and 4 show speedup on the different architectures for a few different grid sizes. Speedup is formally defined as the execution time of the fastest serial algorithm on one processor divided by the execution time of the parallel algorithm [7], and it is a measure of the number of processors that can be efficiently used to solve a given problem. The results shown were obtained by comparison to a modified version of the parallel code, in which all the communication has been removed. The timings were measured over 100 time steps using `MPI_Wtime`. Realistic parameters were used and initial data was chosen such that the fastest transients had died out. The results should be representative for a typical simulation.

Table 3. Speedup on an IBM SP2 for different grid sizes.

	Grid size			
Nodes	200 × 200	400 × 400	600 × 600	800 × 800
2	1.92	1.89	4.27	4.73
4	3.56	3.63	8.14	9.15
8	6.56	6.34	15.30	17.22
16	11.25	11.38	26.24	31.45
32	12.80	17.36	44.10	51.37

Table 4. Speedup on clusters of Sun workstations (Ultra 5) for different grid sizes.

	Grid size		
Computers	200 × 200	400 × 400	600 × 600
2	1.91	2.61	4.09
4	3.83	5.17	31.03
8	6.14	10.00	60.60
16	7.34	16.49	93.78

The speedup for the larger grids is more than optimal. For these simulations, the numerical parameters used for the one-processor job are different from the ones used on several processors, due to RAM memory limitations. In most cases, the dimensions of the Krylov subspaces have been decreased (compare Table 2). For the 600 × 600–grid on one Sun workstation, finite difference approximations were used to evaluate the action of the Jacobian.

The results are encouraging even for the smaller grids. All simulations on eight processors show a parallel efficiency, defined as the ratio of speedup to the number of processors, exceeding 75%. On the IBM SP2, the parallel efficiency is more than 70% for all grids on 16 processors.

Another scalability metric is the scaled speedup [7], which is a measure of how long a given parallel program would have taken to run on a hypothetical serial processor with the same amount of RAM memory as the parallel machine. The motivation for this metric is that a larger machine allows the solution of larger problems, which is not reflected in the fixed-size speedup. By increasing the problem size linearly with the number of processors, memory and surface-to-volume effects remain constant; and usually the efficiency is higher than for the fixed-size case, which is limited by Amdahl's law [7].

Because the considered implementation is based on iterative methods, it is not obvious how to increase the problem size. Parameters such as grid resolution, time step, and tolerances all affect execution time and performance. Furthermore, the convergence depends on data to a large extent. For these reasons, two different limits have been investigated: increased size of the physical domain with all numerical parameters kept fixed, and decreased grid spacing with the physical problem kept fixed. In both cases the problem size was increased such that approximately 400^2 and 200^2 gridpoints reside on each processor on the IBM SP2 and Sun workstations respectively. The results are shown in Table 5.

Table 5. Scaled speedup for increased size of the physical domain with numerical parameters kept fixed (case 1), and decreased grid spacing with all other parameters kept fixed (case 2).

	IBM SP2		Sun workstations	
Nodes	Case 1	Case 2	Case 1	Case 2
2	1.99	1.92	1.92	1.67
4	3.97	3.63	3.72	3.31
8	7.94	7.60	6.78	6.14
16	15.70	12.90	12.71	10.04

The scaled speedup is closer to the optimal than the fixed-size speedup, as expected. When the size of the physical domain is increased, there is only a slight decrease in efficiency due to the global reductions. The results are not as good when the grid is refined, because more iterations are required for GMRES to converge and thus more global reductions are needed. It is possible that a better efficiency can be obtained by changing the time step as well as the grid spacing.

4.2 Computed Dendrites

Results from two different computer simulations are given in this section. For both cases, four-fold anisotropy has been used with the x-axis taken as reference direction, which corresponds to $k = 4$ and $\theta_0 = 0$ in (4).

The first example is a reproduction of results presented by Braun and Murray [1]. The physical parameters used for the simulation were $\alpha = 400$, $m = 0.035$, $\epsilon = 0.005$, $\gamma = 0.015$, and $\Delta = 0.25$. The initial shape of the solid was given by (5), with $r_0(\psi) = 0.05\,(1 + 0.5\cos(\psi))$, and the initial temperature was taken to be $u(x, y, 0) = -\phi(x, y, 0)$. Computations were carried out on a 601×601–point discretization of $[0, 3] \times [0, 3]$, with constant time step $\Delta t = 0.001$. The shape of the solid in Color Plate 19 on page 296 agrees with the results presented in [1, Fig. 1].

The parameters for the second simulation were chosen to model pure nickel, $\alpha = 400$, $m = 0.05$, $\epsilon = 0.005$, $\gamma = 0.025$ and $\Delta = 0.5$. It should be noted that the exact value of γ is unknown, but the choice made here is well within the range of values chosen by other scientists (e.g., [9,11,19]). The undercooling Δ is larger than in the simulation above, causing the dendrite to grow faster. To further stimulate side-arm growth, a random perturbation of 1% amplitude was added to the non-linear function $F(\phi, u)$ in (15), which can be interpreted as thermal noise at the interface [19].

The computation was initiated with a circular seed of radius 0.02, and a uniform temperature equal to -1. Results obtained from calculations on a 451×451–point discretization of $[0, 4.5] \times [0, 4.5]$ with $\Delta t = 0.0005$ are shown in Color Plate 20 on page 296.

In both simulations presented here, smaller time steps have been chosen than motivated by stability and the convergence radius of the Newton iteration. Empirical choices were made in an attempt to decrease the wall clock time needed for the entire simulation. For larger time steps, the initial guess for the Newton iteration (taken to be the solution at the old time level) will be farther from the root, and convergence in the backtracking line search framework is only linear[2]. The time needed to solve the arising linear systems will also increase. Adaptive BDF methods provide a more systematic way of selecting time steps.

Newton iteration was performed until the discrete Euclidean norm of the residual (as given by (15)) was smaller than 10^{-4}. Convergence was typically reached in 3–5 iterations, apart from the first time steps where there are still

[2] Without the backtracking line search modification, Newton's method may diverge for these cases.

fast transients coming from initial data. The largest part of the residual is located in the vicinity of the solid-melt interface, but the residual in each component of the solution remains much smaller than 10^{-4}. The error tolerance can be decreased at the expense of larger computational time so that only the discretization error remains. More sophisticated convergence criteria can also be implemented with a small amount of work.

Solutions obtained by the first-order method described in Sect. 2.3 have been inserted in (15), and the norm of the residual for the second-order discretization has been computed in order to get a feeling for how large the errors due to the splitting are. With parameters from the simulation in Color Plate 20 on page 296, and initial data taken from various time levels in the second-order accurate simulations, the norm of (15) after one time step is approximately 0.03–0.04 for $\Delta t = 10^{-4}$, and approximately 0.001–0.002 for $\Delta t = 10^{-5}$. It is not clear at this point how this affects the global solution, but it is a topic for further investigation.

The wall clock time required to compute Color Plate 20 on page 296 was approximately 24 hours on four Sun workstations. This particular computation has not been performed with the first-order scheme, but numerical experiments indicate that the wall clock time required would be about 36 hours running on a single Sun workstation with $\Delta t = 10^{-5}$. The increase in computational cost for using the second-order method would be about 2.7 if the wall clock time is scaled by the number of computers, which seems to be a justifiable expense for the increased accuracy.

5 Conclusions

A second-order accurate numerical method for simulating dendritic solidification has been introduced. This method is different from the first-order method chosen by many other scientists, and it is computationally more expensive per time step. It was shown how the scheme can be implemented on parallel architectures to allow computation on larger domains and/or with finer grid resolution.

The code scales well with the number of processors, both on parallel computers and networks of workstations. Performance is reasonable for sparse matrix computations. Numerical experiments indicate that the increased amount of RAM memory made available by the parallelization can be exploited to allow accurate simulation within an acceptable increase of run time. A more careful investigation of the errors in the two numerical methods is needed to confirm this conclusion.

The uniform discretization of space and constant time stepping is expected to be replaced by adaptive methods in the future.

References

1. R. J. Braun and B. T. Murray. Adaptive phase-field computations of dendritic crystal growth. *J. Crystal Growth*, pages 41–53, 1997.
2. P. N. Brown and Y. Saad. Hybrid Krylov methods for nonlinear systems of equations. *SIAM J. Stat. Comp*, 11:450–481, 1990.
3. G. Caginalp and X. Chen. Phase field equations in the singular limit of sharp interface problems. In M. E. Gurtin and G. B. McFadden, editors, *On the evolution of phase boundaries*. Springer-Verlag, 1992.
4. S. Chen, B. Merriman, S. Osher, and P. Smereka. A simple level set method for solving Stefan problems. *J. Comp. Phys*, 135:8–29, 1997.
5. J. E. Dennis and R. B. Schnabel. *Numerical methods for unconstrained optimization*. Prentice Hall, 1983.
6. W. Gropp, E. Lusk, and A. Skjellum. *Using MPI: Portable Parallel Programming with the Message Passing Interface*. The MIT Press, 1994.
7. J. L. Gustafson, G. R. Montry, and R.E. Benner. Development of parallel methods for a 1024-processor hypercube. *Siam J. Sci. Stat. Comp.*, 9:609–638, 1988.
8. E. Hairer, S. P. Nørsett, and G. Wanner. *Solving Ordinary Differential Equations I: Nonstiff Problems*. Springer-Verlag, 1987.
9. A. Karma and W.-J. Rappel. Phase-field methods for computationally efficient modeling of solidification with arbitrary interface kinetics. *Phys. Rev. E*, 53, 1996.
10. S. K. Kim and A. T. Chronopoulos. A class of Lanczos-like algorithms implemented on parallel computers. *Parallel Comput.*, 17:763–778, 1991.
11. R. Kobyashi. A numerical approach to three-dimensional dendritic solidification. *Experimental Mathematics*, 3:59–80, 1994.
12. Y. Saad. *Iterative methods for sparse linear systems*. PWS Publishing Company, 1996.
13. Y. Saad, G.-C. Lo, and S. Kuznetsov. *PSPARSLIB Users Manual*. `<http://www.cs.umn.edu/Research/darpa/p_sparslib/psp-abs.html>`.
14. Y. Saad and M. Schultz. GMRES: A generalized minimal residual algorithm for solving non-symmetric linear systems. *SIAM J. Stat. Comp.*, 7:856–869, 1980.
15. A. Schmidt. Computation of three dimensional dendrites with finite elements. *J. Comp. Phys.*, 125:293–312, 1996.
16. R. Tönhardt. *Convective effects on dendritic solidification*. PhD thesis, Royal Institute of Technology, 1998.
17. S.-L. Wang, R. F. Sekerka, A. A. Wheeler, B. T. Murray, S. R. Coriell, R. J. Braun, and G. B. McFadden. Thermodynamically consistent phase-field models for solidification. *Physica D*, 69:189–200, 1993.
18. J. A. Warren and W. J Boettinger. Prediction of dendritic growth and microsegregation patterns in a binary alloy using the phase-field method. *Acta metall. mater.*, 43:689–703, 1995.
19. A. A. Wheeler, B. T. Murray, and R. J. Schaefer. Computation of dendrites using a phase field model. *Physica D*, 66:243–262, 1993.

Astrophysical MHD Simulation and Visualization

Bertil Dorch

Stockholm Observatory; The Royal Swedish Academy of Sciences; SE–133 36 Saltsjöbaden; Sweden; <dorch@astro.su.se>

Abstract. The combination of numerical simulations, interactive data-mining, and visualization has proven to be very powerful when attempting to understand complex astrophysical systems. The present contribution aims at illustrating this by discussing examples from numerical magneto-hydrodynamical simulations within the context of astrophysical dynamos. Two qualitatively different simulations are discussed: A study of the magnetic field topology in a kinematic dynamo model, and a model of the buoyant rise of a twisted magnetic flux rope through a stellar convection zone.

1 Introduction

Magnetic fields are ubiquitous in the Universe. Astrophysical systems possessing magnetic fields range from planets and stars to accretion discs, active galactic nuclei, and clusters of galaxies. In most of these astrophysical systems, the magnetic field is thought to be maintained by a magneto-hydrodynamical (MHD) dynamo process, a process whereby the magnetic field gains energy from the kinetic energy of fluid motions. However, in none of these systems is the dynamo process well understood, if even recognized. The most accessible and best studied example of an astrophysical dynamo is the solar dynamo that gives rise to the 11-year sunspot cycle and the associated occasional violent fluctuations in the Earth's magnetic field, for example, Northern Lights and satellite surges.

The full radiative-MHD (RMHD) solar dynamo problem, which includes the entire volume of the Sun, is not a feasible task using any current computer. This is not so much because of the difficulty of the physics involved as because of the range of scales. From the bottom of the Sun's outer atmosphere (the "convection zone") to its surface—a stretch of 200,000 km—the pressure drops by 10 orders of magnitude. The spatial scales range from of the order of 100 km—the size of the smallest observationally resolved magnetic features on the Sun—to the scale of the Sun itself, which is also the scale of the largest magnetic structures and flow patterns. The time scales range from a millisecond of sudden energy and particle bursts (flares) to several times 11 years, which is the scale of the generation and maintenance of the solar magnetic field. Because the full problem is not feasible, it has to be broken down into conceptual pieces that may be studied in detail in their own right, at the same time taking into consideration their place in the global picture.

The following simulations were performed with the solar dynamo in mind, but the results may easily be extrapolated to other types of astrophysical dynamos as well.

- The principles of dynamo action: Kinematic ABC flow dynamos, see Archontis and Dorch ([1] and [11])
- Dynamics of magnetic flux ropes in a stratified "convection zone," see Dorch and Nordlund ([10] and [13])

Below, I describe first the method used to perform the simulations and then turn to visualizations of the resulting data.

2 Simulation

In some cases, first-principle physics may be extracted from simplified 1D (or even 2D) models, but in most cases where magnetic fields are involved, their dynamics and topology demands 3D (or at least 2.5D). On the one hand, this forces the use of fast supercomputers to handle the associated enormous number of operations; on the other hand, the inherent complexity and intermittency of astrophysical magnetic fields forces the use of advanced tools to visualize the simulation results. For example, data browsing through magnetic field lines might lead to progress that cannot be achieved otherwise.

2.1 MHD Code

The quantitatively different simulations discussed in this paper were performed using slightly different versions of the same multi-dimensional MHD code based on the finite difference Fortran code developed by Galsgaard, Nordlund, Stein, and others (e.g., [22] and [25]). The code was originally designed for stellar (magneto) convection studies, but it has been used for a variety of astrophysical simulations during the last decade, including solar magneto-convection, magneto-turbulence, the heating of the solar corona, shocks in the interstellar medium, magnetic accretion disks, and various studies of galaxies.

The MHD equations are essentially a set of partial differential equations consisting of the conservation equations for mass (the continuity equation, Eq. 1), momentum (the equation of motion, Eq. 2), and energy (Eq. 3), solved together with Maxwell's equations, which are often combined into the induction equation for the magnetic field (Eq. 4).

$$\frac{\partial \rho}{\partial t} = -\nabla \cdot \rho \mathbf{u} \ , \tag{1}$$

$$\frac{\partial \rho \mathbf{u}}{\partial t} = -\nabla \cdot (\rho \mathbf{u}\mathbf{u} - \tau) - \nabla P + \mathbf{F}_g + \mathbf{F}_{\text{Lorentz}} \ , \tag{2}$$

$$\frac{\partial \rho e}{\partial t} = -\nabla \cdot (h \rho \mathbf{u}) + (\mathbf{u} \cdot \nabla) P + Q \ , \tag{3}$$

$$\frac{\partial \mathbf{B}}{\partial t} = \nabla \times (\mathbf{u} \times \mathbf{B}) + \eta \nabla^2 \mathbf{B} \ , \tag{4}$$

where ρ is the mass density, $\mathbf{u}$ the velocity, τ the viscous stress tensor, P the gas pressure, $\mathbf{B}$ the magnetic field density, and η the magnetic diffusivity. In the momentum equation (Eq. 2), $\mathbf{F}_g = \rho \ \mathbf{g}$ is the gravitational force, and $\mathbf{F}_{\text{Lorentz}} = \mathbf{j} \times \mathbf{B}$ is the Lorentz force. Furthermore, e is the internal thermal energy, h is the enthalpy (i.e., $h\rho\mathbf{u}$ is the convective flux), and Q is the net heating (the sum of viscous heating, Joule heating, and radiative heating or cooling).

The equations are solved using a staggered grid method with 6th-order staggering operators, 5th-order centering operators, and a 3rd-order Hyman (predictor-corrector) time-stepping routine. In most cases the grid is regular, but experiments are being made with irregular grids by Rögnvaldsson (see [26]).

Viscous and magnetic diffusion are in general a combination of numerical, artificial, and physical diffusion. In the present scheme, the diffusion terms are quenched by the convergence of the flow across the magnetic field, to reduce the diffusion of well-resolved structures. Typical magnetic Reynolds numbers ($\text{Re}_\text{m} = u\ell/\eta$, where ℓ is a characteristic length scale) are in the range from a few times 10^2 up to 10^5 in regions with smooth flows. In a few simulations, the quenching is turned off in order to compare the results to simulations from the literature done with constant and uniform diffusion. In these cases the magnetic Reynolds number typically range from 10 to 10^3.

2.2 Computers

The basic code routines have been compiled and run on several different hardware architectures (e.g., Sun Sparc, Connection Machine (CM), Silicon Graphics (SGI), Cray, IBM SP, and Fujitsu VX), and versions of the code exist for compilers such as F77, F90, CMF, HPF, and even for the graphically oriented IDL from Research Systems.

The simulations discussed here ran on several different machines, including supercomputers at UNI-C in Denmark (a CM200, an SGI Origin 2000, and an IBM SP2), a CM5 supercomputer in France, and at various SGI machines at the Niels Bohr Institute at Copenhagen University. Currently a version is also running on a fast DEC Alpha EV-6 workstation at the Royal Swedish Academy's Research Station for Astrophysics located at the Stockholm Observatory in Saltsjöbaden.

3 Visualization

The data from the simulations were visualized with both IDL and NAG Explorer to render isosurfaces of magnetic field strength, flow speed, gas pressure, and mass density, as well as magnetic field lines and stream lines. The visualizations were mainly done using various SGI machines, and discoveries were made that would have been hard without interactive 3D data browsing. Furthermore, 3D visualization readily permits easy communication of the results through VR by-products such as VRML and Inventor (iv) format files, which can be made accessible to the scientific community via the World Wide Web (see `<http://www.astro.su.se/~dorch/>`).

Below, I describe the results of two conceptually different approaches to the astrophysical (solar) dynamo problem. The first—ABC flow dynamos—concerns the exponential amplification of a weak seed magnetic field by a prescribed velocity field, while the second—magnetic flux ropes—deals with the dynamics of a single magnetic flux structure embedded in a solar convection zone–like environment.

3.1 ABC Flow Dynamos

The class of steady velocity flows called ABC flows (named after Arnold, Beltrami, and Childress; see [2], [3], and [8]) is an example of how complex flows may be able to amplify weak seed magnetic fields. This rather special type of flow is interesting to dynamo theory, because on the one hand it constitutes a situation so simple that it is possible to understand the details of the amplification process, while on the other hand, it displays behavior that in some respect resembles that of astrophysical dynamos such as the Sun. The amplification is related to the stretching, twisting, and folding of magnetic field lines, and the role played by the finite magnetic diffusivity is as essential to the amplification process as it is believed to be for the Sun.

The ABC flow is a 3D periodic, incompressible, and steady flow given by the sum of three parameterized Beltrami waves (see [3]):

$$\mathbf{u} = A(0, \sin kx, \cos kx) + B(\cos ky, 0, \sin ky) + C(\sin kz, \cos kz, 0) \ . \quad (5)$$

The approach is purely kinematic, so that the back-reaction in the equation of motion Eq. 2 by the Lorentz force is ignored in the treatment. Here, only results from cases with constant and uniform diffusion are presented (i.e., with a fixed $\mathrm{Re_m}$).

A special case is the so-called *normal* ABC flow with $A = B = C = 1$, hence also called the "111" flow. The normal ABC flow has two regimes of dynamo action: one at comparatively low magnetic Reynolds numbers $\mathrm{Re_m} = 8.5$–17.5 (Arnold and Korkina in [2]), and one starting at $\mathrm{Re_m} \approx 27$ (Galloway and Frisch in [19,20]). A main theme of kinematic dynamo theory is whether the exponential growth rate of the magnetic energy remains finite in the limit of zero magnetic diffusivity (i.e., of infinite Re_m)—whether the dynamo is a so-called *fast* dynamo. Hence, we shall discuss here only results for the second window of dynamo action.

The normal "111" ABC flow has eight stagnation points and eight points where the velocity is maximum (see top left panel in Color Plate 21 on page 297). There are two different types of stagnation points for the normal ABC flow:

- α-type stagnation points, where stream lines are diverging along an axis through the stagnation point and converging in the plane perpendicular to the axis.
- β-type stagnation point, where stream lines are converging along the axis and diverging in the plane.

While the diverging stream lines in the plane of a β-type stagnation point connect to the converging stream lines of three α-type stagnation points the reverse is true for an α-point. In itself, it is not so important that there happen to be stagnation points in the normal ABC flow. After all, many interesting flows do not possess stagnation points (such as other types of ABC flows and turbulent flows in general), and in any case a stagnation point may be removed by a simple translation of the coordinate system. Rather, it is the stretching ability of the flow that is relevant for the dynamo action: stretching implies an exponential growth of the magnetic field magnitude. In the high degree of symmetry of the normal ABC flow, the stagnation points coincide with local extrema of the stretching rate. Hence, the two types of stagnation points may be seen as convenient markers of these regions.

Evolving the induction equation (Eq. 4) from a randomly chosen weak seed magnetic field, an oscillating behavior is associated with the exponential growth (as also seen by Galloway et al. in [20] and Galanti et al. in [17,18]). As seen in Fig. 1, there is an initial transition from the seed field state to the fastest growing mode of the magnetic field (see the discussion below of the structure of this mode).

As mentioned by Galloway and O'Brian (see [21], as well as Childress and Gilbert in [9]), it is possible to set the initial condition of the seed magnetic field such that only one mode is present in the calculations; the special initial condition for which that is possible is given by an eigenmode of the induction equation in the case of zero diffusivity $\eta = 0$. In that case, the magnetic field is not amplified in the second window of Re_m as mentioned by Galloway and O'Brian (see [21]), but only in the first window. In [9], however, Childress and Gilbert used this initial condition together with a so-called flux conjecture to try to deduce the growth rate of the ABC flow in the limit of infinite Re_m (i.e., in the second Re_m window). The growth rate derived does not, however, agree with the asymptotic growth rates found by Galloway and O'Brian (see [21]). One should indeed not expect to be able to recover the growth rate of a completely different (exponential growing) mode by studying the stretching of field lines in another (secularly decaying) mode.

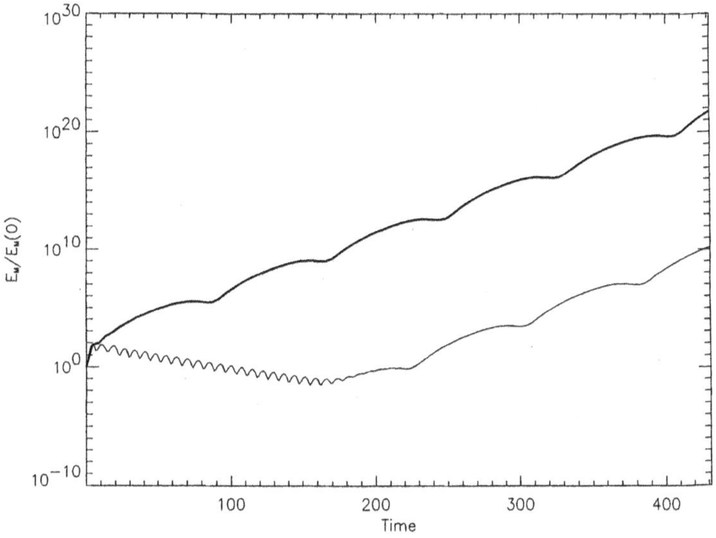

Fig. 1. The total magnetic energy E_M as a function of time on a logarithmic scale for two experiments ($\mathrm{Re_m}$=200), with uniform initial condition (thick line) and the special initial condition discussed in the text (thin line). Time is in arbitrary units.

Figure 1 (the lower curve) shows that the mode, in the case of the special initial condition mentioned above, is an exponentially decaying mode. The decay is associated with an oscillation with a period about 10 times smaller than that of the growing mode. Because of numerical round-off errors, the amplitude of the growing mode is not identically equal to zero in the initial condition, and eventually its inevitable growth combined with the decay of the initialized mode results in a transition from decay to growth of the total magnetic energy (see Fig. 1). Galloway and O'Brian (see [21]) found no growing solution, and they concluded that "something odd is going on."

When the induction equation is evolved from a uniform weak seed field, flux "cigars" rapidly arise at four of the eight stagnation points—the α points (see top right panel in Color Plate 21 on page 297). These are the regions in the flow where the magnetic field is most rapidly advected by a converging flow in two of the three dimensions. The flux cigars are aligned along the axis of divergence through the α-type stagnation points and point directly to the β-type stagnation points. The regions around the β type stagnation points are not unstable to flux cigar generation because the flow there is diverging in two directions.

The four flux cigars seen in Color Plate 21 on page 297 (top right panel, remember the periodicity of the computational box) correspond to the fastest growing eigenmode in the diffusion-less case $\eta = 0$. Initially several modes may be present, but eventually the one with the largest exponential growth rate will be the only one that remains in the solution.

Galloway and O'Brian (see [21]) performed a calculation at a relatively high magnetic Reynolds number and found that the flux cigars were double structures. They offer the remark that the double cigars "obscure the physics of what is happening." In our experiments we do indeed find that a second set of flux cigars is formed next to the four primary cigars, but we conclude that rather than obscuring the physics, they are absolutely essential. These secondary cigars have the opposite polarity of the neighboring cigars, and they are connected to them by reconnecting field lines.

The secondary flux cigars arise after an initial transient phase. These cigars begin to form at the α-point at the separator line next to the primary cigar (four lower panels of Color Plate 21 on page 297). As the secondary cigar forms, the primary cigar moves slightly away from the center of the stagnation point. In the simulation, as the energy increases at the beginning of a "cycle," the secondary cigar increases in size and field strength until the two cigars become equal in size midway through the cycle. The primary cigar now becomes smaller and actually vanishes at the end of the cycle, at which time the growth of the energy slows down, and even reverses briefly. At that point, the flux cigar that was previously the secondary cigar moves into the center of the stagnation point (see the evolution in the four lower panels of Color Plate 21 on page 297).

In fact, it is possible to sketch a dynamical picture of the entire amplification cycle, by carefully data-mining the evolution of the flux cigars along with the motion and reconnection of magnetic field lines (as illustrated by Dorch in [10]). From this analysis it is found, for example, that the magnetic energy grows through most of the cycle but that the main growth takes place while the two cigars have about equal sizes. This is the time at which the most rapid reconnection take place and thus the time where the largest amount of magnetic flux is released down along the axis of divergence through the α points.

The replenishing of the field near the β points is crucial for the operation of the dynamo. In these regions, a complicated folding of field lines takes place that is accompanied by a discontinuity of the directions of field lines across the plane. The latter result would have been hard to discover without the use of modern visualization technics that allow direct "eye contact" when data browsing.

Certain properties of the magnetic transport are nearly invariant as $\mathrm{Re_m}$ is increased. The size of the regions where diffusion is important scale as $\mathrm{Re_m}^{-\frac{1}{2}}$, but reconnection still takes place and the field in the crucial β regions continues to be replenished. In the bulk of the flow where the stretching takes place, the decrease of the diffusivity is unimportant, because the field lines there are not influenced by diffusion. They tend to obtain a certain alignment with the flow topology given by the stretching that is an invariant property and hence the rate of increase of the field strength remains nearly the same. It thus appears very unlikely that the double-cigar mode should go away in the limit of infinite $\mathrm{Re_m}$ and that the normal ABC flow would not be a fast dynamo.

3.2 Magnetic Flux Ropes

Conceptually, the solar dynamo is thought to be a consequence of the combined reconnection, stretching, twisting, and folding of the magnetic field by convection and rotation; that is, it is governed by a process not dissimilar to the one described in the previous section. The current paradigm states that the solar differential rotation (the Sun's outer layers rotate faster at the equator than at the poles) winds up weak magnetic field lines to produce a strong longitudinal (or azimuthal) field. This strong field is intermittent in the form of ropes of magnetic field lines. The magnetic field provides a pressure in addition to the gas pressure, which makes the magnetic flux ropes buoyant, by reducing the mass density inside them. Convection is capable of keeping the flux ropes down near the bottom of the convection zone until they become buoyant enough to begin an ascent towards the solar surface, where they break through to produce bipolar magnetic regions, eventually developing into sunspots.

Numerical simulations of buoyant magnetic flux ropes carried out within the so-called "thin flux tube approximation" (e.g., Spruit in [28] and Moreno-Insertis in [23]) are consistent with the observations of the latitudes of emergence of bipolar magnetic regions on the surface of the Sun, for example (see Fan et al. in [16], and Caligari et al. in [4]). However, more general numerical simulations, not relying on the assumption that the flux tubes are thin, have shown that cylindrical flux tubes are quickly disrupted by a magnetic Rayleigh-Taylor instability, whereby characteristic "mushroom" structures are created, and the flux tubes lose their buoyancy. Examples include the work of Schüssler in [27], of Cattaneo et al. in [6] and [7], of Emonet and Moreno-Insertis in [14], [15], and [24], and of Dorch and Nordlund in [12]. This is a result of the simple topology assumed for the magnetic field of the flux tubes; that is, parallel field lines have no inhibiting effect on the Rayleigh-Taylor instability.

Tension among magnetic field lines resulting from a twist of the magnetic field, for example, may suppress the Rayleigh-Taylor instability and hence prevent the flux ropes from disintegrating. This has been demonstrated in numerical 2D simulations (e.g., Emonet and Moreno-Insertis in [14] and [15]). In 3D, several types of instabilities are possible, such as the kinking instability (resulting from the twist of the rope) and curvature along the flux rope. Because these and other 3D effects cannot be represented in 2D, numerical 3D simulations are called for.

We present here an illustration of results from simulations of flux ropes with an initial entropy balance between the interior of the rope and the external medium. This type of initial condition was also recently used by Moreno-Insertis and Emonet (see [15] and [24]), and it has the advantage that the relatively high buoyancy of this initial condition results in a significantly shorter run-time than, for example, evolving the flux ropes from an initial perturbed state of mechanical equilibrium. The initial twist of the flux ropes are given by a simple expression also applied in 2D simulations. It has been shown that for these cases, the Rayleigh-Taylor instability is inhibited if the degree of twist is sufficiently high (Emonet and Moreno-Insertis in [14], [15], and [24]).

The vertical, essentially 1D, stratification in the computational box is two-fold consisting of a lower, sub-adiabatically (i.e., stable) stratified layer and an upper, adiabatically stratified "convection zone." The purpose of the sub-adiabatic layer is to model the layer of undershooting plasma below the convection zone of the Sun. Both layers are initially in hydrostatic equilibrium, and convection is absent during the simulations.

A 3D simulation of a twisted undulant flux rope was performed with $150 \times 80 \times 200$ grid points. Color Plate 22 on page 298 shows a snapshot from a well-developed stage. While the part of the rope that is in the adiabatic zone starts to ascend, the feet remain anchored in the lower stable layer throughout the simulation, because of the strong sub-adiabatic stratification. The structure of the magnetic field near the apex of the flux rope is similar to what was obtained in 2D simulations. The weak magnetic field line shown has a very high-pitch angle in front of the rope but is almost parallel behind it. The effect of this weak field is the same as in the case of straight but twisted 2D flux ropes, that is, it helps to suppress the Rayleigh-Taylor instability that otherwise threatens to break up the ropes.

After the initial acceleration due to the rope's buoyancy, its apex enters a quasi-stationary regime, where its speed approaches a terminal value in an oscillatory manner. In 2D, this behavior is a result of the competition between buoyancy and aerodynamic drag (Emonet and Moreno-Insertis, see [15]), but in 3D the primary competitor to buoyancy is the magnetic force associated with the main curvature of the rope. As a whole, the flux rope has twisted isosurfaces and looks similar to a twisted rubber band (Color Plate 22 on page 298), but it does not develop a full-blown kink. As the apex of the flux rope approaches the upper boundary of the box, the open boundary begins to influence the evolution of the rope. Color Plate 23 on page 298 shows horizontal cross sections of the magnetic field as the rope emerges through the boundary. A "bipolar region" is formed by the emerging legs, and the "spots" formed by the flux rope's legs separate as the rope rises through the surface (Color Plate 23 on page 298, left panel). At some stages of the emergence, the magnetic field has a characteristic S-shape (see Color Plate 23 on page 298, third row in right panel) similar to what is observed in soft X-ray images of solar magnetic surface structures (e.g., the Yohkoh satellite, see Canfield et al. in [5]).

Magnetic tension is present in topologically complex fields, such as fields generated by turbulent motions producing a chaotic mapping of field lines. It is likely that the dynamo-generated field stored in the undershoot layer has such a complex topology. The origin of a net-twist such as the one assumed above is non-trivial. It might be a consequence of radial shear at some depth in the Sun, acting on flux structures connected across the equator, but this has not yet been proven.

In real solar and stellar convection zones, buoyant flux structures are constantly interacting with the surrounding turbulent convection, convective downdrafts and updrafts, and neighboring flux structures. The question remains whether the quasi-steady state topology that the flux ropes reach in the later phase of their rise is stable towards perturbations from the surroundings, and whether the results reported here, for 3D flux ropes moving in a 1D static stratification, carry over to the more realistic case. Simulations of 3D flux ropes ascending in a dynamic 3D convection zone are under way.

4 Conclusions

The solar dynamo process is at the same time a very complex process and an example of a process that is generic in astrophysics. Turbulent MHD and dynamo processes are of key importance in many astrophysical contexts, and the solar dynamo is unique only in so far as it is the only astrophysical dynamo that we can observe with good resolution in both time and space. Even so, we are severely limited, by only being able to observe the surface manifestations of the solar dynamo.

Access to supercomputers with ever-increasing computing power, and the rapid development of hardware and software for 3D graphics rendering, have offered new tools that may partially compensate for the limited instrumental vision; with supercomputer simulations and 3D visualization, it is possible to "look under the surface" of the Sun. Current computer capacity is not large enough to make models that cover all the relevant length and time scales, but it is possible to make idealized experiments that reveal the basic principles at work. We may then use the insight gained from such experiments as pieces of a puzzle, and try to piece together a larger and more coherent picture.

The experiments reported on here have provided some key pieces and principles and have allowed at least a partial picture of the solar dynamo process. Future more-refined experiments will reveal the extent to which the picture is correct and the extent to which there are pieces still missing in the current paradigm.

Acknowledgment

The author acknowledges support through an EC-TMR grant to the European Solar Magnetometry Network.

References

1. V. Archontis and S. B. F. Dorch. Numerical simulations of dynamos associated with ABC flows. In *Stellar Dynamos: Nonlinearity and Chaotic Flows*, volume 178 of *ASP Conf. Series*, pages 1–11, 1999.
2. V. Arnold and E. Korkina. The growth of a magnetic field in a steady incompressible flow. *Vest. Mosk. Un. Ta. Ser. 1, Matem. Mekh.*, 3:43–46, 1983.
3. E. Beltrami. *Opere Matematiche*, volume 4. Ulrico Hoepli, Milano, 1920.
4. P. Caligari, F. Moreno-Insertis, and M. Schüssler. Emerging flux tubes in the solar convection zone I: Asymmetry, tilt, and emergence latitude. *Ap.J.*, 441:886–902, 1995.
5. R. C. Canfield, H. S. Hudson, and D. E. McKenzie. Sigmoidal morphology and eruptive solar activity. *Geoph. Res. Letters*, 26(6):627, 1999.
6. F. Cattaneo, T. Chiueh, and D. Hughes. A new twist to the solar cycle. *Journal of Fluid Mechanics*, 219:1, 1990.
7. F. Cattaneo and D. Hughes. The nonlinear breakup of a magnetic layer: Instability to interchange modes. *Journal of Fluid Mechanics*, 1969:323–344, 1988.
8. S. Childress. New solutions of the kinematic dynamo problem. *Journ. Math. Phys.*, 11:3063–3076, 1970.
9. S. Childress and A. Gilbert. *Stretch, Twist, Fold: The Fast Dynamo.* Springer-Verlag, 1995.
10. S. B. F. Dorch. *The Solar Dynamo.* PhD thesis, Niels Bohr Institute at Copenhagen University, 1998.
11. S. B. F. Dorch. New results on kinematic dynamo action by ABC flows. Submitted to *Physica Scripta*, 1999.
12. S. B. F. Dorch and Å. Nordlund. 3 D numerical simulations of buoyant magnetic flux tubes. *A&A*, 338:329–339, 1998.
13. S. B. F. Dorch and Å. Nordlund. On the transport of magnetic fields by solar-like stratified convection. In preparation, 1999.
14. T. Emonet and F. Moreno-Insertis. Equilibrium of twisted horizontal magnetic flux tubes. *Ap.J.*, 458:783–801, 1996.
15. T. Emonet and F. Moreno-Insertis. The physics of twisted magnetic tubes rising in a stratified medium: Two-dimensional results. *Ap.J.*, 492:804–821, 1998.
16. Y. Fan, G. H. Fisher, and A. N. McClymont. Dynamics of emerging active region flux loops. *Ap.J.*, 436:907–928, 1994.
17. B. Galanti, A. Pouquet, and P. L. Sulem. *Theory of Solar and Planetary Dynamos*, chapter Influence of the period of an ABC flow on its dynamo action, page 99. Cambridge University Press, 1993.
18. B. Galanti, P. L. Sulem, and A. Pouquet. Linear and non-linear dynamos associated with ABC flows. *Geophy. & Astroph. Fluid Dyn.*, 66:183, 1992.
19. D. Galloway and U. Frisch. A numerical investigation of magnetic field generation in a flow with chaotic streamlines. *Geophy. & Astroph. Fluid Dyn.*, 29:13, 1984.
20. D. J. Galloway and U. Frisch. Dynamo action in a family of flows with chaotic streamlines. *Geophy. & Astroph. Fluid Dyn.*, 36:53–83, 1986.
21. D. J. Galloway and N. R. O'Brian. *Solar and Planetary Dynamos*, chapter Numerical calculations of dynamos for ABC and related flows, pages 105–113. Cambridge University Press, 1992.

22. K. Galsgaard and Å. Nordlund. The heating and activity of the solar corona II: Kink instability in a flux tube. *Journ. Geophysical Res.*, 102:219–235, 1997.
23. F. Moreno-Insertis. Nonlinear time-evolution of kink-unstable magnetic flux tubes in the convective zone of the sun. *A&A*, 166:291–305, 1986.
24. F. Moreno-Insertis and T. Emonet. The rise of twisted magnetic tubes in a stratified medium. *Ap.J.*, 472:L53–L57, 1996.
25. Å. Nordlund, K. Galsgaard, and R. F. Stein. Magnetoconvection and magnetoturbulence. In *Solar Surface Magnetism*, volume 433 of *NATO ASI Series*, pages 471–498, 1994.
26. Ö. E. Rögnvaldsson. *Magnetized Cooling Flows.* PhD thesis, Niels Bohr Institute at Copenhagen University, 1999.
27. M. Schüssler. Magnetic buoyancy revisited: Analytical and numerical results for rising flux tubes. *A&A*, 71:79–91, 1979.
28. H. C. Spruit. Motion of magnetic flux tubes in the solar convection zone and chromosphere. *A&A*, 98:155–160, 1981.

On Grid Partitioning for a High-Performance Groundwater Simulation Software

Erik Elmroth*

NERSC, Lawrence Berkeley National Laboratory; University of California; Berkeley, CA; USA.

Abstract. The partitioning of a discretized computational domain is fundamental for solving many application problems on parallel high-performance computer systems. This is also the key to workload balancing and minimizing communication overhead. This contribution focuses on the partitioning of real application problems for the parallel TOUGH2 software. TOUGH2 is a widely used reservoir simulator for solving subsurface flow-related problems. It solves a set of coupled mass and energy balance equations using a finite-volume method. The application problems considered are from the Yucca Mountain study for a potential repository of civilian nuclear waste. The quality of the partitions produced by three algorithms in the METIS package are analyzed. Performance results are presented for up to 512 processors on a Cray T3E-900.

1 Introduction

TOUGH2 is a widely used reservoir simulator for solving subsurface flow-related problems, such as nuclear waste geologic isolation, environmental remediation of soil and groundwater contamination, and geothermal reservoir engineering. It solves a set of coupled mass and energy balance equations using a finite volume method.

The serial version of TOUGH2 (Transport Of Unsaturated Groundwater and Heat version 2) is now being used by more than 150 organizations in more than 20 countries (see [9] for some examples). It is one of the official codes used in the United States Department of Energy's civilian nuclear waste management for the evaluation of the Yucca Mountain site as a repository for nuclear wastes. In this context arises the largest and most demanding applications for TOUGH2 so far. Scientists at Lawrence Berkeley National Laboratory are currently developing a 3D flow model of the Yucca Mountain site, involving computational grids of 10^5 to 10^6 grid blocks, and related coupled equations of water and gas flow, heat transfer, and radionuclide migration in subsurface [1]. Considerably larger and more difficult applications are anticipated in the near future, with the analysis of solute transport, with ever-increasing demands on spatial resolution, and with a comprehensive description of complex geological, physical, and chemical processes. High-performance capability of the

---

* Present address: Department of Computing Science and High Performance Computing Center North (HPC2N); Umeå University; SE–901 86 Umeå; Sweden.

TOUGH2 code is essential for these applications. A parallel version has recently been presented and shown to give very good parallel performance with parallel speedups of up to 489 on 512 processors on a Cray T3E-900 [2,3].

This contribution focuses on the partitioning of the discretized computational geometry. The partitioning is fundamental for the parallel implementation. It is also the key to workload balancing and minimizing the communication overhead.

2 Parallel TOUGH2

During the initialization phase, input data specifying the problem is read, the discretized computational geometry is partitioned among the processors, and the initial data are distributed. This phase is partly serial, but the computations that follow are completely performed in parallel with communication required for exchanging subdomain border elements and convergence tests.

The time stepping, with dynamically adjusted step length, is performed as an outer loop. For each time step, a non-linear system of equations is solved using Newton's method. For each Newton iteration, a Jacobian matrix is generated by numeric differentiation, and a linear system is solved using a preconditioned BICGSTAB implemented in Aztec [4].

2.1 Grid Partitioning and Grid Block Reordering

Given a computational geometry, space is discretized into many small volume blocks describing a computational mesh or grid. In the following, we consider the dual grid, obtained by representing each *block* (or volume element) by its centroid and by representing the interfaces between blocks by *connections.* (The nomenclature "blocks" and "connections" is used in consistency with the original TOUGH2 documentation [8].) The physical properties for blocks and their interfaces are represented by data associated with the blocks and connections, respectively.

In TOUGH2, the computational domain is defined by the set of all connections given as input data. From this information, an adjacency matrix is constructed, that is, a matrix with a non-zero entry for each element (i, j), where there is a connection between blocks i and j. In the current implementation, the value 1 is always used for non-zero elements, but different weights may be used. The adjacency matrix is stored in a compressed-row format, called CSR format, which is a slight modification of the Harwell-Boeing format.

After partitioning the grid on the processors, the blocks (or more specifically, the vector elements and matrix rows associated with the blocks) are reordered by each processor to a local ordering. The blocks for which a processor computes the results are denoted the *update* set of that processor. The update set can be further partitioned into the *internal* set and the *border* set.

The border set consists of blocks with an edge to a block assigned to another processor, and the internal set consists of all other blocks in the update set. Blocks not included in the update set but needed (read only) during the computations define the *external* set.

2.2 Partitioning Algorithms

We will analyze the quality of the partitionings produced by these algorithms implemented in the METIS package: METIS_PartGraphKway, METIS_PartGraphVKway, and METIS_PartGraphRecursive [5–7]. The algorithms are here named *K-way*, *VK-way*, and *Recursive*, respectively. First, we give brief presentations of the algorithms, which all partition an irregular graph $G = (V, E)$, where V is the set of vertices and E is the set of edges.

All three algorithms are based on a multi-level scheme. The graph is initially coarsened down to a few hundred vertices by grouping subsets of vertices into multi-nodes. The actual coarsening may be performed using a number of different algorithms, and here this is done using a Sorted Heavy Edge Matching algorithm in all three cases. The coarse graph is then partitioned into k parts. This is done by a traditional k-way algorithm for K-way and VK-way and by a recursive bisection algorithm for Recursive. The final step is the uncoarsening phase, in which the multi-nodes are expanded until the original graph is obtained. The partitioning is refined during the uncoarsening phase, in order to take advantage of the additional degrees of freedom obtained by expanding the multi-nodes.

K-way and Recursive minimize the edge cut, that is, the number of edges that straddle partitions (which is as an approximation of the communication volume). VK-way minimizes the true communication volume. All algorithms, however, are based on heuristics and none of them is guaranteed to find the global minimum of its object function.

We note that the edge cut is only an approximation of the true communication volume. Since several edges cut may correspond to the same element in the actual communication taking place, the edge cut overestimates the true communication volume. (We count each bidirectional edge cut as 2.) As a consequence, a partitioning that minimizes the edge cut may not minimize the true communication volume, as illustrated in the example in Fig. 1.

The partitioning $\{A\ B\ C \mid D\ E\ F\}$ minimizes the edge cut with 6 cut edges, that is, 3 bidirectional edges, whereas the true communication volume is 5. The total communication volume, however, is only 4 for the partitioning $\{A\ B\ D \mid C\ E\ F\}$, even though the edge cut is 8 for that partitioning.

3 Evaluation Criteria

The overall objective for a partitioning algorithm in this application is to divide the computational domain into subdomains for the available processors so that the computations can be performed in parallel efficiently.

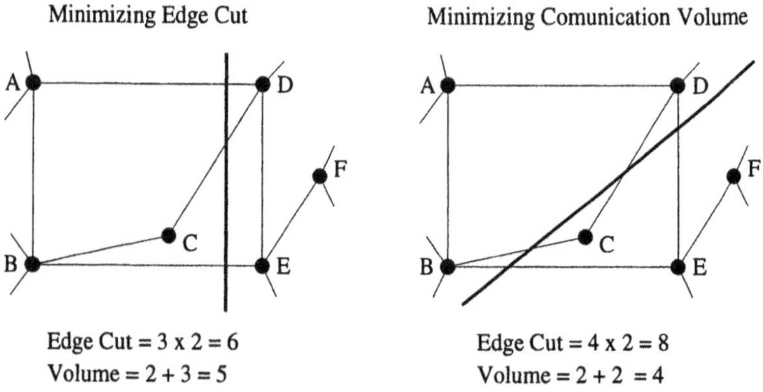

Fig. 1. An example illustrating that a partitioning that minimizes the edge cut does not necessarily minimize the communication volume.

So, what characterizes a partitioning that leads to efficient parallel computations? The final answer is the best trade-off when optimizing for a number of objectives; and the best compromise is not only dependent on the problem and algorithm but also dependent on the characteristics of the parallel computer system.

These issues include both load balancing and minimization of the total amount of overhead. Here we analyze computational load balance, communication volume, and number of messages to be sent. For each of these, we investigate both the total (or average) load per processor and the difference between the maximum load at any processor and the average, as this may be the processor all other processors will be have to wait for during the computations. In summary, we focus on how well the algorithms perform in the following aspects:

Computation load balance Minimize the difference between the maximum number of elements for any processor and the average number of elements per processor.

Average (or total) communication volume Minimize the average number of external elements per processor.

Communication volume load balance Minimize the difference between the maximum number of external elements for any processor and the average number of external elements per processor.

Average (or total) number of messages Minimize the average number of neighboring processors per processor.

Load balance in number of messages Minimize the difference between the maximum number of neighbors for any processor and the average number of neighbors per processor.

The partitioning algorithms considered here are not explicitly designed to minimize all these criteria. They are, however, considered state-of-the-art algorithms for partitioning irregular graphs, and our main interest is to find out how well they are doing in all these aspects.

3.1 Test Problems

Evaluation of the partitionings and the parallel performance of TOUGH2 using these partitionings have been performed for a 2D and a 3D real application problem arising in the Yucca Mountain nuclear waste site study. Results have been obtained for up to 512 processors of the Cray T3E-900 at NERSC, Lawrence Berkeley National Laboratory.

The linear systems have been solved using BICGSTAB with 3×3 Block Jacobi scaling and a domain decomposition based preconditioner with the ILUT incomplete LU factorization. Different levels of overlapping have been tried for this procedure, though all results presented are for non-overlapping tests, which have given good performance in general.

The 2D problem consists of 17,584 blocks (with 3 components per block) and 43,815 connections between blocks, giving a total of 52,752 equations. The Jacobian matrix in the linear systems to be solved for each Newton step is of size $52{,}752 \times 52{,}752$ with 946,926 non-zero elements.

The 3D problem consists of 97,976 blocks (with 3 components per block) and 396,770 connections between blocks, giving in total 293,928 equations. The Jacobian matrix in the linear systems to be solved for each Newton step is of size $293{,}928 \times 293{,}928$ with 8,023,644 non-zero elements.

4 Grid Partitioning Analysis

Table 1 displays the maximum number of elements in any processor's partition for the three algorithms. We note that Recursive clearly outperforms the other two algorithms in terms of averaging the load for all tests from 2 to 128 processors on the 2D problem and 16 to 512 processors for the 3D problem. Only for 256 processors on the 2D problem is VK-way producing a slightly better partition in this aspect.

The difference between the maximum and average number of elements is illustrated in Fig. 2. We also note that VK-way performs slightly better than K-way.

Table 2 shows that VK-way clearly outperforms the other two algorithms in minimizing the maximum and average communication volume in terms of number of external elements required per processor. This supports our discussion about differences between minimizing the edge cut and the true communication volume. K-way appears to give slightly lower communication volume than Recursive.

The number of external elements required per processor decreases as the number of processors increases; and of course, the number of elements per partition does the same. However, the ratio of the number of external elements

Table 1. Maximum and average number of elements per processor.

2D PROBLEM

	Maximum for any processor			*Average per processor*
# Processors	*K-way*	*VK-way*	*Recursive*	*All algorithms*
2	8817	*8793*	*8793*	8792.00
4	4514	4400	*4397*	4396.00
8	2260	2208	*2199*	2198.00
16	1130	1131	*1100*	1099.00
32	565	563	*551*	549.50
64	284	282	*276*	274.75
128	142	141	*138*	137.38
256	72	*70*	71	68.69

3D PROBLEM

	Maximum for any processor			*Average per processor*
# Processors	*K-way*	*VK-way*	*Recursive*	*All algorithms*
16	6308	6295	*6124*	6123.50
32	3154	3153	*3063*	3061.75
64	1579	1576	*1532*	1530.88
128	792	789	*767*	765.44
256	396	395	*384*	382.72
512	198	198	*193*	191.36

2D PROBLEM

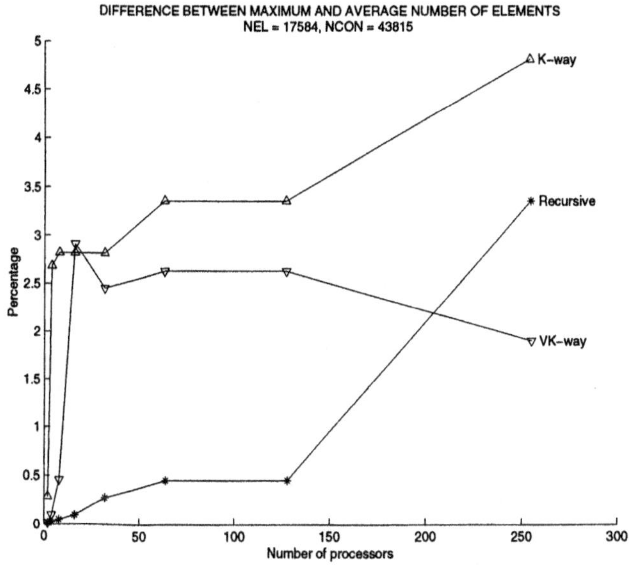

3D PROBLEM

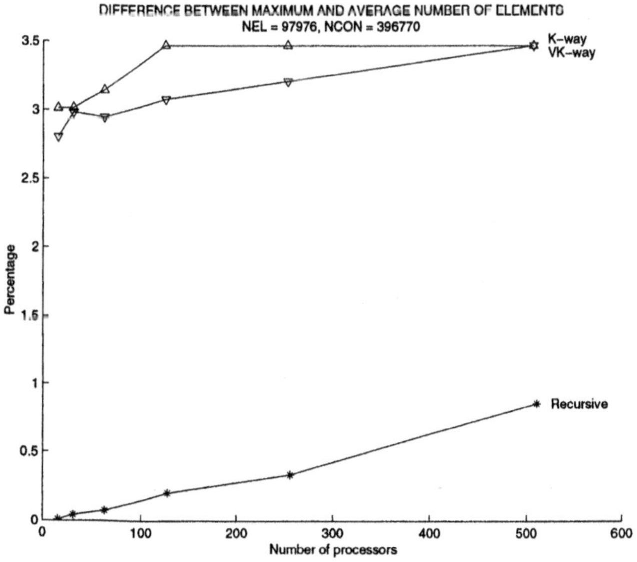

Fig. 2. Computation load imbalance. Difference between maximum and average number of elements per processor (in percentage).

Table 2. Maximum and average number of external elements per processor.

2D PROBLEM

	Maximum for any processor			*Average per processor*		
# Processors	*K-way*	*VK-way*	*Recursive*	*K-way*	*VK-way*	*Recursive*
2	134	*122*	150	134.00	*121.00*	149.00
4	266	*238*	254	175.25	*162.75*	191.00
8	212	*198*	220	173.75	*153.50*	171.00
16	198	*176*	206	146.88	*126.06*	142.75
32	150	*128*	158	110.62	*95.53*	113.28
64	110	*100*	120	83.20	*71.89*	85.58
128	80	*69*	84	61.02	*52.92*	62.83
256	64	*64*	74	44.65	*40.17*	45.30

3D PROBLEM

	Maximum for any processor			*Average per processor*		
# Processors	*K-way*	*VK-way*	*Recursive*	*K-way*	*VK-way*	*Recursive*
16	2172	*1762*	2299	1624.50	*1360.94*	1639.25
32	1536	*1265*	1468	1109.78	*949.19*	1106.12
64	1062	*908*	1129	768.69	*664.25*	770.27
128	746	*670*	736	533.48	*455.76*	526.59
256	529	*472*	521	355.00	*308.94*	352.21
512	340	*301*	374	232.89	*206.64*	234.12

to the number of elements per partition increases. We note that the number of external elements required (206.64) actually is larger than the number of elements to be updated (191.36) by the average processor for the 3D problem on 512 processors.

We also observe a significant imbalance in the communication volume between different processors for all three algorithms. The difference in percentage between maximum and average in Fig. 3 is between 30% and 65% for large numbers of processors for all algorithms on both problems.

We remark that the number of external elements affects not only the communication volume, but also the computation load, as these elements also are used in computations. As the number of elements per partition becomes small and the imbalance between the number of external elements per processor becomes large, this actually becomes a significant factor for load balance. We note that the imbalances caused by this are smaller for VK-way than for the other two algorithms, which then may show an advantage for VK-way on large numbers of processors.

Finally, the number of messages to be sent or received by each processor is proportional to the number of neighboring processors. Hence, the average and maximum number of neighbors per processor should be small. Table 3 shows that the three algorithms produce partitions where the average number of neighbors is roughly the same. The maximum number of neighbors for any processor is, however, slightly lower for K-way for large number of processors.

We conclude that none of the algorithms is superior in all aspects analyzed. Recursive gives most even distribution of elements, that is, computation workload; VK-way gives least communication volume; and K-way gives the smallest number of messages for the "worst case" processor. How these differences affect the parallel performance in practice is analyzed in the following section.

5 Performance Analysis

The two test problems normally require simulations of 10^4–10^5 simulated years. During such long simulations, one would normally observe variations in numbers of time steps, Newton steps, and iterations in the linear solver required when the problem is solved on different numbers of processors. These issues are further discussed in [2]. In order to analyze the performance differences when using different partitioning algorithms, we have kept the simulation fixed to 1 time step, 1 Newton step, and 8 iterations in the linear solver. This is indeed a tiny simulation but still a valid instrument for analyzing the different partitioning, because these steps are repeated over and over again when performing longer simulations.

The execution times in seconds for both problems are presented in Table 4.

The general trend is that Recursive shows best performance for small numbers of processors and that VK-way gets better as the number of pro-

2D PROBLEM

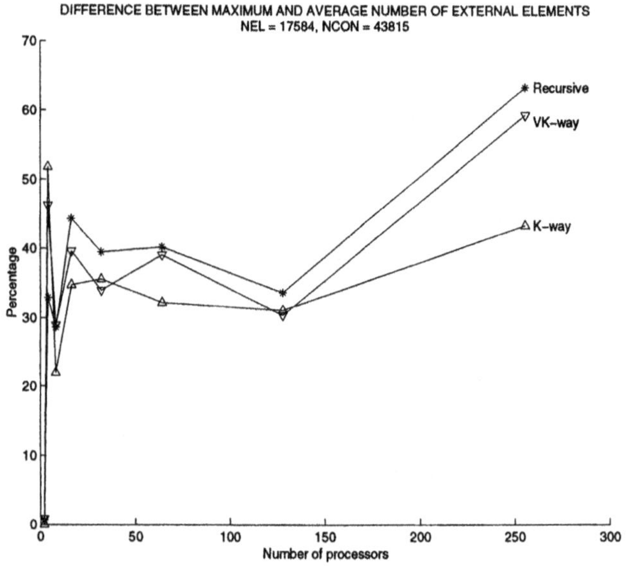

3D PROBLEM

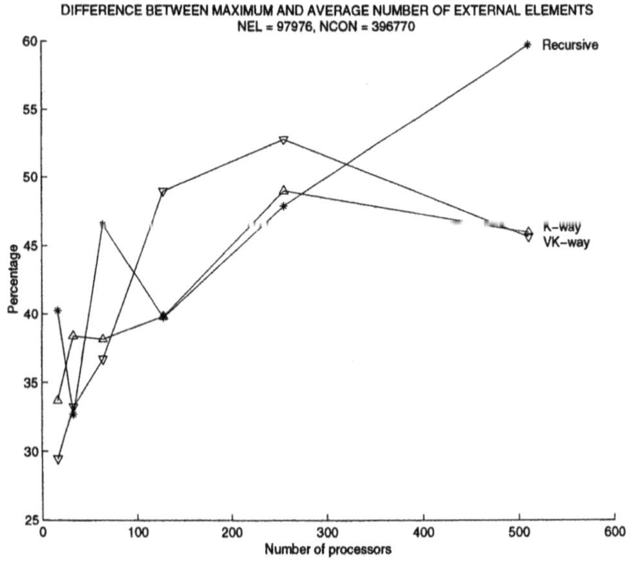

Fig. 3. Communication volume imbalance. Difference between maximum and average number of external elements per processor (in percentage).

Table 3. Maximum and average number of neighbors per processor.

2D PROBLEM

# Processors	*Maximum for any processor* K-way	VK-way	Recursive	*Average per processor* K-way	VK-way	Recursive
2	*1*	*1*	*1*	*1.00*	*1.00*	*1.00*
4	*2*	*2*	3	*1.50*	*1.50*	2.00
8	*4*	5	*4*	*3.00*	*3.00*	3.25
16	*6*	*6*	*6*	4.00	4.00	*3.88*
32	*7*	*7*	*7*	4.50	4.44	*4.38*
64	*7*	*7*	8	4.94	4.91	*4.78*
128	*7*	8	8	*5.11*	5.19	5.22
256	*7*	9	9	*5.33*	5.36	5.40

3D PROBLEM

# Processors	*Maximum for any processor* K-way	VK-way	Recursive	*Average per processor* K-way	VK-way	Recursive
16	*9*	11	11	7.25	7.12	*7.00*
32	*12*	15	13	8.62	*8.31*	8.87
64	*14*	17	19	9.88	*9.47*	10.12
128	*15*	22	23	10.70	*10.58*	11.20
256	*18*	23	24	11.90	*11.65*	11.74
512	*19*	22	25	12.59	*12.21*	12.59

Table 4. Execution time for 1 time step, 1 Newton step, and 8 iterations in the linear solver.

2D PROBLEM

# *Processors*	*K-way*	*VK-way*	*Recursive*
2	11.787	11.799	*11.507*
4	5.188	5.103	*5.010*
8	2.441	2.423	*2.378*
16	1.179	1.187	*1.178*
32	0.618	0.613	*0.611*
64	0.357	0.352	*0.347*
128	0.260	0.261	*0.251*
256	0.244	*0.240*	0.255

3D PROBLEM

# *Processors*	*K-way*	*VK-way*	*Recursive*
16	15.234	16.384	*15.143*
32	5.709	5.991	*5.636*
64	2.541	2.589	*2.525*
128	1.325	*1.300*	1.307
256	0.910	0.886	*0.866*
512	0.834	*0.821*	0.881

cessors increases. This follows naturally from the increasing importance of low communication volume as the number of processors becomes large and the fact that computation load balance is the most critical factor for small numbers of processors.

A closer look at the results reveals the following. For the 2D problem, the best result is always obtained for the partitioning that gives the best load balance, as in fact the VK-way algorithm actually gives better load balance than Recursive for 256 processors. For the 3D problem, Recursive gives better load balance for all tests, even though the difference becomes smaller as the number of processors increases. This, together with increased importance of low communication volume, allows VK-way to perform better than Recursive for 512 processors, slightly better on 128 processors, and almost as well on 256 processors. The differences on 128 and 256 are too small to be significant, but the trend as the number of processor increases is clear.

6 Conclusions

We conclude that the three partitioning algorithms clearly produce partitions that are beneficial from different aspects. Recursive performs the best workload distribution up to a large number of processors, VK-way is best in minimizing the communication volume, and K-way is slightly better than the other two in minimizing the number of messages to be sent during the computations. Which is the best choice depends on the characteristics of the computer system (communication bandwidth, start-up time for sending a message, etc.), and the characteristics of the problem (i.e., the number and size of messages to be sent).

For the TOUGH2 software and our two application problems from the Yucca Mountains study on up to 512 processors on Cray T3E, we find Recursive to be the best choice up to around 128 processors and VK-way for very large numbers of processors. However, in these tests we never find very large differences in execution times using different partitioning algorithms, which shows that this software and problem combination is not very sensitive to the differences provided by the three algorithms. Finally, we note that all three algorithms produce partitions with large imbalances in the number of external elements per processor, that is, large imbalances in communication volume.

Acknowledgment

We thank Chris Ding for helpful discussions and Yu-Shu Wu for providing the test problems.

This work is supported by the Director, Office of Science, Office of Laboratory Policy and Infrastructure, of the U.S. Department of Energy under contract number DE-AC03-76SF00098. This research uses resources of the National Energy Research Scientific Computing Center, which is supported by the Office of Science of the U.S. Department of Energy.

References

1. G. S. Bodvarsson, T. M. Bandurraga, and Y.-S. Wu. The site-scale unsaturated zone model of Yucca Mountain, Nevada, for the viability assessment. Yucca Mountain Characterization Project Report LBNL-40376, UC-814, Earth Sciences Division, Lawrence Berkeley National Laboratory, Berkeley, CA, 1997.
2. E. Elmroth, C. Ding, and Y.-S. Wu. High performance computations for large scale simulations of subsurface multiphase fluid and heat flow. Report, NERSC, Lawrence Berkeley National Laboratory, Berkeley, CA, 1999. Submitted to *The Journal of Supercomputing.*
3. E. Elmroth, C. Ding, Y.-S. Wu, and K. Pruess. A parallel implementation of the TOUGH2 software package for large scale multiphase fluid and heat flow simulations. In *Supercomputing '99 Proceedings.* ACM, 1999.
4. S. A. Hutchinson, L. V. Prevost, J. N. Shadid, C. Tong, and R. S. Tuminaro. Aztec user's guide. Version 2.0. Technical report, Massively Parallel Computing Research Center, Sandia National Laboratories, Albuquerque, NM, 1998.
5. G. Karypsis and V. Kumar. A fast and high quality multilevel scheme for partitioning irregular graphs. Report 95-035, Department of Computer Science,

University of Minnesota, 1995, Revised 1998. To appear in *SIAM Journal of Scientific Computing.*
6. G. Karypsis and V. Kumar. Multilevel K-way partitioning scheme for irregular graphs. Report 95-064, Department of Computer Science, University of Minnesota, 1995, Revised 1998. To appear in *Journal of Parallel and Distributed Computing.*
7. G. Karypsis and V. Kumar. METIS: A software package for partitioning unstructured graphs, partitioning meshes, and computing fill-reducing orderings of sparse matrices. Version 4.0. Technical report, Department of Computer Science, University of Minnesota, 1998.
8. K. Pruess. TOUGH2: A general-purpose numerical simulator for multiphase fluid and heat flow. Technical Report LBL-29400, UC-251, Lawrence Berkeley Laboratory, University of California, Berkeley, CA, 1991.
9. K. Pruess, editor. *TOUGH Workshop '98 Proceedings*, LBNL-41995, Conf-980559, Lawrence Berkeley National Laboratory, Berkeley, CA, 1998.

Visualization of Multi-Scale Data Sets in a Self-Organized Criticality Sandpile Model

Bogdan Hnat and Sandra C. Chapman

Space and Astrophysics Group; Physics Department; University of Warwick; Coventry CV4 7AJ; UK

Abstract. Generic methods for the description of a complex system are sandpile (avalanche models), cellular automata, and coupled map lattice (CML). The emergent phenomenology of these is broadband, but it is represented on a finite grid. To study the three-dimensional evolution of a CML model of a sandpile as a function of control parameters, a semi-immersive VR environment has been used. Sandpile profiles are calculated in real time and are superimposed over the color grid that represents one of the control parameters. This represents a generic problem of visualizing multi-scale structures in data sets derived from nonlinear systems. The software modules were developed in OpenGL and Glut library and will ultimately be implemented in semi-immersive VR on an ImmersaDesk. To achieve sufficient speed in real-time display, the full data set is represented graphically in a compact form.

1 Introduction

All multi-scale physical systems—such as Computational Fluid Dynamics, CML, or CA—typically yield gridded data that are inherently multi-scale [1]. Among these avalanche models, sandpile provides a relatively straightforward test case for other multi-scale systems where grids might be in addition irregular and nonuniform. To study the dynamics of the sandpile in space and time requires simultaneous display of structures differing in scale by several orders of magnitude. The challenge in visualization of the sandpile model data comes not only from their multi-scale character, but also their size and complexity.

Most visualization techniques deal, to some extent, with the handling of large datasets. Algorithms such as visibility culling, triangle meshes, or polygonal surface simplification are available to reduce the complexity of the graphical representation of the model [6]. Many of those techniques, however, have serious limitations when used for scientific applications. Details of the structure are often neglected or compromised to minimize the size of the dataset [3]. This might lead to false conclusions about the details of the investigated system. Therefore, it is critical that the full three-dimensional surface is retained with all the information required for understanding a decimated dataset without a priori knowledge of the region of interest. Another problem that most visualization applications must overcome is real-time managing and rendering of the data. For applications that perform data storage, some optimization and compression of the geometry can be performed to speed up future rendering. Most scientific applications, however, must respond to user's interrupts at a speed that makes any disk storage rather impractical to use. A scientific visualization of the multi-scale datasets must therefore balance the amount of data required to study the system and amount of time needed to compute, encode, and display those data.

The goal of this project was to develop a tool for display and analysis of the large, multi-scale structures, that is, with generic features on different scales. To achieve that, one has to be able to select interesting regions of the structure and focus on its detailed topology as well as gain an appreciation of trends in the data over their entire dynamic range. Here we discuss developing, in OpenGL, rendering techniques for real-time display of the multi-scale data. We also attempt to combine two datasets, using OpenGL blending algorithms to study the impact of the changing parameters on the sandpile evolution. Information of detailed structure in the sandpile has also been coded into the sound, which can be played simultaneously by the user to further enhance data analysis.

Because the future goal is to port the code to an ImmersaDesk tracked semi-immersive VR system as a real-time computational steering application, the benefits and drawbacks of the different implementations will be discussed in some detail.

1.1 Terminology

Figure 1 shows a two-dimensional (length vs. height) time slice of the sandpile surface. The sandpile length is discretized (slots), each part with one unit of length. The sand is always added to slot one, positioned at the origin of the graph. Dynamic rules applied to the model cause the distribution of the sand from one slot to another. These *avalanches* can move from left to right only. Avalanche length is defined as the number of slots involved in a one-slope relaxation process. Let $y_{i,t}$ describe a parameter used in the dynamic rule for slot i and time t. The value of the parameter is selected from two available values for each slot i involved in the relaxation process at time t. The system is open at the end, allowing the sand to leave the system when the boundary is reached.

2 Structure Visualization

The easiest method of summarizing the evolution of a complex system over its entire bandwidth is to use a state variable graph. Figure 2 (left) shows a time evolution of the total energy of the CML system. The simplicity of the plot allows display of the entire evolution of the system and at the same time presents the researcher with essential information—the existence and starting position of the equilibrium state. This is crucial, because it indicates

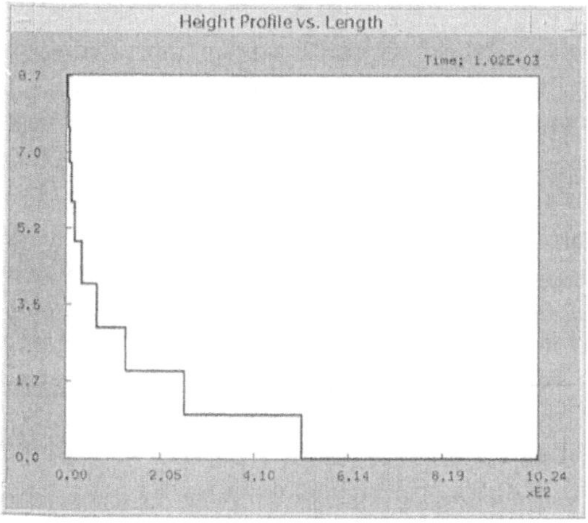

Fig. 1. Time slice of the sandpile evolution surface.

the interval where the time series is stationary. The multi-scale character of the system is also revealed. In the equilibrium state of the system, the energy plot has a clearly self-repeating pattern. Different scales of these oscillations allow us to expect the existence of similar features in the generated 3D image.

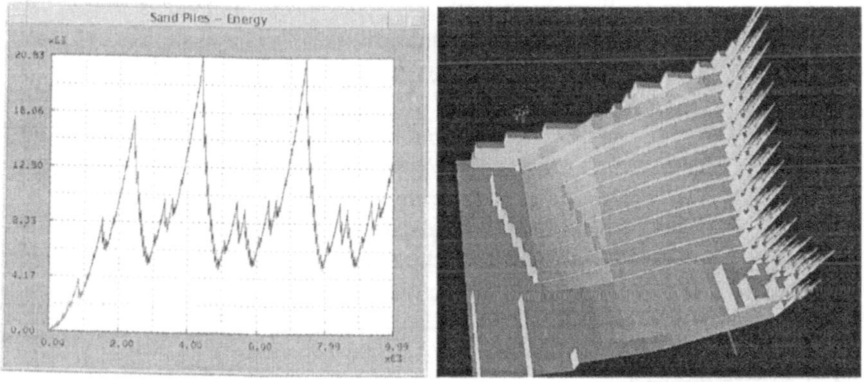

Fig. 2. State variable (total energy) of the sandpile in time (left). 3D surface obtained from part of one cycle in the steady state of the system.

Figure 2 (right) presents the surface of the sandpile profile at a time that corresponds to this steady state. A polygonal surface with the OpenGL lighting model was used to generate the image. The 3D surface can be rotated, translated, and viewed in a close-up mode by the user. The multi-scale character of the structure is very obvious in Fig. 2. The topology of the surface varies widely with distance across the system and time—it is an inverse cascade from small to large scales. Over a short length range, one can observe a plethora of short peaks oscillating in time. The structure contains complex patterns the user needs to extract. If we look on increasing length scales, larger, simpler "step-like" structures are apparent.

Such a large dynamic range in the structures is hard to manage. Because different regions of the surface must be studied in a close-up mode, we cannot use any one simple optimization technique for coarse graining of the entire surface [2]. Different regions need to be optimized differently, based on the details of their topology. The simplest regions are the ones that create flat facets. Successive vertex removal can be used for these areas as described in [5]. The evolution surface contains step-like features up to the smallest resolution. This is due to the self-similar nature of the sandpile. One could further reduce the number of polygons in the mesh by replacing non-flat, monotonically changing areas with smooth surfaces. Normal vector direction changes can be used to estimate the amount of information loss during the surface coarse graining process. No simplification is possible for the most rapidly changing areas (peaks).

The image in Color Plate 24 on page 299 (top) presents a special case in our studies. An exact analytical solution exists for a sandpile model with the nonlocal redistribution rule operating over the entire length of the system and with a single value for the control parameter in the dynamic rules [4]. Because this special case has simple self-similarity, information on all scales is of equal importance. This allowed us to test the efficiency of our visualization techniques against the existing mathematical model. The next simplest case is when the dynamic rules include a control parameter that can assume only two values (with equal probability).

To study the correlation of this dynamic parameter with the behavior of the system, the two datasets are visualized simultaneously in a 3D view (see Color Plate 24 on page 299). A flat color mesh representing two values of the dynamic parameter is drawn under the translucent, 3D sandpile profile surface. In 3D, one can see that the edges of the surface facets tend to align with an area of change of the dynamic parameter.

3 Sonification

Quantitative sonification was also used to enhance the user's experience of the explored system. In our case we used sound to improve the analysis of the most complex areas of the surface. The sound's frequency, volume, and duration are all based on the sequence of avalanche lengths in time. The choice of the parameter was based on the analytical solution presented in Fig. 2. In the special case of the sandpile with dynamic rule acting over the length of the system, avalanche lengths are shown only to exist with lengths

2^N cells (where N is an integer) [4]. For other values of the range of the redistribution rule, the small avalanches cluster around values of 2^N. Therefore, the avalanche lengths are approximately always linear on the logarithmic scale and are therefore tones in the octave. Combined with a close-up view of the structure, sonification is an additional tool for visualizing the multi-scale topology. In particular, it flags to the user patterns in the time series that would be difficult to see on the visual representation of the sandpile. We should mention, however, that the large disadvantage of sonification is that various parameters describing the sound can influence each other (e.g., the volume and the frequency of the sound), giving false perceptions of the changes within the system.

4 Dataset Reduction and Surface Optimization

The storage requirements for the presented problem are quite stringent. A typical run of the simulation generates between 20,000 to 250,000 records. Each record contains energy data, avalanche data, and a height profile. The evolution surface may contain between 500 and 15,000 polygons. Table 1 presents the size of the datasets before and after the optimization process. Additionally, the size of the dataset increases if the dynamic parameter color mesh is displayed together with the height profile. If the performance of the application is not to be compromised, one has to find a simple yet efficient method to reasonably reduce the original datasets.

Table 1. Dataset size before and after optimization.

System Length	Algorithm Applied	Output Size	Number of Polygons
256	None	69 MB	132608
256	Run-length Encod.	8.0 MB	8000
256	Vertex Removal	6.4 MB	2500

Closer analysis of the sandpile output structures suggested that the form of the run length encoding could be used for data size reduction. The encoding is done in two steps. First, after each relaxation process, the heights are scanned. If the heights in the neighboring sites are the same, only one record is used to record the heights together with its start and end position. Later, when the polygonal mesh is created, the same procedure is used to combine flat areas for different relaxation records. A doubly linked list is used to store the mesh allowing immediate release of the purged polygons. This reduces the number of polygons from about 10,000 to 2,000 on average (see Table 1). It should be stressed that this method does not eliminate any details from the original dataset, and thus the image always reflects the true evolution of the system.

The color mesh of the dynamic parameter is stored in a very compact form. No value of the parameter is stored, but only an index to a matrix where the two possible values of the parameter are stored. Since the index can be only 0 or 1, we pack eight indexes in one character type variable (8 bits long). The run length encoding algorithm is run against the mesh to further reduce the number of polygons.

5 Further Development

The ultimate goal for this development process is to port the simulation into an ImmersaDesk semi-immersive virtual reality environment. An eight processor Onyx2 will be used to calculate multiple instances of the sandpile running under different control parameters. The CAVE library will be used to develop rendering routines. Since CAVE parallelizes the application, differentiating between computational and display parts, data storage methods should be reviewed to avoid duplication of the datasets. The user interface must be modified to implement pop-up panels to allow the user to access parameters with a wand, as there is no keyboard access. Selection of the time interval of the data and the length interval of the sandpile, currently done through the state variable window, may be then changed to a cube selection directly from the surface model. Higher computational capability would allow us to view multiple surfaces and compare their structures. The number of records displayed in three dimensions could be increased also (currently, a maximum of 2048 records can be drawn).

Acknowledgment

S. C. Chapman was supported by PPARC, and B. Hnat was supported by HEFCE for this work.

References

1. P. Bak, C. Tang, and K. Weisenfeld. Self-organised criticality: An explanation of $1/f$ noise. *Phys. Rev. Lett.*, 59(4):381–384, 1987.
2. M. F. Barnsley, R. L. Devaney, B. B. Mandelbrot, H. O. Peitgen, D. Saupe, and R. F. Voss. *The Science of Fractal Images.* Springer-Verlag, 1988.
3. M. M. Chow. Optimized geometry compression for real-time rendering. In *8th IEEE Visualization '97 Conference Proceedings*, Phoenix, AZ, 1997. IEEE Computer Society Press.

4. P. Helander, S. C. Chapman, R. O. Dendy, G. Rowlands, and N. W. Watkins. Exactly solvable sandpile with fractal avalanching. *Phys. Rev. E*, 59:6356–6360, 1999.
5. P. Lindstrom and G. Turk. Fast and memory efficient polygonal simplification. In *IEEE Visualization '98 Conference Proceedings*, pages 279–286, Research Triangle Park, NC, 1998. IEEE Computer Society Press.
6. Alan Watt and Mark Watt. *Advanced Animation and Rendering Techniques.* Addison-Wesley, New York, 1998.

Simulation and Visualization of Climate Scenarios on a Distributed Memory Platform

Martin Kücken[1], Ulrich Schättler[2], Friedrich-Wilhelm Gerstengarbe[3], and Peter Werner[3]

[1] Center for Nonlinear Dynamics; Institute of Physics; University of Potsdam; P.O. Box 601553; D-14415 Potsdam; Germany
[2] Deutscher Wetterdienst; P.O. Box 10 04 65; D-63004 Offenbach; Germany
[3] Climate System Department; Potsdam Institute for Climate Impact Research; P.O. Box 601203; D-14412 Potsdam; Germany

Abstract. A parallel version of a local grid point model (LM) of the German Weather Service (DWD) has been implemented on the IBM SP computer system of the Potsdam Institute for Climate Impact Research for the computation of various regional climate scenarios. Multi-variate statistical techniques (pattern recognition, cluster analysis) are used to optimize the validation of different climate variables in different climate scenarios. Our visualization approach incorporates both domain and presentation knowledge in the form of different visualization techniques. The need for a standardized tool to configure any system will become more and more important. The first aim of this work was to develop some basic concepts for a Graphical User Interface (GUI) based on a Motif application.

1 Introduction

Since the summer of 1998, a parallel version of the non-hydrostatic local grid-point model (LM) of the German Weather Service (DWD) has been available. It runs on configurations with up to several hundred processors. The LM will be used by DWD for weather forecasts over Germany. A version of LM has been implemented on the IBM SP computer system of the Potsdam Institute for Climate Impact Research. We try to answer the question of what way this model might be used efficiently for the computation of various regional climate scenarios. In particular, we are concerned with the following problems.

- For the simulation and validation of climate scenarios, there has been an enormous increase in computational demands concerning processor time, processor memory, and the amount of external data storage.
- The evaluation of complex climate datasets requires additional applications in a wide range field, including multi-variate statistical methods and interactive high-end scientific visualization tools.
- It was necessary to develop a graphical user interface (GUI) to efficiently consider the different software tools with the parallel hardware platform.

2 Description and Parallelization of the LM

2.1 Equations, Algorithms, and Grid Structure

The LM is a nonhydrostatic model that has been developed at the DWD as a next-generation limited-area prediction model. It is designed as a flexible tool to be used for operational numerical weather predictions (NWP) on a meso-β and meso-γ scale as well as for the evaluation of the local climate and for various scientific applications covering a wide range of spatial scales (down to grid spacings of about 100 meters). After one year of pre-operational testing, the LM will be the main forecast tool for DWD, starting in December 1999.

The model is based on primitive hydro-thermodynamical equations describing compressible nonhydrostatic flows in a moist atmosphere without any scale approximations. A basic state is subtracted from the equations to reduce numerical errors associated with the calculation of the pressure gradient force in case of sloping coordinate surfaces. The continuity equation is replaced by an equation for the perturbation pressure, which becomes a prognostic variable in addition to the three velocity components: temperature, water vapor, and cloud water. The set of model equations is formulated in rotated geographical coordinates and a generalized terrain-following vertical coordinate. As the Jacobian matrix of the transformation is evaluated numerically, the vertical coordinate and grid stretching can easily be changed, depending on the problem or application.

Spatial discretization is realized by standard second-order finite difference schemes on a staggered Arakawa-C-/Lorenz-grid. The leapfrog-method with a variant of the Klemp and Wilhelmson [10] time-splitting technique is employed for time integration. The extensions proposed by Skamarock and Klemp [15] are also included, whereby horizontally propagating waves are treated explicitly with a forward-backward scheme, and buoyancy and acoustic vertical oscillations are both integrated with an implicit Crank-Nicolson method. A weak 3D divergence damping is applied to control unstable acoustic modes. Additional numerical smoothing is realized by Asselin time filtering and by linear fourth-order computational mixing. To suppress gravity wave reflection at the upper boundary, a Rayleigh-type wave-absorbing layer can optionally be included.

The present physics package of LM is adapted from the operational hydrostatic model DM. The lower boundary conditions for temperature and humidity are provided by a two-layer soil model after Jacobsen and Heise [7], which includes interception storage and snow. Climate values are prescribed in a third layer. Short- and long-wave radiative energy fluxes are calculated by a revised version of the Ritter and Geleyn [12] δ-two-stream radiation scheme, where subgrid-scale cloudiness is diagnosed from relative humidity. Vertical diffusion is treated with the Louis [11] scheme for the surface layer and with a diagnostic second-order level-2 scheme for the Prandtl-layer and the free atmosphere. The parameterization of moist convection is realized by the Tiedtke [19] mass-flux scheme. Grid-scale clouds and precipitation are represented by parameterized cloud microphysics, including water vapor, cloud water, rain, and snow as interacting water categories. Saturation adjustment is applied to calculate condensation and evaporation of cloud water. Rain and snow are treated diagnostically by assuming instantaneous fallouts of particles.

A more detailed description of the LM is found in Doms and Schättler [3].

2.2 Parallelization

To meet the computational requirements for operational NWP and for climate simulations, the program of the LM has been coded in Standard Fortran 90 and parallelized using the Message Passing Interface library (MPI) for explicit message passing on distributed memory machines. Thus, it is portable and can be run on any parallel computer providing MPI. Moreover, it can still be executed on conventional scalar and vector systems where MPI is not available.

The model area of the LM is represented as a 2D regular latitude-longitude grid with up to 50 vertical levels. The parallelization strategy is the 2D domain or data decomposition (grid partitioning), which is well suited for grid-point models using finite differences. This strategy is also used and described by several other authors [2,5,9,14]. Each processor gets an appropriate part of the data to solve the model equations on its own subdomain. These subdomains are arranged in a 2D array of rectangular tiles. The local data structure of every processor contains additional rows and columns to store the values of grid points belonging to neighboring processors (see Fig. 1).

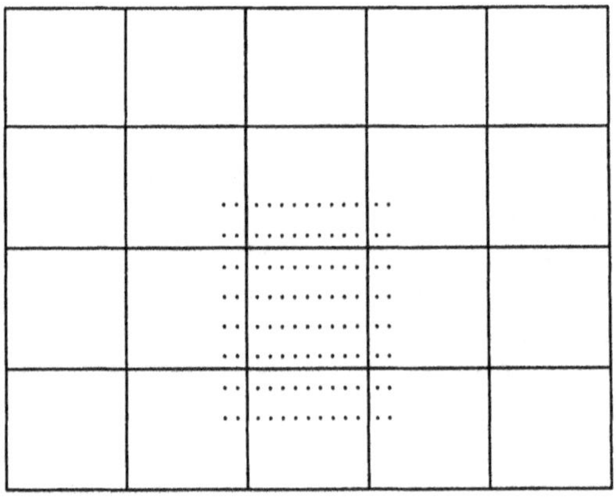

Fig. 1. 2D domain decomposition with local data structure.

During the integration step, each processor updates the values of its local subdomain; grid points on the edges are exchanged using explicit message passing. For the algorithms described above, only a nearest-neighbor exchange (i.e., local communication) is necessary. Figure 2 shows the wall-clock time for a forecast of one hour from full-physics test runs with different domain sizes on a SGI/Cray T3E. The wall-clock times are illustrated on a log-log-scale, where perfect scalability is indicated by a straight line. From this figure it can be seen that the model scales well to hundreds of processors for appropriate domain sizes.

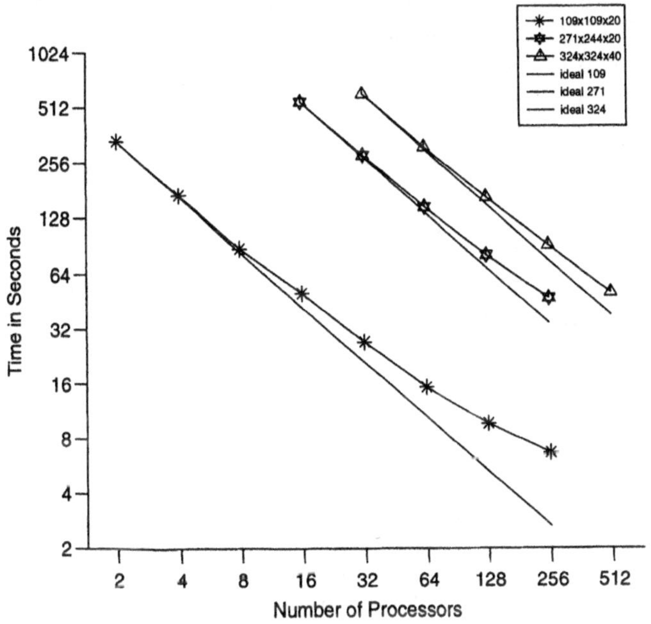

Fig. 2. Wall-clock time for different model sizes.

3 Multi-variate Statistical Techniques

Multi-variate statistical techniques (pattern recognition, cluster analysis) are used to optimize the validation of several variables in different climate scenarios. The advantage of this technique is that the dataset used for the evaluation is not constrained by a single variable limitation, but instead represents the true environmental conditions that occur during the study period. The code of the cluster analysis is useful for structured (model data) or unstructured (observation data) grids. Despite the fact that the code itself does not work in parallel, we can use a massively parallel system by submitting jobs to a batch queuing system of the IBM SP. In this way, all jobs that are required for the model evaluation can work in parallel.

The aim of the cluster analysis is the separation of several elements into homogeneous groups. Two main techniques are possible. Using hierarchical methods (e.g., Bacher [1]), different sequences of groups on different levels may be constructed. The result is a hierarchy of clusters in a tree structure. This method is commonly used in most of the existing statistical software tools (e.g., [13,17,16]). The disadvantage of this technique is based on the fact that an exchange of elements between the groups is impossible after the "tree structure" was built up. With non-hierarchical methods, this disadvantage vanishes because the elements are simultaneously partitioned into a given number of clusters (see Steinhausen and Langer [18] or StatSoft [17]). For independent samples, Jahnke [8] showed that the minimum-distance methods fulfill the consistence criterion of statistical estimation procedures, which therefore appear to be an ideal tool for model output validation.

In the following analysis, we apply the minimum-distance method (Forgy [4]) of the non-hierarchical cluster analysis. The starting condition is to attribute an equal number L of elements e_i from a total of M to the initial number of K_0 clusters (initial partition) so that each cluster receives $L = M/K_0$ elements as follows:

$$\begin{aligned} &e_1, \ldots, e_L \in c_1 \\ &e_{L+1}, \ldots, e_{2L} \in c_2 \\ &\vdots \\ &e_{(k-1)L+1}, \ldots, e_{kL} \in c_k \ , \end{aligned} \tag{1}$$

where c_i, $i = 1, \ldots, k$ represents the cluster. A so-called group centroid $\bar{e}_k$ is then calculated for each k of the K_0 clusters:

$$\bar{e}_k = \frac{1}{L} \sum_{i=(k-1)L+1}^{kl} e_i \ . \tag{2}$$

The Euclidean distance between the elements and the group centroid $\bar{e}_k$ defines the following target function $a(g)$ at each grouping step g:

$$a(g) = \sum_{k=1}^{K} \sum_{i \in k} |e_i - \bar{e}_k|^2 \ . \tag{3}$$

In this sense, each grouping step can be seen as a displacement of the element e_i into that cluster of the nearest centroid. Thus, the target function can be minimized:

$$a(g) \forall g \longrightarrow min \ . \tag{4}$$

This procedure is repeated until a local minimum of the target function is reached. The initial and final number of clusters are the same and subjectively defined when applying the "standard" non-hierarchical cluster analysis algorithm. If, for example, the initial number of clusters is too small, the number of elements within a single cluster will be relatively large. Consequently, possible internal structures of an initially identified cluster cannot be considered any further. A solution for the problems associated with this procedure (e.g., the subjectively defined number of clusters or the unknown statistical significance of the cluster separation) has been suggested by Gerstengarbe and Werner [6].

For the validation of the model results, one can now proceed as follows. The test sample (e.g., grid-point values of the model output of one or more meteorological parameters) has to be compared with the training sample (e.g., grid-point values of the analysis of the same meteorological parameters). Therefore, a spatial clustering of the training sample has to be carried out with the described method. After this, the Euclidean distance has to be calculated between each grid point of the test sample and the nearest group centroid of the training sample clusters. The number of corresponding grid points between test and training sample as to their affiliation to the clusters serves then to measure the quality of the model outputs.

4 Graphical User Interface for Large-Scale Climate Simulations

4.1 General Principles

In climate simulation, any simulated system or pre/post-processing subsystem involves many elementary functions. Furthermore, the use of high- and low-level simulation requires the ability to merge different architectures from different users. It seems to be important to have not only a normalized library of components to simulate and evaluate any system but also a normalized tool to assemble those components at different levels. The need to have a standardized tool for configuring any system will become more and more important. The first aim of this work is to develop some basic concepts for a GUI based on a Motif application. In the present case, the fields of interest can be summarized as follows.

- Parameterization of a simulation run (serial and parallel platforms).
- Parameterization of a cluster analysis run (serial and parallel platforms).
- Data collection and preparation (serial and parallel platforms).
- Any field of any object can be displayed in any graphical representation (only serial platforms).

In the case of Unix-based platforms, any application can be created and managed as a child process of the GUI (by using the UNIX system calls `fork` and `execlp` to control user processes).

4.2 Main System Features

1. Definition of input data for a simulation program:
   - An ASCII-file that contains the namelist variables, which control the execution of the LM (i.e., the domain and the size of the grid parameters for the model run and variables for controlling the input/output of a domain).
   - GRIB-Code files (GRIded Binary) for initial and boundary values.

   Figure 3 shows the configuration window for preparing boundary files. In the first step, a parameter file is chosen with the help of a file-selection box or by building a new parameter file. Then the parameter file itself is available for manipulating control statements. The preparation run can, at least, be initialized either by submitting a job to the IBM SP or by running this job directly on its own host.
2. Definition of input data for the cluster analysis:
   - An ASCII-file that contains the namelist variables that control the execution of the cluster analysis (i.e., the domain and the size of the grid parameters for the model run and variables for controlling the input/output of a domain).
   - Binary data files that contain time series of prognostic variables.
3. Transformation of compressed binary GRIB-Code files to uncompressed binary files.
4. Creation of an executable LM, the following four options are available:
   - sequential LM with debug option,
   - sequential LM with optimized code,
   - parallel LM with debug option,
   - parallel LM with optimized code.
5. Execution of a chosen LM-version:
   After starting an executable program including debug informations, the debugger will be automatically activated and will provide a convenient GUI. Simulation runs without a debug option will be submitted to the IBM SP.
6. Off-line visualization of the simulation results.

4.3 Object-Oriented High-End Visualization Tools

For large-scale climate simulations, an interactive visualizer is the only reasonable method to view large datasets, to interpret them from different perspectives, to identify special areas of interest, to recognize anomalies, and to detect errors that might not otherwise be apparent.

A three-month climate simulation (166 x 242 grid points and 20 vertical levels) requires approximately 17 days on an IBM SP computer (25 processors of type Power2 66 MHz thin; each processor equipped with 128 MB memory). All model and intermediate data have been stored on disk space

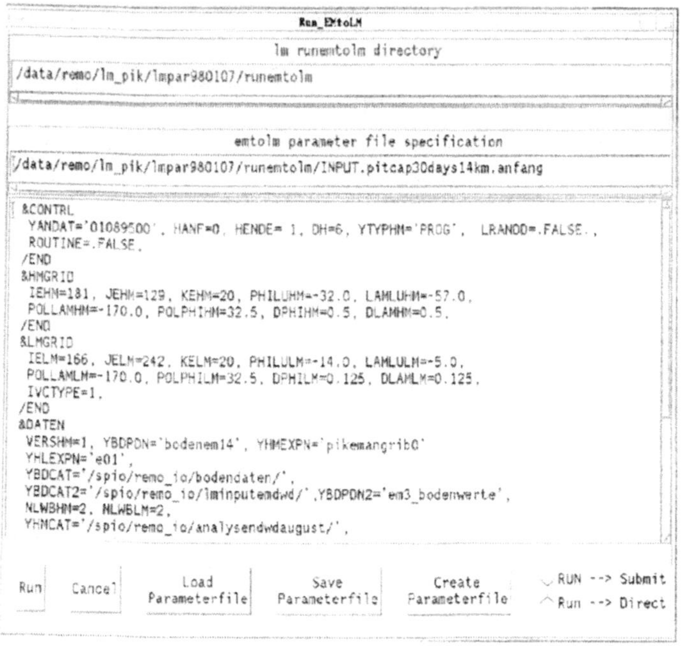

Fig. 3. Graphical user interface window for preparing the LM boundary files.

requiring more than 10 GB for each three months of generated data. These large amounts of data are required for a thorough statistical analysis.

We use a Motif environment with XRT-Widgets (KL-Group) for the basis visualization of time series and 3D model grids. The widgets display 3D data in surface, contour, and bar chart representations. The user-interaction features of XRT (rotation, translation, scaling, zooming, picking) are fully supported. The XRT family of widgets allows developers to quickly add highly functional and flexible GUI components to their OSF/Motif applications.

For our applications, we use the following widget specifications.

XRT/graph A complete and powerful 2D Motif graph widget.
XRT/3d A sophisticated 3D surface, bar, and contour widget.

It is possible to generate PostScript, CGM, XWD, PNG, JPEG, and other bitmap formats of graphs, which makes it easy for end-users to print what they see on the screen. Furthermore, we use object-oriented high-end visualization tools like the IBM Data Explorer.

These applications allow the user to interact with the data and allow an interactive development of the visualization program itself. Both of these facts enable users to change design parameters and to visualize the result rapidly. Currently, our visualization tools do not use the parallel possibilities of a massively parallel platform.

After experimenting with other tools, we focused on IBM Data Explorer. The Data Explorer GUI allows end users to perform tasks at various levels of sophistication. For example, this interface can be used to apply data and adjust input values to an existing visualization process. It is possible to construct a new visualization process, called a visual program, by connecting a network of Data Explorer's modules. It is also possible to create new modules with C or Fortran to be used with the system modules. The second feature is very helpful for the visualization of a given model area, including a choice of different special geographical projections.

Besides the user interface, Data Explorer also supports a scripting language interface for users who want to build their own visualization functions in a more traditional programming style.

Data Explorer's GUI provides an integrated online help facility. This makes available online access to the Data Explorer user manuals as well as to context-sensitive help information. In addition, it allows users to document various aspects of particular visual programs. This program-specific documentation can then be accessed online by other users.

Data Explorer provides an extensive set of modules that can be used to visualize data. The `Isosurface` and `AutoColor` modules, for example, perform the standard visualization functions of creating constant-value surfaces and/or coloring objects based on a data value. This method is shown in Fig. 4.

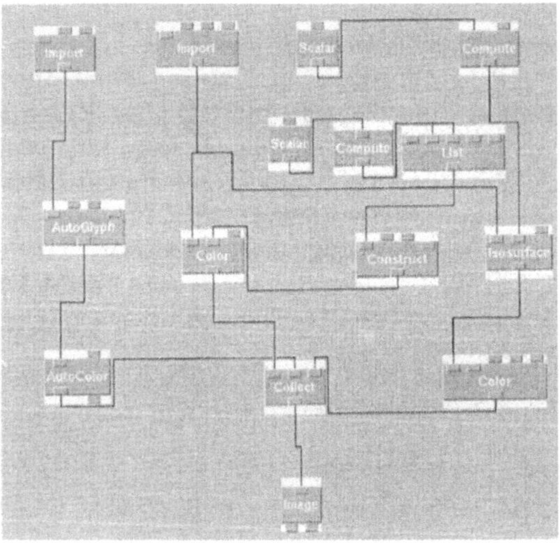

Fig. 4. Data Explorer modules for the visualization of a cluster analysis.

Here, the Data Explorer program uses the `Import` module to read two external files (a file that contains the cluster distribution, and a file for the description of the domain orography). Then this program shows a way to visualize a distribution of clusters in a given model area with embedded model orography. In addition to these expected functions, Data Explorer also provides tools to perform a more sophisticated manipulation of data. The `Compute` module can perform arithmetic or trigonometric operations point-by-point not only with its own data but also on the grid itself. Thus, changing the contour style of a grid, for example, is a simple process of entering an expression.

Even standard tools, such as `Isosurface`, operate on multiple types of input grids. For example, if the input field to `Isosurface` is 2D, the module will automatically create contour lines. A proven method for the visualization of meteorological data fields is based on these `Isosurface` properties. In this way, it is possible to visualize cluster distributions of both irregular and regular grids with the same visual program.

Due to the underlying data model, the modules are efficient and interoperable. The modules are able to describe a wide variety of types of input data. Because the data itself are self-describing, modules can be flexible in the type of data they accept and can perform their actions appropriately based on their input.

Color Plate 25 on page 300 illustrates the graphical output of the Data Explorer program shown in Fig. 4.

5 Conclusions

From this assessment, it can be concluded that the graphical user interface developed so far has fulfilled almost all of the functions that were set down at various stages of the project.

There are time problems when working interactively with the visualized data.

Furthermore, the preparation time for visualization is quite long, if the datasets are too large. In some cases we solved the time problems with a coarse grid, but this can also lead to a loss of information.

The choice of the XRT family of widgets to facilitate visual interactivity with large datasets is perhaps not an ideal one, but it makes it possible to add these widgets to an OSF/Motif application.

Future work will focus on interactive data analysis of large-grid datasets such as those produced by numerical climate models.

Acknowledgments

We would like to thank our colleagues Detlef Hauffe at the Potsdam Institute for Climate Impact Research and Uwe Böhm at the University of Potsdam, Institute of Physics, for hints and help in carrying out parts of the practical implementation.

References

1. J. Bacher. *Clusteranalyse.* Oldenbourg, München, Germany, 1996.
2. A. Dickinson, P. Burton, J. Parker, and R. Baxter. Implementation and initial results from a parallel version of the meteorological office atmosphere prediction model. In G.-R. Hoffmann and N. Kreitz, editors, *Coming of Age (Sixth ECMWF Workshop on the Use of Parallel Processors in Meteorology Proceedings)*, pages 177–194, Singapore, 1995. World Scientific.
3. G. Doms and U. Schättler. The nonhydrostatic limited-area model LM (lokal modell) of DWD, part I: Scientific documentation. Technical report, DWD, March 1997.
4. E.W. Forgy. Cluster analysis of multivariate data: Efficiency versus interpretability of classifications. *Biometrics*, 21:768, 1965.
5. I. Foster and J. Michalakes. MPMM: A massively parallel mesoscale model. In G.-R. Hoffmann and T. Kauranne, editors, *Parallel Supercomputing in Atmospheric Science (Fifth ECMWF Workshop on the Use of Parallel Processors in Meteorology Proceedings)*, pages 354–363, Singapore, 1993. World Scientific.
6. F.-W. Gerstengarbe and P.C. Werner. Applying non-hierarchical cluster analysis algorithms to climate classification: Some problems and their solution. In press in *Theor. Appl. Climatol.*, 1999.
7. I. Jacobsen and E. Heise. A new economic method for the computation of the surface temperature in numerical models. *Contr. Atmos. Phys.*, 55:128–141, 1982.
8. H. Jahnke. *On a graphic consistency conception for cluster analysis procedure*, chapter Cluster analysis as a procedure in inferential statistics, page 168. Vandenhoek & Ruprecht, Göttingen, Germany, 1988.
9. T. Kauranne, J. Oinonen, S. Saarinen, O. Serimaa, and J. Hietaniemi. The operational HIRLAM 2 model on parallel computers. In G.-R. Hoffmann and N. Kreitz, editors, *Coming of Age (Sixth ECMWF Workshop on the Use of Parallel Processors in Meteorology Proceedings)*, pages 63–74, Singapore, 1995. World Scientific.
10. J. B. Klemp and R. B. Wilhelmson. The simulation of three-dimensional convective storm dynamics. *Journal of the Atmospheric Sciences*, 35:1070–1096, 1978.
11. J.-F. Louis. A parametric model of vertical eddy fluxes in the atmosphere. *Bound. Layer Meteor.*, 17:187–202, 1979.
12. B. Ritter and J.-F. Geleyn. A comprehensive radiation scheme for numerical weather prediction models with potential applications in climate simulations. *Monthly Weather Review*, 120:303–325, 1992.
13. SAS Institute Inc., Cary, NC. *SAS/STAT User's Guide: Version 6*, fourth edition, 1990.
14. U. Schättler and E. Krenzien. The parallel "Deutschland-Modell": A message-passing version for distributed memory computers. *Parallel Computing*, 23:2215–2226, 1997.
15. W. Skamarock and J. B. Klemp. The stability of time-splitting methods for the hydrostatic and nonhydrostatic elastic systems. *Monthly Weather Review*, 120:2109–2127, 1992.
16. SPSS Inc., Chicago, IL. *SPSS for Windows manual*, 1999.
17. StatSoft, Tulsa, OK. *Statistica, Vol. III: Statistics II*, 1994.

18. D. Steinhausen and K. Langer. *Clusteranalyse: Einführung in Methoden und Verfahren der automatischen Klassifikation.* Walter de Gruyter, Berlin, 1977.
19. M. Tiedtke. A comprehensive mass flux scheme for cumulus parameterization in large-scale models. *Monthly Weather Review*, 117:1779–1799, 1989.

The Grid: What's Really Going On?

Maxine D. Brown[1], Tom DeFanti[1], Carl Kesselman[2], Jesper Oppelstrup (Master of Ceremonies)[3], Thierry Priol[4], and Karl-Einar Sjödin[5]

[1] Electronic Visualization Laboratory/University of Illinois at Chicago; USA
[2] University of Southern California, Information Sciences Institute; USA
[3] Royal Institute of Technology; SE-100 44 Stockholm; Sweden
[4] IRISA/INRIA; Campus de Beaulieu; F-35042 Rennes Cedex; France
[5] NUTEK and KTH Network Operation Centre; Stockholm; Sweden

Abstract. This panel discussion features experts in networking and applications on the Grid. They address successes as well as pitfalls of the high-performance computing Grid, now and in the next millennium. The following is a paraphrase of participants' remarks.

1 Introduction

MC **Jesper Oppelstrup**: The topic for today's panel discussion is "The Grid: What's Really Going On?" I would like to open up the discussion by putting up a slide containing a quote from Dr. Sheryl Handler, then president of Thinking Machines Corporation, given in 1989 during a U.S. House Subcommittee hearing on the U.S. supercomputer industry.[1]

Fig. 1. Jesper Oppelstrup

> It's hard to understand an ocean because it is so large.
> It's hard to understand a molecule because it is so small.
> It's hard to understand nuclear physics because it's so fast.
> And it's hard to understand the Greenhouse Effect because it is so slow.
> Supercomputers break these barriers to understanding.
> They, in effect, shrink oceans, zoom in on molecules,
> slow down physics, and fast-forward climates.
> Clearly, a scientist who can see natural phenomena
> at the right size and at the right speed
> learns more than one who is faced with a blur.[2]

[1] During the panel discussion, this quote was misattributed to Al Gore in a 1992 presidential initiative on high-performance computing and networking.

With this quote in mind, if one were to make a quick assessment of the presentations during this year's PDC conference, one might conclude that many of the presentations were about interaction; not just about seeing, but interacting—with the computer, the simulation, and other human beings. Is this the way things are going? From scientists watching a simulation at the right speed and size, to people interacting with the simulation and others; interacting with the Grid, as it were?

I might also mention that the Grid is getting some popular attention currently in the form of the movie *The Matrix.*

I'd like to ask each of the panel members to make a few opening remarks to provoke comments from the audience.

2 Panel Members' Opening Remarks

PANEL MEMBER **Tom DeFanti**: Between the talk Maxine gave and the one I gave, you probably figured out that our goal is 3D phone calls. We want to call up someone or something else and really be there. It's a nice general goal, but it's fraught with all sorts of problems. We would like to feel things in these environments. I've thought about putting everything from a car wash to sun lamps in the CAVE so you could get more tactile feeling. It would be nicc to be able to feel things. It would also be nice to smell them. Some work is going on with computational smells.

Fig. 2. Tom DeFanti

It will take the rest of our lives for such environments to be better than being there. They're already better than getting there. Most computer experiences are better than most airport experiences. It's gotten to a place that is interesting. We just need to turn up the quality now, in the best way we can.

PANEL MEMBER **Carl Kesselman**: I would actually argue that the idea behind the [Sheryl Handler] quote is wrong. Supercomputers in and of themselves don't break any barriers to anything and don't let you see anything. The great potential of the type of Grid environments that we've been talking about is by looking at science as an end-to-end activity, not just producing numbers. We actually can produce these types of systems that help us see what's going on. It really is a system-level activity. What Tom talks about is part of that.

Fig. 3. Carl Kesselman

The current mode of viewing a supercomputer as something that stands off in the corner and we essentially submit a card deck to it and wait for an answer to come back is not the most effective way to use these machines.

I would furthermore argue that in a place like Sweden, with a limited budget, and centers spread geographically throughout the country, Grid technology will really be essential for maximizing the interaction going on at Chalmers, Umeå and other universities, and to collaborate with industry. Also, industry in Sweden is multinational. I would argue that it is essential, especially now as you're reevaluating the role of supercomputing centers in Swedish science, that Grid technology is considered and embraced within the science community—as to some extent it has been in the scientific community in the United States.

PANEL MEMBER **Karl-Einar Sjödin**: I came here not really knowing where Grid computing is going. I have had some experience with high-performance computing and networking, in European Union research programs in high-performance computing and networking, but I was somewhat unfamiliar with Grid technology until recently. My expectations of what it is were somewhat wrong. The two speakers before me on this panel have talked about things that I have been involved in recently, but I didn't think those things were Grid computing. I thought Grid computing was a more narrow concept.

Fig. 4. Karl-Einar Sjödin

There is a problem here: There should be a definition of Grid computing that makes it distinguishable from other things that are happening. For example, in the European Union program, there has been—over the last several months—a lot of effort in reworking the content of the Information Society Technologies [IST] program. The rework started out with a new concept named "ambient intelligence." Nobody knew initially, and maybe not even now, what ambient intelligence is. It means that the computers will disappear; and that your environment is aware of you, in some strange way; and without thinking about it, you will interact with a lot of information technology. It means that you will do computing without realizing it.

Now I realize that is Grid computing. As I said, it's too broad for me. In the work for the IST program there have been a lot of key words added, like openness and standardization. There is a lot of work going on toward clarifying the term in various organizations like the Grid Forum. I still find it difficult to say what is *not* Grid computing in the future world of computing.

PANEL MEMBER **Maxine Brown**: In the two statements you have up there on the overhead, there is one major difference between what [Handler] was talking about in [1989] and what we're talking about today. That major difference is networking. In [1989], graphics displays were tightly coupled to supercomputers under one roof. People felt that if they just bought more or bigger computers, they could solve more problems. But problems are becoming complex. There are scientists at NCSA that are trying to simulate an entire factory, not just chemical spillage, for example. It's no longer good enough to just have a bigger machine. You must have a collection of computers and a collection of scientists with different expertise that can all work on solving these complex problems. So the network facilitates not only more resources, but more resources for a more diverse set of scientists to solve a more complicated type of problem.

Fig. 5. Maxine D. Brown

PANEL MEMBER **Thierry Priol**: I have more questions about the Grid than answers. I will make some provocative statements. I will take the first three sentences of the overhead.

Fig. 6. Thierry Priol

It is hard to program a PC because you need to learn languages and all the software environments.
It is very hard to program a supercomputer because you have to learn new techniques.
It is hard to program a Grid system.

Of course, all of this is for someone who is trying to solve a problem. What we need really soon now is some problem-solving environments on top of the Grid that will hide all the big machines and big networks, and we need some programming models that will help in programming these Grid systems. We have to discuss this issue during the panel.

3 How Will the Grid Improve the Quality of Life?

MC **Jesper Oppelstrup**: How will Grid technologies improve the quality of life for the average individual? If we succeed in our efforts of building the infrastructure, where will we see the benefits in the next five years? Telemedicine? Being able to see Grandma when we make an IP phone call? You talk about a 3D phone call, and that sort of ties everything together. But there might be other things. So, where do you see the benefit to the citizen-at-large?

PANEL MEMBER **Carl Kesselman**: I don't consider the examples you gave as Grid technologies. The Grid is more than just a network. It's the integration of collections of resources to solve end-to-end problems. I don't view a 3D phone call, necessarily, as being a Grid technology. If you take that and integrate with a simulation environment and a couple of petabytes of data, then you're starting to look at a problem that requires Grid technology.

Going back to your questions, I have a direct and an indirect answer. The indirect answer is that it's bound to have a profound impact on the way that we do design in engineering and manufacturing technology. I was talking to some people in the Netherlands about something very mundane: producing soap for a washing machine. I mean, this is the most boring thing you could possibly think of, right? They said, "You don't understand, it's *tablets*." (We don't have that in the U.S.) What's the big deal about soap tablets? They said, "This is an engineered delivery vehicle for soap. We can have different densities in the tablets, and we can put conditioner in tablets."

It turns out that the design of soap is very competitive. If you can get your product to market just a few months before your competitor, you get the shelf space. You end up with a market share and they don't.

If you look at the large manufacturers of soap in Europe, like Unilever, they have their chemists in England and their people who understand soap's interaction with clothes in the Netherlands. They're running very complex, multi-resolution simulations from molecular dynamics all the way down to the structural mechanics of cotton.

You listen to the problem; they say, "We have a lot of codes, we have to do work-flow management, we're doing large-scale simulations, we have to do international collaboration—and we want to tie all of this in with the marketing process. The way it feels and smells is going to affect whether people will even want to touch our tablets." This is a big Grid problem. It's a coupling of multidisciplinary multiresolution simulations with visualization, databases, and marketing. This will have a profound effect on the way people will be cleaning their clothes.

That's an indirect effect. Grid technology has a lot to offer in problems like that. It applies to many other fields: aircraft design, etc. You can turn to almost any engineering process, and it will look like this.

Regarding the alternative question as to what more direct impact you could have: that is a little more difficult to answer. You could imagine situations where, let's say, as we get more and more electronic devices, why shouldn't you in your house be able to create an animated greeting card with just as good animation quality as *Toy Story*? It's basically a Grid problem. We should be able to buy that kind of service on the Web, and have some software on our desktop that interacts with that service. We should be able to get this sort of service in our homes. I think this will have a direct impact on peoples lives, but I feel it is longer range than the indirect engineering problems previously mentioned.

PANEL MEMBER **Tom DeFanti**: There is another dynamic going on here. It's not a particular surprise that a lot of us are in academia. What we really do in academia is train students. If any of this is ever going to get done for a soap company, *we* aren't the ones who are going to do this; it's our students, or someone else's students. To a certain extent, what we're doing is trying to find answers. Things that don't work are almost as interesting as things to do work—as long as we can definitively find out why. We hope our students get to make all the mistakes they can when they're with us, so when they go out into the real world, they make fewer mistakes and they can take this technology and solve some of these problems. To me, that's what we're doing. We're creating an environment in which a whole new generation of people can think and operate differently.

PANEL MEMBER **Maxine Brown**: That environment that we're creating is coming from two directions. It's top down. I agree with Carl that the concept of the Grid will facilitate engineering. It will also facilitate drug design as diseases get more complex and more problematic. It all adds ultimately to the quality of life. There will be smart homes. Your coffee maker will go on when your alarm goes off. We'll have more of a democratization of information from the Web. We will have more entertainment and "edu-tainment" that comes right to the home from our Internet port when we plug our smart appliances into the wall. But while we're driving it from the top—and while we have students that we're training—there are children out there that are growing up with today's media. They are accustomed to real-time interaction video games. At least in the U.S., kids wear holes in the palms of their hands from playing Nintendo. When these kids grow up, they are not going to be satisfied with keyboards and monitors in their faces. They are going to want a very different world, and so I think this process will accelerate as the game market and the entertainment market raise our children as well.

4 The Intelligent Pill

MC **Jesper Oppelstrup**: There is a comment that I would like to make, at least for the benefit of our Swedish audience. You mentioned drug design; do you think it is fair to say that there will be "intelligent" drugs on the market in a couple of years?

[LAUGHTER]

PANEL MEMBER **Maxine Brown**: I hope so.

MC **Jesper Oppelstrup**: Recently, a TV commercial was forbidden because it said that a particular pill is intelligent. So, they're just out of synch with what's going on, I guess.

PANEL MEMBER **Maxine Brown**: Well, *Star Trek* has been telling us this for years. They stick little nano-things down you and cure you. So, I hope so.

MC **Jesper Oppelstrup**: But it's illegal to say that a drug is intelligent right now in a Swedish TV ad. Karl?

PANEL MEMBER **Karl-Einar Sjödin**: That's illegal because it isn't true. In a few years, when it is true, you can re-run the commercial, but you will have to prove it by submitting it to an IQ test or something.

MC **Jesper Oppelstrup**: There will be a Turing test for pills.

PANEL MEMBER **Carl Kesselman**: We'll just call it ambient intelligent chemistry, right?

MC **Jesper Oppelstrup**: Then nobody will understand anyway what it is.

5 Does the Grid Face a Software Barrier?

MC **Jesper Oppelstrup**: What about software? We've seen great examples here at the conference, and everyone is very enthusiastic that this will happen. There is the matter of software that Thierry has brought up: it's hard to program a PC, it's hard to program anything. Is there a possible barrier to the Grid in programming and how to develop software systems? How do we try to overcome that? Do we have the right angle of attack and the right ideas at the moment?

PANEL MEMBER **Thierry Priol**: I think so. I would say high-performance computing is not very different from other computing. We have some programming models available, like component programming. I think that is a good candidate to program some Grid systems. Of course, we have this requirement of high performance. We have to adapt this model to be able to communicate very efficiently between components. We have to investigate which modifications need to be made to this model. You have some ongoing work on this in the U.S., for example, you have the CCA forum. When you are adapting a component model, you have to keep in mind that you have this problem of interoperability between components. For instance, if you have to solve a problem and you have to use services available in different places, you have to combine all these services together using a problem-solving environment, for example. You have to be sure everything is interoperable, both in terms of data exchange and control exchange between the components.

MC **Jesper Oppelstrup**: Is this coming about? Is everything becoming interoperable? Is everything plug-and-play at the moment, or when will it ever be?

PANEL MEMBER **Thierry Priol**: I think that's a good question. If you look at what is happening with component programming, we're at the early stage. I don't have any example of a large-scale application based on component programming distributed on a network. It is quite difficult to know when this technology will be used everywhere. I don't have any answer, of course.

6 Does the Grid Face a Lack of Programmers?

PANEL MEMBER **Maxine Brown**: We have a worse problem than programs. The problem is programmers: there aren't enough. We have problems in the schools attracting good students and keeping them, because industry hires them off. Industry wants to hire more and more talent. Our students have no problem getting jobs, and then moving from job to job. As we become more automated—although it will take many lifetimes to build the Grid—it will take competent programmers. I think that is a real problem.

MC **Jesper Oppelstrup**: And you are inferring, then, that the young people who go out into industry will not contribute to Grid programming? Because one might think that they could do it at Boeing just as well as they could do it at KTH.

PANEL MEMBER **Maxine Brown**: No, I'm saying they'll contribute; I just don't think there are enough. I think that with all the talent, there aren't enough people to do the jobs that need to be done.

MC **Jesper Oppelstrup**: That might be a very serious thing then: if there is, at all, enough talent or not to create what we want to create. Anybody else want to comment on the availability of enough talent? This might be a serious barrier.

PANEL MEMBER **Carl Kesselman**: No. We have all the talent we need right here.

[CARL PUTS HIS ARM AROUND TOM]

MC **Jesper Oppelstrup**: You indicate a volume of programmers. Why can't we make systems that make a single programmer more productive? Perhaps more automatic programming? Higher-level languages? That's been going on for decades and decades, so that will surely play a role also here.

Another comment is that just five years ago, all these gizmos—PCs, and projectors, and video, and everything—would not have worked together in this room. They did now, even when a projector failed. Partly thanks to human talent. So, I think we can rely on people being interested in making things work in the future.

PANEL MEMBER **Carl Kesselman**: I was just going to comment. If we had all those thousands of people who are writing device drivers for Linux, and got them doing something else, we would have plenty of programmers.

[LAUGHTER]

7 Human Nature, Funding, and the Grid

AUDIENCE MEMBER **Andrew Chien**: I think that using high-level programming tools and making things easier is absolutely an important thing to do. But it's been my experience that whenever you make something like this easier, people come up with ten times as many applications and things they want to do in that space. I think we are always going to be in this situation of shortage, because there are many more things that people want to do in this space than we actually have time or resources to do.

Fig. 7. Andrew A. Chien

MC **Jesper Oppelstrup**: I had a list here of possible barriers: lack of creativity, lack of computer cycles, lack of infrastructure, lack of software programmers. I'm inclined to agree with the audience member that we probably feel a lack of all of these things at any given time.

PANEL MEMBER **Carl Kesselman**: These are general problems that really have nothing to do with Grids versus anything else. There's always a lack of everything, and we could sit here and wring our hands, and say "Oh, we don't have the perfect programming model." There is a wide range of applications which we're very successful at, right now. There are classes of programming models that we know work fairly well. We have some ideas of how to generate new classes of programming models. There are research activities going on—for example, that Lennart [Johnsson] and I are engaged in—that are attempting to develop some new programming tools to allow us to take steps forward. There will always be more work to do, and wherever you go, there you are.

[LAUGHTER]

PANEL MEMBER **Tom DeFanti**: Okay, I want to pick on you a little bit there. Even though we have a global society, our individual countries make decisions about what kinds of things to support, and how much to support it, and how much infrastructure to support. I'll give you an example. I haven't noticed any professors of video game construction around here, and there aren't any in the United States, to my knowledge—even given the fact that a good deal of our balance-of-trade problem with Japan tends to be about video games.

PANEL MEMBER **Carl Kesselman**: There's a Nintendo chair somewhere isn't there?

PANEL MEMBER **Tom DeFanti**: Yeah, okay, but Singapore doesn't have these problems. They see a market, and they go after it. If you go to the supercomputer center in Singapore, there are all sorts of people worried about video games, because they see that as a market that they can go after. So, we have to see what we can do to keep from falling behind other countries that are very aggressive. I think there is a relative nature here, besides the global nature of just wringing your hands.

PANEL MEMBER **Carl Kesselman**: Just to make a snide remark: you didn't see the lack of a fine programming model for video games holding Nintendo back.

8 Industry and Academics

AUDIENCE MEMBER **Per Öster**: Multinational companies on a global scale have been able to monitor all their production plants, centralize their payrolls, and communicate globally for years. So what's so difficult for us in the scientific community that we spend these conferences and all of these hours discussing how to do these things? What's the difference for us?

Fig. 8. Per Öster

PANEL MEMBER **Karl-Einar Sjödin**: One difference might be that the industrial environment is more focused. In the academic environment, there are so many possibilities that you might get lost instead of focusing. For example, if you want to improve production, you have very strong targets. This world of opportunities that is in the future with Grid computing is creating some of the problems.

Let me say that I think the big enabler here that will help solve the problems is the infrastructure. The key thing that we could do is improve the infrastructure, the networks if you like. That will make the other things happen. It will open up for new creativity, and it will involve more people—maybe even in helping to address the programming problem by giving many more people a chance to be involved. It will open up the marketing possibility by making a global market available for the ones with new ideas in new ways. There are many things we should do—improve education, improve programming tools—but improving the infrastructure, I think, is at the base.

MC **Jesper Oppelstrup**: Karl, you are speaking as our sponsor. Are you willing to write the check so we can go out and do this?

PANEL MEMBER **Karl-Einar Sjödin**: I'm talking in a very generic mode here.

[LAUGHTER]

PANEL MEMBER **Carl Kesselman**: Now I'll argue with both of you. I don't have deep knowledge of industry, but based on what I've seen, I don't believe that industry is addressing the types of collaboration that we're talking about. Why do they do what universities don't do? Well, because they go out and buy Lotus Notes, and they use that so they can collaborate. So, there are a bunch of stupid things they do that we don't do. I do not believe that they have in place this kind of widely distributed sharing and collaborative engineering and design that we're talking about.

My second comment is that I'm not actually worried about infrastructure. The infrastructure is going to happen. The technology marches on, and sooner or later, the networks and so on are going to be there. What I'm more concerned about is how that infrastructure will be used, and can we influence how it will be used? A big divergence of what we're calling Grid computing, from the normal evolution, is this attempt to shift away from a focus of information sharing to information creation and sharing in the engineering process. I'm glad people are investing in the infrastructure, but that's not really what I'm worried about. I think that's just going to happen due to the normal demands. What's going to drive it is that as things globalize, and as time-to-market becomes more important, there are going to be economic pressures. It will be economically impossible for Boeing not to do this because Airbus is going to do it, and vice versa.

PANEL MEMBER **Maxine Brown**: Back to the comment of Per Öster. Per, I'm not sure if you used the word *monitoring*—that industry is *monitoring* their business with their current collaborative tools—but I think their goal is to *run* their companies with their tools, and not just monitor them. A couple of years ago, a colleague in Chicago, where McDonald's is headquartered, said that McDonald's was very interested in the kind of infrastructure that we were talking about. In my naïveté, I asked, "Why would they care?" Well, first, McDonald's wanted to look—real-time—at weather forecasts, so they would know that if it were going to snow, they wouldn't get a lot of customers in a certain area. This affects their suppliers, so they can tell the people they buy food from how much food they're going to order. Then, that affects the trucking companies: How many trucks do they deploy to pick up the food to deliver it? This also affects the number of employees coming in to work. So one things leads to another. With all the real-time inventory in Sears Roebuck and other major department stores, it affects their suppliers, and the suppliers' suppliers—so they can order the parts or order the goods to stock the shelves in real time. Industry is interested—not just in engineering processes, but also in service to the consumer—in the kinds of services to allow this real-time resource sharing and real-time accessibility.

MC **Jesper Oppelstrup**: This is what IBM calls deep computing: this type of problem that pervades many layers and many tiers of activity; the real-time aspect is also very important in that.

9 Continental Differences

MC **Jesper Oppelstrup**: We have heard about the U.S., Europe, Asia. Is there a significant difference in the focus—if there is a focus—of the U.S., of the European Union, or of Asia in approaching the growth of the information technology society?

PANEL MEMBER **Thierry Priol**: I could comment on this. One of the problems is related to funding. We are all driven by funding sometimes. In Europe, the funding to do research activities on the Grid is mainly driven by industrial requirements. If you want to work on the Grid, you will apply for funding to the European Commission. You will have to prove to the European Commission that what you do will have an impact on European industry. And this has side effects on the way we do research in Europe.

In the U.S., you're probably running some large applications, grand-challenge applications. It is quite different in Europe. We are running industrial applications. To my mind, that makes a difference between what is done in the U.S. and what is done in Europe.

MC **Jesper Oppelstrup**: What about Asia? What's going on there? Japan? Singapore? Southeast Asia?

PANEL MEMBER **Carl Kesselman**: I know that in Japan they make similar funding complaints, but they are trying to organize some Grid opportunities there. In fact, Andrew [Chien] and I will be in a workshop in March that's being organized specifically to discuss these types of issues.

MC **Jesper Oppelstrup**: So that's very well in-synch with the U.S. approach.

PANEL MEMBER **Carl Kesselman**: I wouldn't say in-synch. But there is certainly a group of researchers in Japan who recognize this as interesting and valid technology, and who want to make sure they don't fall too far behind in that.

PANEL MEMBER **Thierry Priol**: I have some relationship with a Japanese institution. They have the Earth Simulation Project in Japan. Such a project could motivate people to start some research on the Grid, using this project.

PANEL MEMBER **Karl-Einar Sjödin**: One aspect of the European model is mobile communications. The networking in Europe, from a European perspective, is going to be the convergence of a fixed network and mobile networking—with an emphasis on on the mobile networking because it is seen as a strength, particularly here in northern Europe. This of course leads to the idea of mobile Grid computing, whatever that is.

By the way, in the new work program of the European IST program, there is an action line dealing with end-to-end applications, calling for large-scale work involving middleware and end-to-end applications using special terminals. The man in the loop is a key ingredient, and he is probably on the move.

10 Back to the Lack of Programmers

MC **Jesper Oppelstrup**: One of the problems that one must address here is to attract enough students and people to work on making these dreams or plans come true. How do you go about that? We do see people playing video games, so the interest seems to be there. What is lacking if we don't produce enough programmers who can make it happen?

Panel member **Carl Kesselman**: I don't see how you make the correlation between playing video games and having an interest in doing computer science.

MC **Jesper Oppelstrup**: I think it's a very clear correlation.

Panel member **Carl Kesselman**: The analogue of that is that all these people who like to watch TV should be actors?

MC **Jesper Oppelstrup**: No, I think that's a false analogy. Everybody can be a programmer, because you have a PC at home. You can write your own games, and that's what people who are interested try to do.

Panel member **Carl Kesselman**: I suppose this is off topic. This is somewhat related to the Linux versus NT discussion. Just because you have a PC at home, that doesn't make you a developer. Most of the users of these technologies don't want to know anything about them. So I guess I don't see a correlation between video games and computer programmers, actually.

MC **Jesper Oppelstrup**: Okay, I'll withdraw the remark, but the question still stands: How do we attract more people to make this Grid computing happen?

Panel member **Tom DeFanti**: In graduate school, of course, you pay them. You basically outbid the other fields. You have assistantships or fellowships in Grid computing instead of chemistry or something, and you find people there. And also by having jobs at the other end. I don't know how you do this in elementary schools and so on. I guess you have movies with more programmer heroes, like *The Matrix*, I really don't know. But having this equipment at home does help, because when the kids start to get interested, they have access to essentially phenomenal tools, to learn.

Audience member **Bernard Pailthorpe**: I have a few things to add. It's really about how much computers are used. It wasn't very long ago that only two percent of humanity had access to computers. Maybe it's four percent now, or ten percent, at the most wild extrapolation. There are somewhat larger issues. We're talking about the high-end here, rather than just computing generally.

Another point is related to what Carl was just saying in relation to computer games and programming and so on. That reminds me of the debate that there's no software industry in Japan. Of course there is; we just don't recognize it as such. What's going on in computing games is a very large-scale software generation effort. It doesn't resonate much with this sort of advanced computing that we're talking about, particularly in the U.S. context.

The third point I wanted to make was that *The Matrix* was made in Sydney.

Fig. 9. Bernard A. Pailthorpe

[LAUGHTER]

PANEL MEMBER **Carl Kesselman**: I think the problem of bringing in students has nothing to do with Grids or anything like that. It's just pure and simple economics. I saw a program on CNN last night lamenting the fact that people were leaving business schools in droves because they could go off and make millions of dollars in start-ups. So, it's just general economics and the problem of how we get students into universities. And then once they're, get them excited and working on *our* project, not some other *boring* project. ;-)

AUDIENCE MEMBER **Per Öster**: Perhaps Carl doesn't believe that this belongs to the Grid discussion. I was thinking about the issue of attracting students. To some extent, a lot of the development that's going on is partly dependent on various open-source initiatives, like Linux. To a very large extent, this kind of development exists because there are a number of people around the world working in government-funded laboratories where people don't care what they are doing during their working hours. Suddenly, the government agencies will start to care about what people are doing. Will this effect this software development that is going on? Also, the things we're discussing here are partly dependent on that kind of software development.

11 What Can Academia Learn from Industry?

AUDIENCE MEMBER **Andrew Chien**: Part of what the Grid is about is developing infrastructure and applying infrastructure that allows people to do complex end-to-end allocations. Are the things that are happening out there in the commercial space—like Sandpiper–Digital Island and Akamai—that are doing end-to-end infrastructure (or Inktomi servers and so on), examples of Grid infrastructure? Or how are they different from the Grid infrastructure that we're talking about?

PANEL MEMBER **Carl Kesselman**: We probably have a lot to learn from Inktomi. I would argue—and this is maybe a self-serving argument—that what distinguishes Inktomi or Akamai is that they've been pretty focused toward delivering a pretty specialized service for a specific application. Akamai, as I understand it, is good at brokering bandwidth for delivering Web pages. I would argue that there are probably things we can learn from that, but that, in and of itself, does not make a pervasive infrastructure that addresses the wide range of applications that one would hope to support.

Clearly those type of commercial activities are important and need to be considered, as are the CORBA activities. (Just to make you happy, Thierry.) Certainly we need, from an academic perspective, to understand where the billions of dollars of commercial money are going and to make sure that we're not duplicating those efforts; but to understand also where there are opportunities which industry doesn't recognize right now. We need to do enough leading-edge development so that they can understand that there is a market here, that this is a relevant technology. Then, they can come rushing in, and develop new and wonderful things, and throw away all the stuff that we've done.

12 Shortening the Product Cycle

PANEL MEMBER **Tom DeFanti**: We've shortened up the cycle a bit between things happening in academia or a sole-inventor's mind. Historically, it took about thirty years for technology to get from the first time you saw it to being in schools. We've tightened that up a little bit, but I can't say that we can look at this stuff and say, "Aha! There's something." Occasionally it happens, but generally speaking it takes a long time before any of this tech happens.

Part of the idea of having these advanced industrial societies is that we can afford things like this, in which we basically come together and throw out ideas and generate some stuff. Fifteen years from now, students will hear about the great Carl Kesselman and wonder how he could have possibly known what was going to happen in the future. Well, we're all just sort of guessing, right? If you make enough guesses, and they're intelligent guesses, some of them come true.

PANEL MEMBER **Carl Kesselman**: Right.

PANEL MEMBER **Tom DeFanti**: I really think that that's a lot of what it's about. Anybody who tries to tell us that we are learning in ways that are really that much different from sampling it, seeing how it tastes, putting it back if it doesn't taste so good, and so on—I think you're wrong. You can put as much intelligence as you want into this, but there's still a certain amount of randomness. And then of course there are all the other random factors in terms of actually succeeding in a market, and things like that. I'm just telling you to be optimistic about this, but not disappointed if you fail a lot, because I just think that makes you more creative as you go on.

PANEL MEMBER **Carl Kesselman**: This reminds me of a wonderful story—your comment about twenty years from now and the closing-up of technology. There is a very good book called *Where Wizards Stay Up Late* [1]. This is a story in there. The first ARPAnet node came up in '69 or '70, and they were doing a demo at a meeting in Washington, D.C., I guess, in '72 or '73, and they had actually set up IMPs [Interface Messaging Processors] in the hotel rooms. Jon Postel was out there, and all the original founders of the ARPAnet. Somebody had tripped over a cable or something, and one of the IMPs crashed. These telephone guys, who were these switching-circuit bigots—they thought this whole notion of packet-circuit switching was just completely absurd. They came there and they saw this little network and it was broken, and they were shaking their heads and laughing at these guys. And here it is twenty years later, and they're running like crazy to deploy packet-switching infrastructure. So, I think this notion of delay is quite important.

MC **Jesper Oppelstrup**: From thirty years to twenty years, is that the message?

13 Closing Comments: Should We Reconvene in Five Years?

MC **Jesper Oppelstrup**: Time is running out, so I would like to see if there are some closing comments, on this note: If we should reconvene here in five years time, and then maybe in ten years in time, what do you think will have happened in the Grid?

PANEL MEMBER **Tom DeFanti**: I guess we're working so that we won't have to reconvene here; we'll convene from where we are. We're probably going to stay up late, or get up early, but we'll just do that. My hope is that we'll only have to travel for vacation.

PANEL MEMBER **Carl Kesselman**: Realistically, five years from now, if we're extremely successful, we'll see some real, persistent, wide-scale infrastructure that will have some basic capability that people are actually using to solve real problems. We're not there yet. Five years is probably a reasonable horizon to try to be there.

MC **Jesper Oppelstrup**: What about ten years from now?

PANEL MEMBER **Carl Kesselman**: In ten years, it should be way cool.

[LAUGHTER]

PANEL MEMBER **Karl-Einar Sjödin**: I'm not sure I'm that optimistic. I think in five years, we'll be in about the same situation. We'll see that the possibilities far outweigh what we have achieved. And there will be so many new challenges: lack of resources, lack of programmers, or whatnot.

Also, I think that some of the deployment of the new possibilities will be slowed down by the capacity of people to take it on. That is in many ways a slowing factor here. The technology is possible. If we just had computers do it all, we could advance much faster; but the human side is going to make it slower. I think five years is probably too short a time. Let's convene in ten years.

MC **Jesper Oppelstrup**: I hope so.

PANEL MEMBER **Maxine Brown**: I think we should convene every year as a touchstone to keep up with what's going on, and to share ideas and share technologies. Let's continue to be optimistic. If today we have the information superhighway, then we have to build an information super-duper highway. If that means jet planes and more landing strips, as opposed to horse-and-buggies on roads, I think we have a long way to go. We're going to have fun getting there!

PANEL MEMBER **Thierry Priol**: I will go back to my first statement about the programming of these machines. I think and I hope that in ten years we'll be able to solve problems—not faster than before, but simpler than before. We'll use the performance of the machine in order to provide some languages or programming models that are easier to use—not only that go faster.

MC **Jesper Oppelstrup**: Somebody spoke about weather prediction. We can think about the weather at this time of year, and Stockholm, and winter, and all that. So why do we really want to convene here? Somebody said the object of all this simulation is not really prediction; it's control. How long do we have to wait to get control of the weather in Stockholm in December?

I want to thank the panel for a very lively discussion, and we hope to see some of you—never mind the weather—about the same time next year.

[APPLAUSE]

References

1. K. Hafner and M. Lyon. *Where Wizards Stay Up Late: The Origins of the Internet.* Simon & Schuster, 1996.
2. Sheryl Handler, president of Thinking Machines. Hearing of the Science, Research, and Technology Subcommittee of the House Science, Space, and Technology Committee. 101st Cong., 1st Sess., June 20 1989.

Presenters

Christer Andersson

Ph.D. Student
Department of Numerical Analysis and Computing Science (NADA), Royal Institute of Technology, Sweden

Christer Andersson received an M.Sc. degree in engineering physics, specializing in applied mathematics, from the Royal Institute of Technology (KTH), Sweden, in 1997. Since then, he has been a Ph.D. student in the Department of Numerical Analysis and Computing Science at KTH `<http://www.kth.se/>`. The position is partly funded by the National Graduate School of Scientific Computing `<http://www.tdb.uu.se/~ngssc/ngssc_home/>`.

You can find out more about Christer Andersson on the Web at `<http://www.nada.kth.se/~christe/>`.

Nicole Bordes, Ph.D.

VisLab Manager
San Diego Supercomputing Center and NPACI, USA

Nicole Bordes has a research background in physics and materials sciences. In the five years she has worked in scientific visualization, she has created a new undergraduate course in that field at the University of Sydney. In the Sydney VisLab `<http://www.vislab.usyd.edu.au/>`, she collaborated on research projects spanning many disciplines, and she provided the essential link between the computer scientists and scientific researchers. She continues these activities in her present role as manager of the San Diego Supercomputer Center Advanced Scientific Visualization Laboratory `<http://vis.sdsc.edu/>`.

Maxine D. Brown

Associate Director, Electronic Visualization Laboratory
Electronic Visualization Laboratory/University of Illinois at Chicago, USA

Maxine D. Brown is an associate director of the Electronic Visualization Laboratory `<http://www.evl.uic.edu/>` (EVL) at the University of Illinois at Chicago (UIC), working with funding, documentation, and promotion of its research activities. She is a co-principal investigator of the U.S. National Science Foundation's (NSF) Science, Technology, and Research Transit Access

Point `<http://www.startap.net/>` (STAR TAP) initiative to provide a persistent infrastructure to facilitate the long-term interconnection and interoperability of advanced international networking. Brown was one of several U.S. technical advisors appointed to the G7 GIBN activity in 1995, the findings of which were presented at the G7 Paris meeting in January 1996. Brown was also responsible for all the logistical operations and application documentation for both the SC'95 I-WAY and SC'98 iGrid (international Grid) research demonstrations. Along with Tom DeFanti and Bruce McCormick of Texas A&M University, she co-edited the landmark NSF report, "Visualization in Scientific Computing."

In recognition of her services to the University and the community at large, Brown was a recipient of the 1990 UIC Chancellor's Academic Professional Excellence (CAPE) award. In 1998 she was the recipient of the first SIGGRAPH Outstanding Service Award. Prior to coming to UIC in 1986, Brown was a consultant in the computer graphics industry, and held several positions in industry, notably at Digital Productions, ISSCO and Hewlett-Packard.

R. Holland Cheng, Ph.D.

Group Leader of Structural Virology Unit
Department of Biosciences Novum, Karolinska Institute, Sweden

R. Holland Cheng received the Ph.D. in structural biology from Purdue University. He is currently an associate professor in structural virology at Karolinska Institute <http://www.cbt.ki.se/> in Sweden. His research includes the structure-function study of macromolecular interactions involved in the virus infection process with special interest is the area of high-resolution microscopy, digital image analysis, and molecular modeling. His extensive efforts have been in the novel development of data acquisition and digital image processing in order to provide new tools for observing the dynamic event and various forms of biological molecules in the virus life cycle. One of his research goals is to extend the knowledge gained from the 3D structures to modulate a rationale drug design against viral infection. Such computer-processed reconstruction can also aid the design of gene-therapy vectors for specific targeting and delivery.

You can find out more about R. Holland Cheng's research group on the Web at <http://www.csb.ki.se/users/kisv/>.

Andrew A. Chien

SAIC Chair Professor
University of California, San Diego, USA

Andrew A. Chien's research involves networks, network interfaces, and the interaction of communication and computation in high-performance systems. Andrew is the Science Applications International Corporation (SAIC) Chair Professor in the Department of Computer Science and Engineering at the University of California, San Diego. He received his undergraduate, master's, and doctoral degrees from the Massachusetts Institute of Technology in 1984, 1987, and 1990 respectively. From 1990 to 1998, Professor Chien was a faculty member in the Department of Computer Science at the University of Illinois and a Senior Research Scientist in the National Center for Supercomputing Applications. He is currently working on large-scale clusters with both NCSA and NPACI. Andrew is also recipient of a 1994 National Science Foundation Young Investigator Award, a 1995 C. W. Gear Outstanding Faculty award, and a 1996 Xerox Outstanding Research Award.

You can find out more about Andrew A. Chien on the Web at `<http://www-csag.ucsd.edu/individual/achien/achien.html>`.

Tom DeFanti

Professor and Director, Electronic Visualization Laboratory
Electronic Visualization Laboratory/University of Illinois at Chicago, USA

Thomas A. DeFanti, Ph.D., is director of the Electronic Visualization Laboratory (EVL), a professor in the department of Electrical Engineering and Computer Science, and director of the Software Technologies Research Center at the University of Illinois at Chicago. DeFanti is an internationally recognized expert in computer graphics. In his 26 years at UIC, he has amassed a number of credits, among them the use of EVL hardware and software for the computer animation in the 1977 *Star Wars* movie. He contributed to and co-edited *Visualization in Scientific Computing*, a 1987 report sponsored by the National Science Foundation (NSF); he received the 1988 ACM Outstanding Contribution Award; he was appointed an ACM Fellow in 1994; and he was one of several U.S. technical advisors appointed to the G7 GIBN activity in 1995. He and EVL director Daniel J. Sandin are recognized for the conception of the CAVE virtual reality theater in 1991.

Currently, DeFanti is chair of the Internet2 Applications Strategy Council, and member of the UCAID Board of Trustees; and he is principal investigator of the NSF Science, Technology, and Research Transit Access Point (STAR TAP) and EuroLink initiatives to provide a persistent infrastructure that will facilitate long-term interconnection and interoperability of advanced international networking. He has also been active in the ACM SIGGRAPH organization and in the ACM/IEEE Supercomputing (SC) conferences, including founding the *SIGGRAPH Video Review* video publication in 1979.

Bertil Dorch, Ph.D.

Research Postdoctoral Fellow
Stockholm Observatory, The Royal Swedish Academy of Sciences, Sweden

Bertil Dorch received a M.S. degree in astronomy from Copenhagen University in 1995 on the subject of buoyant magnetic fields. In 1998 he received the Ph.D. degree in physics from the Niels Bohr Institute at Copenhagen University for his thesis on different aspects of the solar dynamo. Since early 1999 he has been a postdoctoral fellow in the EU-commission collaboration programme *The European Solar Magnetometry Network* at the Stockholm Observatory <http://www.astro.su.se/>, under the Royal Swedish Academy of Sciences. There, he works closely with observers using probably the best solar telescope in the world (the SVST on La Palma). Both his published work and his theses revolve around the subject(s) of magnetic fields in the cosmos. A crucial part of his working time is devoted to code development, numerical simulations, and visualization using a 3D MHD code based on the work by Robert Stein (Michigan) and Åke Nordlund (Copenhagen), among others.

You can find out more about Bertil Dorch on the Web at <http://www.astro.su.se/~dorch/>.

Erik Elmroth

Senior Lecturer / Coordinator at HPC2N
Department of Computing Science and HPC2N, Umeå University, Sweden

Erik Elmroth is a senior lecturer in numerical analysis and parallel computing at the Department of Computing Science at Umeå University. He is also coordinator at the High Performance Computing Center North (HPC2N) <http://www.hpc2n.umu.se/>, and an HPDR/SSF/FRN national HPC lecturer. Current research interests include eigenvalue problems with applications in control theory, parallel linear algebra kernels for HPC systems, and HPC applications. International experiences include positions as postdoctoral fellow at NERSC, Lawrence Berkeley National Laboratory, and visiting scientist at the Department of Mathematics, Massachusetts Institute of Technology (MIT).

You can find out more about Erik Elmroth on the Web at <http://www.cs.umu.se/~elmroth/>.

Björn Engquist

Professor
Department of Numerical Analysis and Computing Science, Royal Institute

of Technology, Sweden
Department of Mathematics, University of California, Los Angeles, USA

Björn Engquist has been a professor of mathematics at the University of California, Los Angeles, since 1978, and he has been a professor of Numerical Analysis and Computing Science at the Royal Institute of Technology in Stockholm (KTH) since 1992. He received his B.S. in 1966 and the Ph.D. in 1975 from Uppsala University, Sweden. Between 1966 and 1985, he held positions at the Courant Institute, the University of California at Berkeley, Stanford University, and Uppsala University. He is the director of the Parallel Scientific Computing Institute (PSCI) and the Center for Parallel Computers (PDC), both at KTH.

You can find out more about Björn Engquist on the Web at `<http://www.math.ucla/people/classes/faculty/engquist.html>` or `<http://www.nada.kth.se/~engquist>`.

Erik Engquist

Master's Student
Department of Numerical Analysis and Computing Science (NADA), Royal Institute of Technology, Sweden

Erik Engquist received his undergraduate degree in computer science from the University of California at Berkeley. He completed his master's degree in scientific computing at the Royal Institute of Technology in Stockholm `<http://www.nada.kth.se/>`.

Henry Fuchs

Federico Gil Professor of Computer Science
Adjunct Professor of Radiation Oncology
University of North Carolina at Chapel Hill, USA

Henry Fuchs has worked with 3D biomedical imaging and graphics since 1969 and with head-mounted displays since 1970. He has been on the faculty of the Department of Computer Science at the University of North Carolina at Chapel Hill since 1978. At present, he is predominantly involved in the field of virtual reality in medicine through his work in ultrasound-guided, head-mounted displays, and in telecollaboration as part of the National Tele-Immersion Initiative. Fuchs is one of the inventors of the Pixel-Planes high-performance graphics engine, several of which, in their day, were judged the world's fastest graphics computers; and he has been a principal investigator for its successors, PixelFlow and ImageFlow. He has more than one hundred publications resulting from his research in computer graphics, particularly interactive, 3D computer graphics.

Fuchs has been a member of the National Research Council Computer Science and Telecommunications Board since 1993. He was an associate editor of *ACM Transactions on Graphics* (1983–1988) and the guest editor of its first issue (Jan., 1982). He was the technical program chair for ACM SIGGRAPH '81 Conference, chair of the 1985 Chapel Hill Conference on Advanced Research in VLSI, chair of the 1986 Chapel Hill Workshop on Interactive 3D Graphics, co-director of the NATO Advanced Research Workshop on 3D Imaging in Medicine (1990), and co-chair of the National Science Foundation Workshop on Research Directions in Virtual Environments (1992). He has served on industrial advisory boards for many years.

You can find out more about Henry Fuchs on the Web at `<http://www.cs.unc.edu/~fuchs/>`.

Christer Fureby, Ph.D.

Senior Researcher
Weapons and Protection Division (FOA), Defence Research Establishment, Sweden

Christer Fureby received an M.Sc. degree in civil engineering in 1989 and the Ph.D. degree in engineering physics in 1995, both from the Lund Institute of Technology. Since 1998, he has been a project manager and senior researcher in computational physics in the Weapons and Protection Division of the Swedish Defence Research Establishment, FOA `<http://www.foa.se/>`. He has been involved in about sixty publications. He works with computational fluid dynamics with LES and code development. His main research areas involve the application of LES to incompressible, compressible, and reacting flows, as well as magnetohydrodynamics, solid-fluid interactions, and multiphase flows.

Dr. Fred G. Gustavson

Research Staff Member
Mathematical Sciences Department, IBM T. J. Watson Research Center, USA

Fred G. Gustavson is a research staff member with IBM at its Mathematical Sciences Department of the Thomas J. Watson Research Center `<http://www.research.ibm.com/mathsci/>`. He is the project manager of the algorithms and architecture method of producing high-performance library software for serial, SMP, and distributed memory processors.

Michael Hammill

Head of Education and Training
Parallelldatorcentrum, Royal Institute of Technology, Sweden

Michael Hammill was the conference coordinator of *Simulation and Visualization on the Grid.* He is currently the head of training and education at PDC. He holds degrees in mathematics, philosophy, and computer science from the University of Iowa. He has been involved in programming, software development, and graphics since 1984, and has worked with high-performance computing since 1990. He was a programmer/analyst and the computer consultant manager from 1984 to 1990 at the Weeg Computing Center, now part of Information Technology Services (ITS) at the University of Iowa. He was a senior computing consultant at ITS from 1990 to 1994, where he served as the academic affiliate for the four NSF-funded supercomputer centers in the United States. In 1992, he was part of a team that won first prize in the MasPar Supercomputing Challenge for research on "Studying Performance and Visualization in Lyapunov's Chaotic Space."

From 1994 to 1998, Hammill worked at the Cornell Theory Center (CTC) as a programmer/consultant and as the director of the Smart Node Program, a national affiliates program of more than 100 academic institutions. He coordinated and chaired annual affiliates' meetings, and provided researchers with technical consulting and training on CTC's computational resources. Beginning in 1996, he served for two year on the SUPER! Steering Committee. He also enrolled in computer science courses at Cornell University.

You can find out more about Michael Hammill on the Web at `<http://www.pdc.kth.se/~mike/>`.

Bogdan Hnat

Research Fellow
Space and Astrophysics Group, Physics Department, University of Warwick, England

Bogdan Hnat joined the Space and Astrophysics Group at the University of Warwick `<http://warwick.ac.uk/>` in January of 1999. He is developing semi-immersive visualization programs for ImmersaDesk VR systems with relevance to astrophysical problems. Previously he worked as a software engineer at the Boeing Company, where he designed avionics software for the C-17 transport aircraft. He earned his M.S. degree in physics at the University of Warsaw in Poland in 1992.

Lennart Johnsson

Hugh Roy and Lillie Cranz Cullen Distinguished Professor of Computer Science, Mathematics, and Electrical and Computer Engineering

University of Houston, TX, USA

Lennart Johnsson received his M.S. degree in engineering physics in 1967 and the Ph.D. degree in Control Engineering in 1970 from Chalmers Institute of Technology, Gothenburg, Sweden. At ASEA AB (now ABB), from 1971 to 1979, he led the development of real-time supervisory data acquisition and control systems for electric utilities and process industries. He was on the faculty of Caltech from 1979 to 1983, where he initiated a course in parallel scientific computation, taught VLSI design, and did research in parallel algorithms and VLSI design. From 1983 to 1990, he was an associate professor of Computer Science and of Electrical Engineering at Yale University, where he introduced courses in parallel scientific computation and also in computer arithmetic. He also led the research effort that resulted in the acquisition of the first Connection Machine at a university after MIT. He was Gordon McKay Professor at Harvard University from 1990 to 1997.

From 1986 to 1995, Johnsson also served as the director of Computational Sciences at Thinking Machines Corporation and was responsible for the development of the Connection Machine Scientific Software Library (CMSSL) as well as parts of the Connection Machine Run-Time System (CMRTS). Since 1995, he has been Cullen Distinguished Professor of Computer Science, Mathematics, and Electrical and Computer Engineering at the University of Houston; an adjunct professor of Computer Science at Rice University; and a visiting professor of Computing Sciences and Numerical Analysis at the Royal Institute of Technology, Stockholm. He also serves as the director of both the Texas Learning and Computation Center at the University of Houston and the Houston Area Computational Science Consortium. Johnsson serves on the executive board of the W. M. Keck Center for Computational Biology in Houston and the Los Alamos Computer Science Institute; he is chair of the PDC External Advisory Committee and the Swedish National Allocations Committee for High-Performance Computation; and he is the editor of seven journals.

You can find out more about Lennart Johnsson on the Web at `<http://www.cs.uh.edu/~johnsson/>`. His e-mail address is `<johnsson@cs.uh.edu>`.

Carl Kesselman

Project Leader and Research Associate Professor in Computer Science
University of Southern California, Information Sciences Institute, USA

Carl Kesselman is a project leader at the University of Southern California's Information Sciences Institute and a research associate professor in Computer Science, also at the University of Southern California. An acknowledged authority in Grid technologies, he has co-led the Globus project `<http://www.globus.org/>`, which is developing core technologies for computational Grid systems in the areas of resource location, resource allocation, security, communication, and data access. Most recently he received international recognition for GUSTO, the world's first high-performance computational Grid. GUSTO pushes the technological envelope by using high-speed networks and software to provide global access to advanced supercomputers and other devices.

Malek O. Khan

Ph.D. Student
Department of Theoretical Chemistry, Lund University, Sweden

Malek O. Khan received his M.Sc. degree in engineering physics at Lund Institute of Technical in 1995. He is currently a Ph.D. student in Theoretical Chemistry at Lund University `<http://www.lu.se/>`, Sweden, and at the National Graduate School of Scientific Computing, `<http://www.tdb.uu.se/~ngssc/ngssc_home/>`. His research interests include statistical mechanical simulations, mainly of biological and colloidal systems. He expects to complete his Ph.D. in the spring of 2001.

You can find out more about Malek O. Khan on the Web at `<http://www.teokem.lu.se/~malek/>`.

Sungye Kim

Master's Student
Graduate School of Advanced Imaging Science, Multimedia, and Film, Chung-Ang University, South Korea

Sungye Kim is a master's student in the computer graphics laboratory at the Chung-Ang University `<http://www.cau.ac.kr/>` in Seoul, South Korea. She received a B.S. degree in computer science from Chung-Ang University in 1998, and she expects to receive an M.S. degree in February of 2000. Her research interests include global illumination, radiosity, and non-photo-realistic rendering.

Anton H. J. Koning, Ph.D.

Consultant
Academic Computing Services Amsterdam (SARA), The Netherlands

Anton Koning received a master's degree in computer science at the University of Amsterdam in 1991 with a thesis on the design and implementation of an integrated robot and sensor simulation environment. This was followed by four years of research on parallel volume visualization at Utrecht University, from which he received his Ph.D. in 1996. In 1997 he started working for the Academic Computing Services Amsterdam (SARA) as a visualization consultant and has since been involved in developing software for their CAVE, which was installed later that year. Beside his visualization work, he also acts as consultant for various parallelization and optimization projects. His special interests include (bio)medical visualization, volume rendering, and massively parallel computing.

You can find out more about SARA on the Web at `<http://www.sara.nl/>`.

Ottmar Krämer-Fuhrmann

Computer Scientist, TTN Manager
Institute for Algorithms and Scientific Computing (SCAI)
GMD—German National Research Center for Information Technology, Germany

Ottmar Krämer-Fuhrmann received his education in computer science and operations research at the University of Bonn. He is currently the head of a research group on integrated simulation environments in distributed systems, as well as a TTN manager `<http://www.gmd.de/SCAI/ttn/index-e.html>`.

You can find out more about Ottmar Krämer-Fuhrmann on the Web at `<http://www.gmd.de/SCAI/people/kraemer.html>`.

Martin Kücken

Research Scientist
Institute of Physics, University of Potsdam, Germany

Martin Kücken graduated in 1973 from the University of Rostock, with a degree in chemistry. Since that time, he developed software tools for mathematical statistics for ten years; he assisted in developing numerical weather forecast models in the software department of the GDR for seven years; and since 1991, he has been at the University of Potsdam `<http://www.uni-potsdam.de/>` and the Potsdam Institute for Climate Impact Research. His main areas of interest are climate analysis and the development of climate models.

Per Linse

Associate Professor
Center for Chemistry and Chemical Engineering, Lund University, Sweden

Per Linse received his Ph.D. degree in 1985. He is now the leader of a group performing statistical mechanical simulations and modeling of colloidal system and liquids covering structural, dynamical, and thermodynamical aspects. He is currently an associate professor in the Center for Chemistry and Chemical Engineering at Lund University, Sweden <http://www.lu.se/>.

André Merzky

Staff Scientist
Konrad-Zuse-Zentrum für Informationstechnik Berlin (ZIB), Germany

André Merzky studied elementary particle physics in Leipzig, Edinburgh, and Berlin. He joined the Konrad-Zuse-Zentrum für Informationstechnik Berlin <http://www.zib.de> (ZIB), Germany, in 1994 as a student, where his interests shifted to distributed computing–in particular to remote steering and visualization. He is currently a staff scientist at ZIB.

You can find out more about ZIB's Department of Scientific Visualization on the Web at <http://www.zib.de/Visual/>.

Jesper Oppelstrup

Assistant Professor
Department of Numerical Analysis and Computing Science (NADA), Royal Institute of Technology, Sweden

Jesper Oppelstrup received an M.S. degree in 1969 in engineering physics, and the Ph.D. degree in numerical analysis in 1976, both at the Royal Institute of Technology (KTH). He was a research project manager at the Swedish Institute of Applied Mathematics from 1977 to 1985, and he has worked in commercial development of finite element software since 1984. He began working at KTH in 1985, and he is currently an assistant professor in numerical analysis. Oppelstrup is engaged in the development of high-performance computing resources and industry/academic collaborative research in high-performance computing and mathematical modeling, especially for fluid dynamics and electromagnetics. He is a member of the national supercomputer resource allocation committee, he is on the Program Board of the Parallel and Scientific Computing Institute at KTH, and he is a member of the Swedish Society of Mathematicians.

You can find out more about Jesper Oppelstrup on the Web at <http://www.nada.kth.se/~jespero/>.

Bernard A. Pailthorpe, Ph.D.

Associate Director, Scientific Visualization
San Diego Supercomputing Center and NPACI, USA

School of Physics, University of Sydney, Australia

Bernard Pailthorpe founded the Sydney VisLab `<http://www.vislab.usyd.edu.au/>`, with Australian Research Council funding, to support computational and visualization research. This laboratory underpins research and teaching innovations in a broad spectrum of disciplines. Its recent work has provided visualization support for high-resolution weather modeling, specifically in the context of the Sydney 2000 Olympics. Currently, he directs the scientific visualization program `<http://vis.sdsc.edu/>` for the National Partnership for Advanced Computational Infrastructure `<http://www.npaci.edu/>` and for the San Diego Supercomputer Center `<http://www.sdsc.edu/>`, at UCSD. The group efforts there are focused on scalable volume visualization. He established and led a research effort at the University of Sydney in computational physics, specifically simulating advanced materials by classical and quantum molecular dynamics. This led to an understanding of the mechanism of diamond formation in carbon thin films. He has wide experience in physics education, including developing new classes in computational physics. He has advised government at senior levels on HPC, including the NSF in the USA; the Australian Prime Minister's Science & Engineering Council, which resulted in a new funding program to establish the Australian Partnership for Advanced Computing in 1998; and the New South Wales State Premier, which also resulted in a new HPC funding program in 1998.

Michael E. Papka

Virtual Environments Engineer
Mathematics and Computer Science (MCS) Division, Argonne National Laboratory, USA

Michael Papka received a B.S. degree in physics from Northern Illinois University in 1990, then continued master's studies in physics for another year while working as a research assistant at the Fermi National Accelerator Laboratory. He received an M.S. in electrical engineering and computer science in 1994 from the University of Illinois at Chicago. Since 1990, he has held various research assistantships within the Mathematics and Computer Science Division of Argonne National Laboratory `<http://www.mcs.anl.gov/>`. He is now a staff member at Argonne doing research in experimental virtual reality, collaborative science, and multimedia as applied to high-performance computing. He is also a Ph.D. student in the Department of Computer Science at the University of Chicago.

You can find out more about Michael E. Papka on the Web at `<http://www.mcs.anl.gov/~papka/>`.

Dr. Thierry Priol

Research Director
IRISA/INRIA, France

Thierry Priol received a master's degree (DEA) in computer science in 1986 from IFSIC at the Université de Rennes 1. He also received the Ph.D. degree in computer science in 1989 and an "habilitation à diriger des recherches" in 1995 from the Université de Rennes 1. Since 1989, he has been a researcher at INRIA, and in 1995 he was promoted to research director in the same institute. His research interests include parallel and distributed systems, metacomputing, distributed numerical simulation, code coupling, distributed objects (CORBA), networks of workstations (NOW), clusters, distributed shared memory (DSM) systems, and high-performance visualization.

You can find out more about Thierry Priol on the Web at `<http://www.irisa.fr/paris/pages-perso/Thierry-Priol/anglais/welcome.html>`.

Faith Short

Public Information Officer
Parallelldatorcentrum, Royal Institute of Technology, Sweden

Faith Short earned her B.A. degree in English from Texas A&M University in 1989. She has been involved in editing and publications for ten years. She was a production editor at Texas A&M University Press and Yale University Press from 1989 to 1994, after which she spent ten terrific months as the managing editor at ILR Press at Cornell University. She worked for the next two years as the communications specialist at the Cornell Theory Center. From 1997 to 1998 she was the publications and publicity coordinator at the Herbert F. Johnson Museum of Art at Cornell University. She has been the public information officer at Parallelldatorcentrum (PDC) at the Royal Institute of Technology (KTH) since June of 1998.

Karl-Einar Sjödin

Senior Program Officer, IT
Swedish National Board for Industrial and Technical Development (NUTEK) and KTH Network Operation Centre (KTHNOC), Sweden

Karl-Einar Sjödin received his M.Sc. from the Royal Institute of Technology (KTH) in 1965. From 1967 to 1992, he held a variety of management positions at Control Data AB. He has been a senior program officer at the Swedish National Board for Industrial and Technical Development (NUTEK) since 1994. He was a member of the Swedish Council for High-Performance Computing from 1994 to 1997. He has been on the program committee of the European Union's IT-research programs, Esprit and IST, since 1994.

Dave Turek

Director, Technical Strategy and Business Opportunity
IBM Server Group, IBM, USA

David Turek was appointed to his current position in January 1999. He is responsible for establishing technical strategies and realizing new business opportunities for IBM's server product lines. He is responsible for emerging technologies, strategic planning, cluster architecture, e-business, deep computing, and Linux technologies. From 1995 to 1998, he was the director of systems development for IBM RS/6000 SP, where he was responsible for the design and development of four generations of node (processor) technology, five operating systems releases, and two generations of switch technology. He joined the RS/6000 Division in 1992 and was responsible for the design and implementation of all RS/6000 SP software. He joined IBM in 1974 as a systems engineer and held a variety of technical and management positions within the IBM Corporation.

Turek has an M.S. degree in mathematics from Trinity College (1976) and a B.A. degree in mathematics and philosophy from the University of Rochester (1972). He was also enrolled in a Ph.D. program in operations research at the University of Pennsylvania (1980).

Color Plates

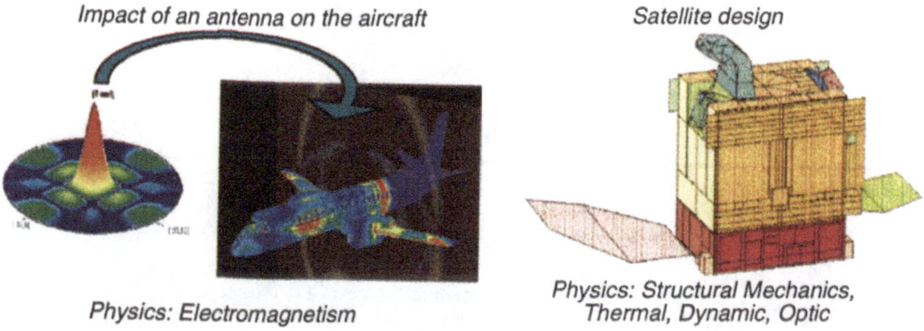

Color Plate 1: Distributed Numerical Simulation Applications. (See Sect. 1 on page 33.)

Color Plate 2: Playback of a large-format movie on the ActiveMural. (See Sect. 3.2 on page 73.)

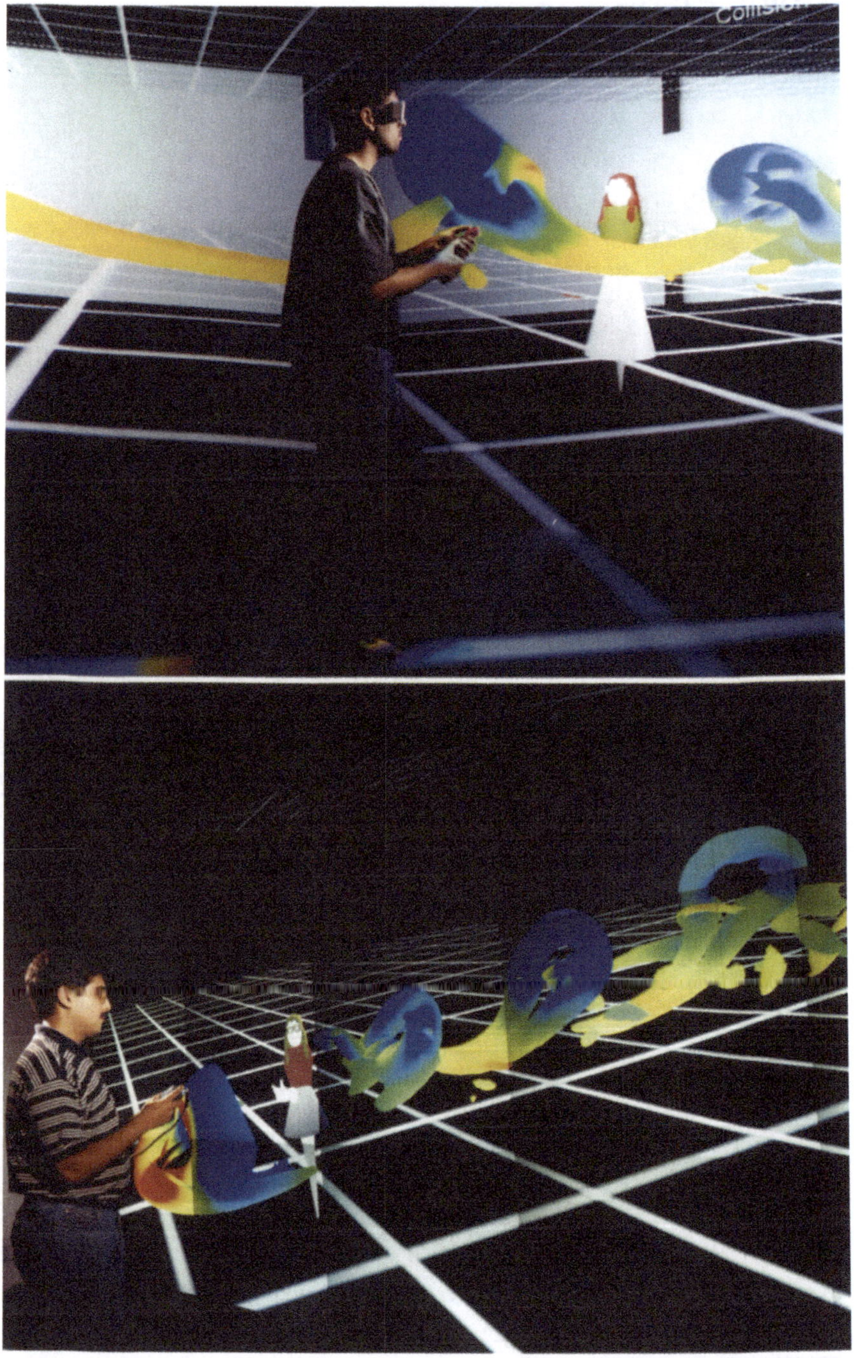

Color Plate 3: A CAVElib-based application running in the CAVE and on the ActiveMural. (See Sect. 3.3 on page 74.)

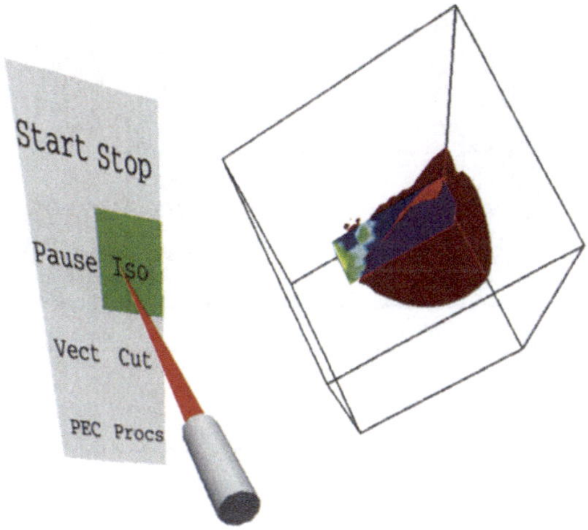

Color Plate 4: Isosurface of electric field strength (red) and surface currents on a conductor. (See Sect. 4.2 on page 89.)

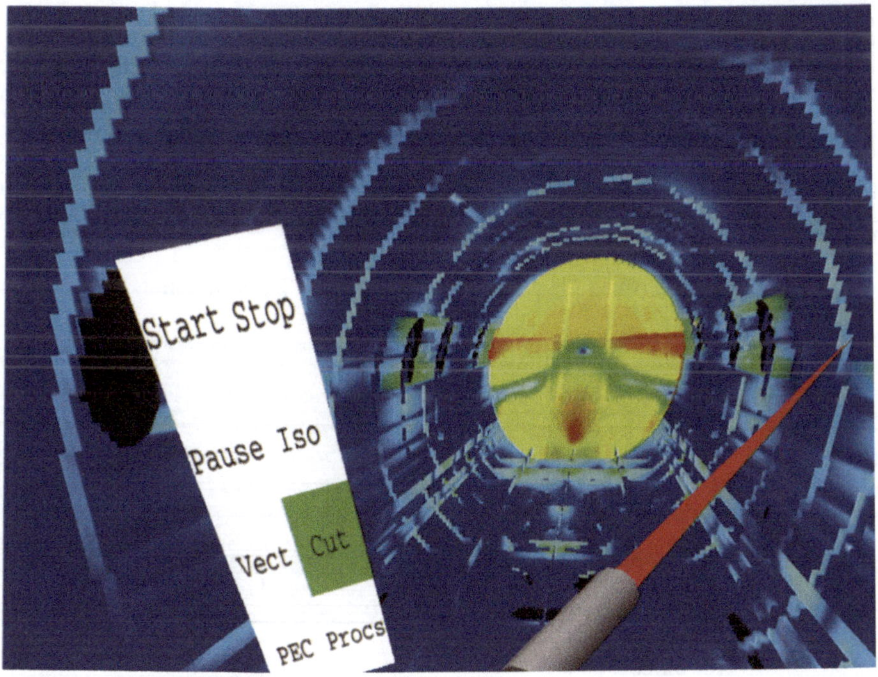

Color Plate 5: Cutting plane inside a large model. (See Sect. 4.2 on page 89.)

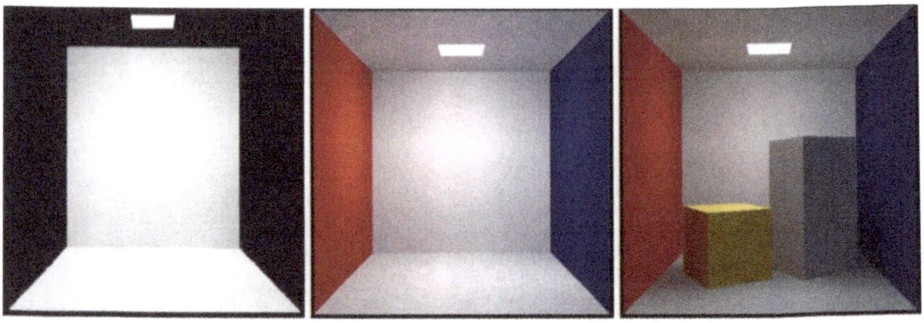

(a) floor and wall (b) cornell room (c) cornell box

Color Plate 6: Result Images 1. (See Sect. 4 on page 109.)

(a) bed room (b) TV and sofa

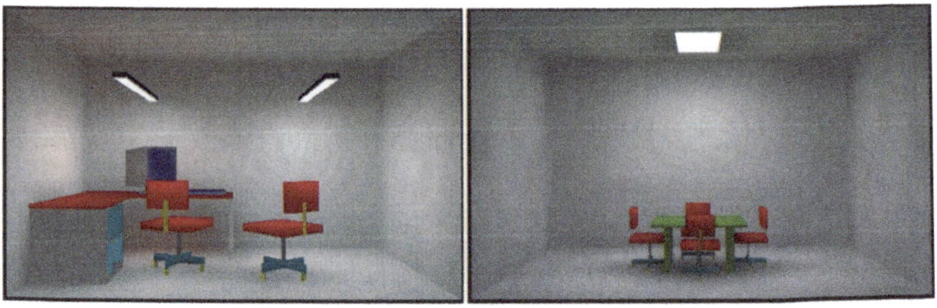

(c) office (d) meeting room

Color Plate 7: Result Images 2. (See Sect. 4 on page 109.)

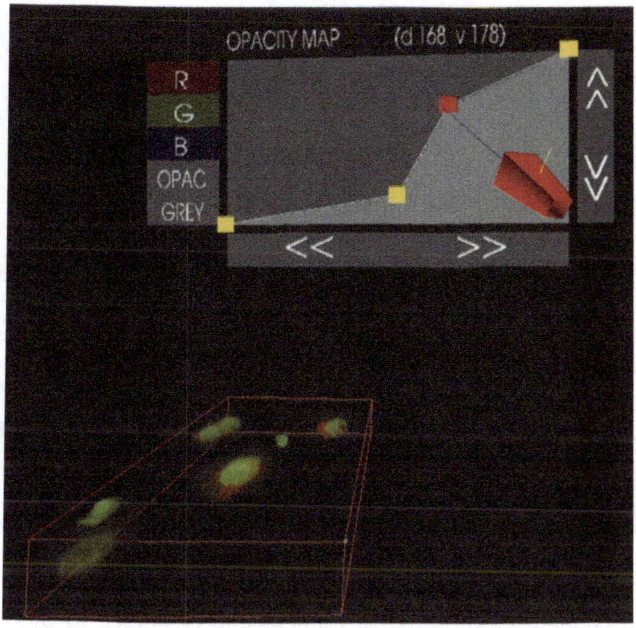

Color Plate 8: The transfer function editor and CLSM data. (See Sect. 2 on page 114.)

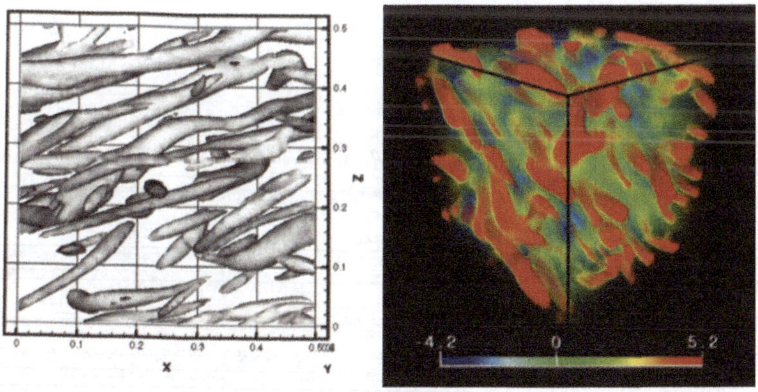

(a) (b)

Color Plate 9: Nomura's CFD of turbulence visualized using (a) isosurfaces at a single value of vorticity (Bailey, after [20]), and (b) rendered volume visualization of the full vorticity distribution. (See Sect. 3.3 on page 129.)

Color Plate 10: One frame snapshot from a fly-through of a 3D model of the Orion Nebula for the Hayden Planetarium (rendered by J. Genetti, D. Nadeau, and G. Johnson, SDSC). (See Sect. 4.1 on page 130.)

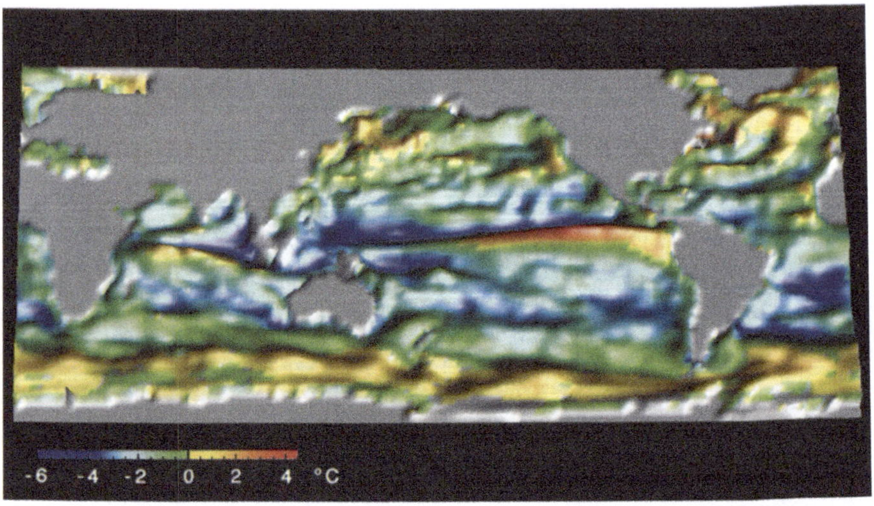

Color Plate 11: Simulation of the 1997 El Niño event, showing relative local warming mapped onto relative sea surface height. (See Sect. 4.2 on page 131.)

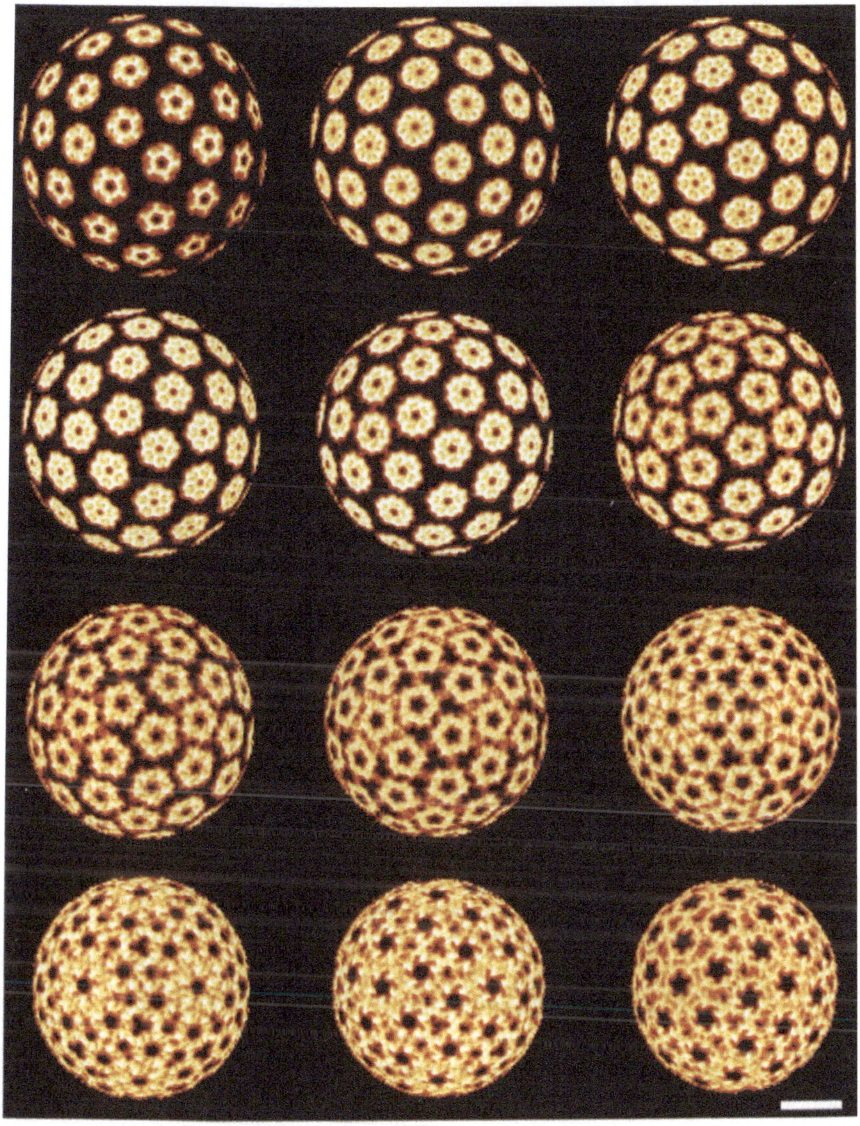

Color Plate 12: Sequential radial shells of human JC virus density map are displayed from 210 Å to 172 Å at 3.5 Å intervals viewed from the twofold axis. Higher density is represented by a lighter color scale. Density points are summed within the radial bins in order to analyze protein-protein interaction among the morphological units. Scale bar = 10 nm. (See Sect. 6 on page 148.)

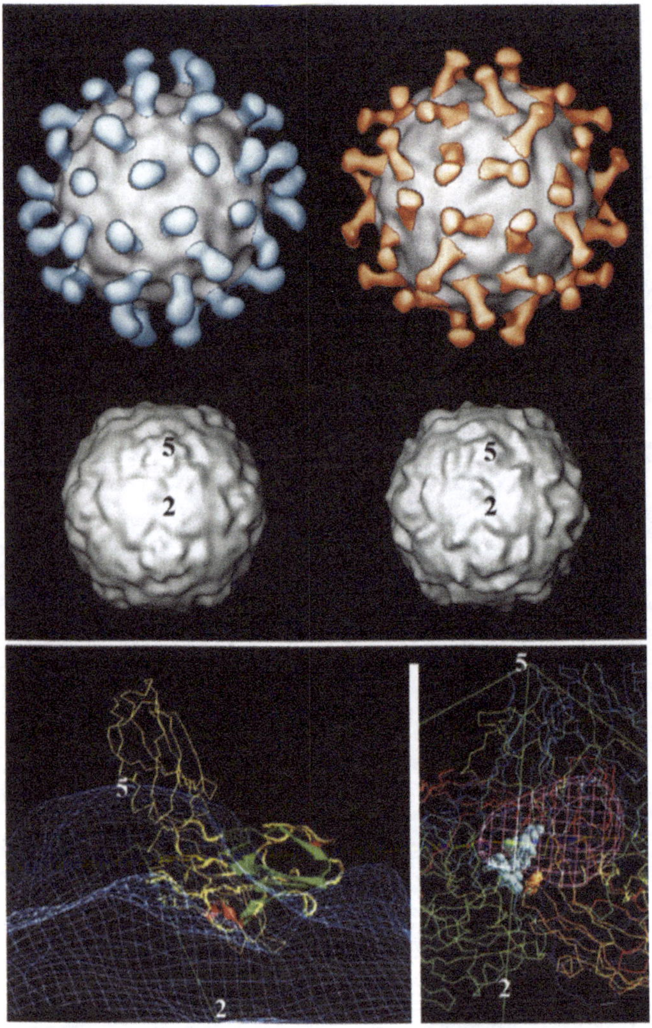

Color Plate 13: The 3D microscopic reconstruction of the receptor-bound poliovirus compared with the receptor-bound human rhinovirus serotype 3 at 25 Å resolution (top panel). The soluble receptors only carry two fragments of the N-terminal domains and bind to the canyon positions (middle panel, circle marked as one binding site) but with different orientation (ribbon drawing represents poliovirus receptor and $C\alpha$ backbone represents human rhinovirus receptor). The detailed molecular contacts could be revealed by the receptor footprint (bottom right). The contour grids show as the virus surface in the bottom left panel and as the receptor footprint in the bottom right panel. The related symmetric axes are shown by the characters. (See Sect. 6 on page 148.)

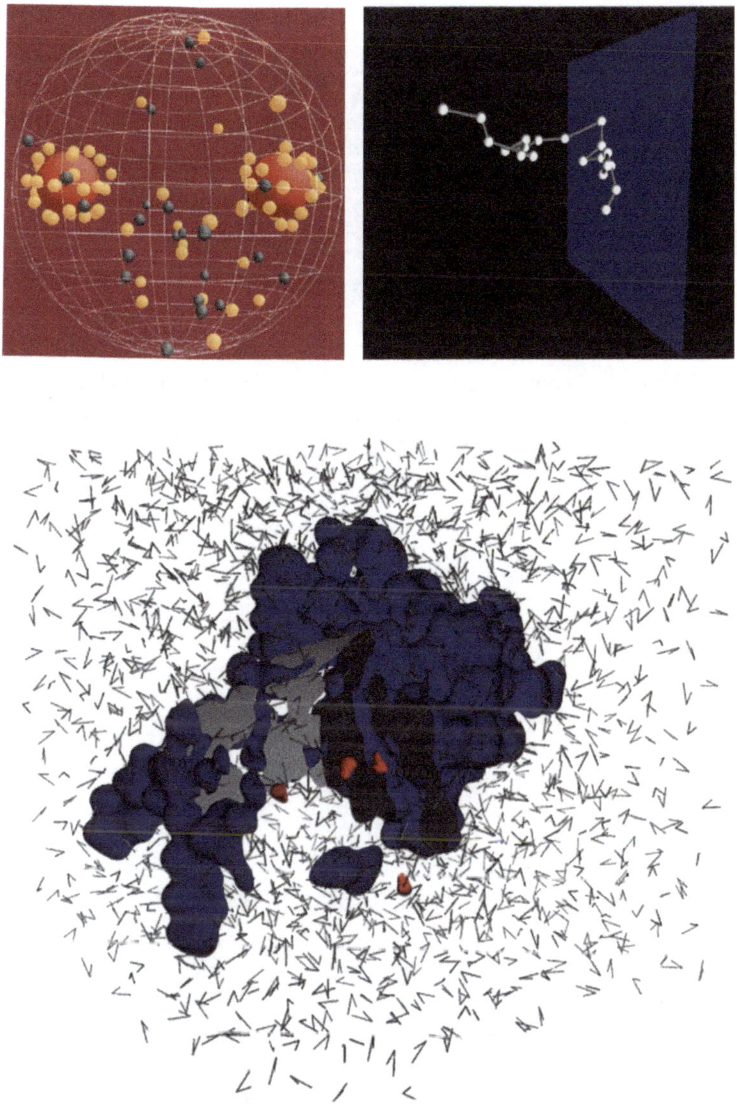

Color Plate 14: (A) Snapshot from a Monte Carlo simulation of two charged spherical aggregates (red). The counterions with opposite charges are shown in yellow, and the co-ions with the same charge as the aggregates are gray. The whole system is enclosed in a spherical cavity indicated by white grid lines. (B) Screenshot of a simulation showing the adsorption behavior of a polymer to a surface. (C) Picture from a molecular dynamics simulation of a 900-particle protein surrounded by 1800 water molecules (shown as Vs). A cylindrical portion of the protein has been cut away to reveal four internal water molecules (colored red) in a "pocket." This image was exported from GISMOS and post-processed using POV-Ray. (See Sect. 3 on page 159.)

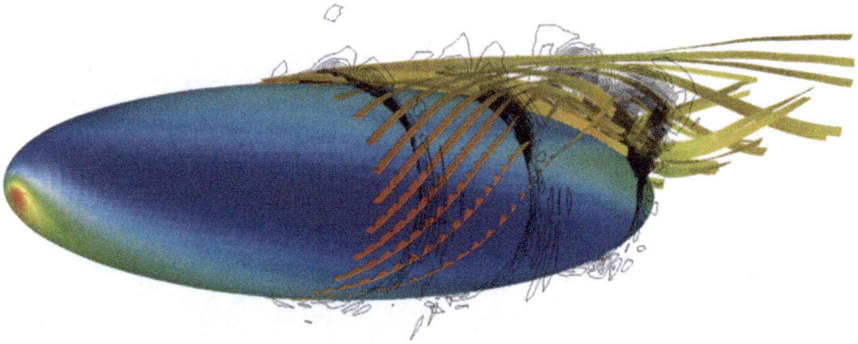

Color Plate 15: Perspective view of the 6:1 prolate spheroid showing stream ribbons, static pressure distribution on the hull, and streamwise vorticity $\overline{\omega}_1$ in three cross sections. (See Sect. 4 on page 185.)

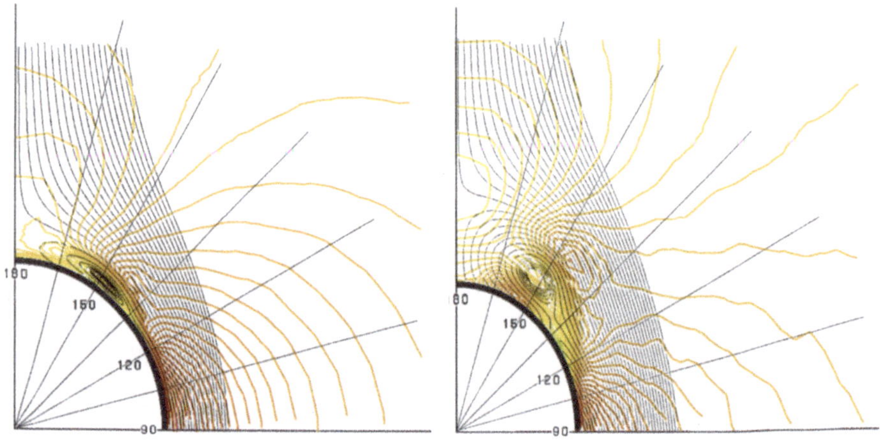

(a) (b)

Color Plate 16: Predicted secondary streamlines and velocity contours (a) at $X/L = 0.600$ and (b) at $X/L = 0.772$. (See Sect. 4 on page 185.)

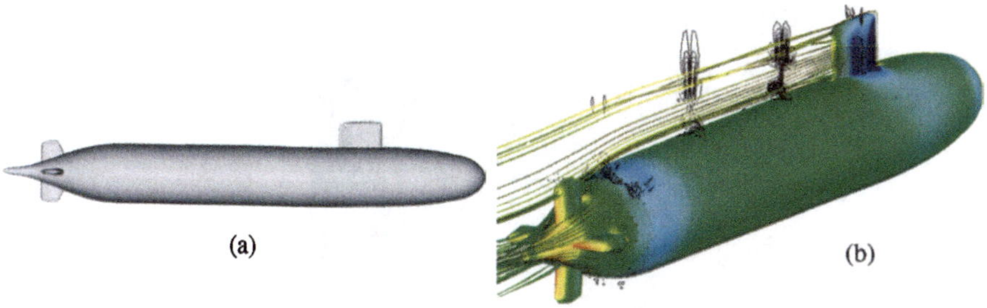

(a) (b)

Color Plate 17: Schematic (a) of the AFF-8 DARPA SUBOFF configuration, and (b) perspective view of the flow at 0° pitch and 0° drift, with the hull colored by the static pressure coefficient and the flow represented by stream ribbons and the streamwise vorticity $\overline{\omega}_1$ at selected cross sections. (See Sect. 5 on page 187.)

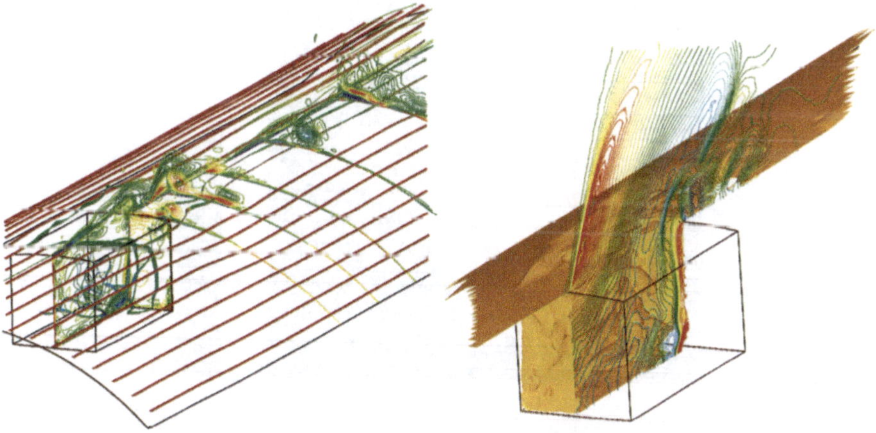

(a) (b)

Color Plate 18: Perspective view of the cavity embedded in the arc-shaped body showing (a) longitudinal vorticity (transversal isolines), transversal vorticity (longitudinal isolines), and streamline patterns; and (b) pressure isolines and an isosurface of the vorticity magnitude. (See Sect. 6 on page 189.)

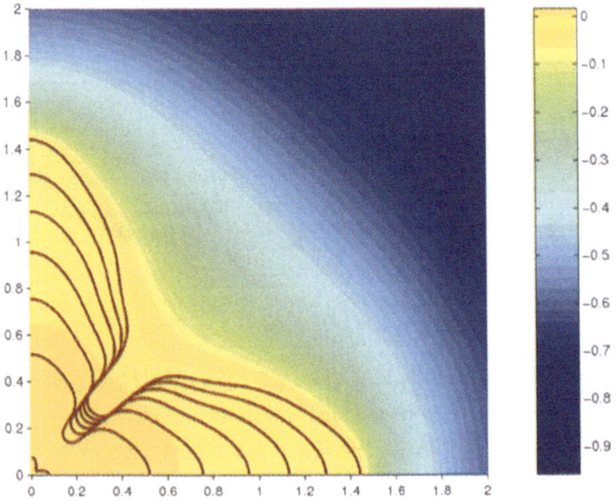

Color Plate 19: Growth of dendrite with physical parameters and initial data as in [1]. The solid lines show the location of the solid-melt interface at the time levels $t = 0.0, 0.25, \ldots, 1.5$. The background color represents the heat distribution at $t = 1.5$. (See Sect. 4.2 on page 206.)

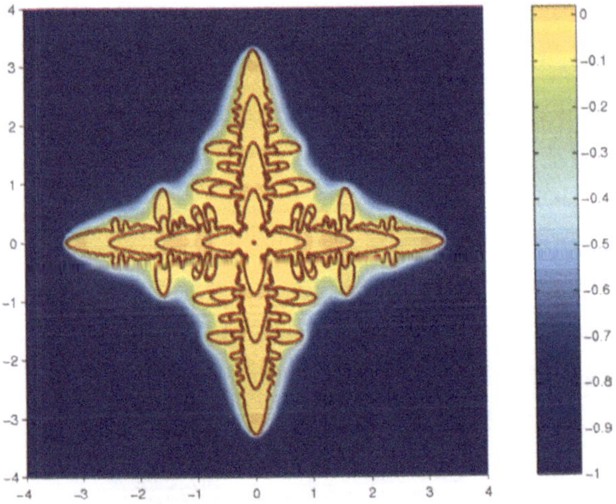

Color Plate 20: Growth of nickel dendrite starting from a small circular initial seed. The solid lines show the location of the solid-melt interface at the time levels $t = 0.0, 0.075, \ldots, 0.3$, and the background color represents the heat distribution at $t = 0.3$. The solution has been computed in one quarter of the domain and reflected around the x- and y-axes. (See Sect. 4.2 on page 206.)

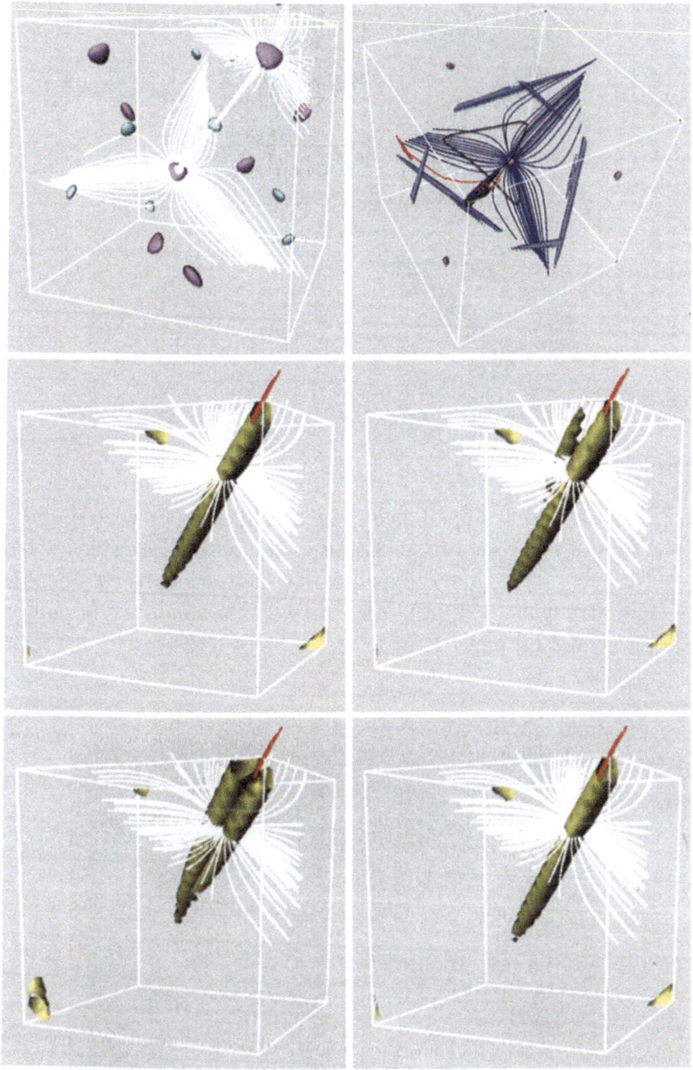

Color Plate 21: Upper left panel: Structure of the normal ABC flow: Stagnation points in purple (α type at upper right corner and β type near center of view), speed maximum in green, and stream lines in white. Upper right panel: Blue flux cigars and stream lines and red-and-black magnetic field lines. Lower four snapshots (only part of the box is shown) showing the evolution of the double flux cigars (yellow). Also shown are converging stream lines in white and the diverging axis through the α point (red). The four pictures correspond to four instants during the oscillation in the magnetic energy. (See Sect. 3.1 on page 212.)

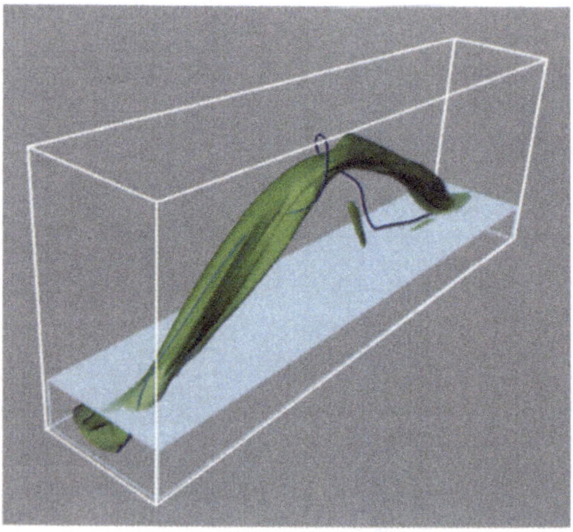

Color Plate 22: A snapshot showing the ascending 3-D twisted undulant flux rope. The feet of the rope are anchored in the stable lower layer. Isosurfaces of magnetic field strength are shown at low field strength (transparent) and at high field strength (opaque). Also shown is a single weak magnetic field line. (See Sect. 3.2 on page 216.)

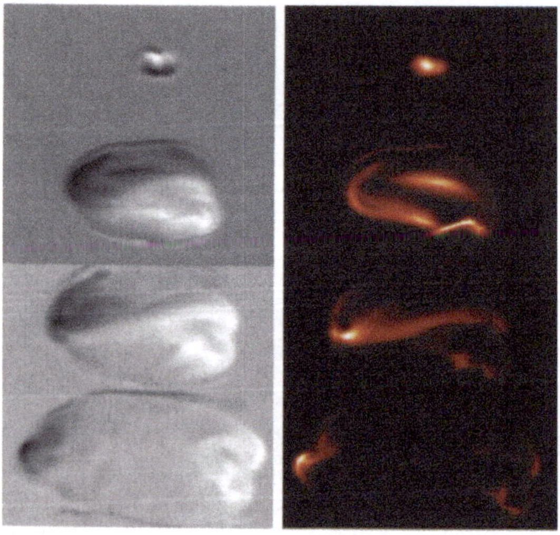

Color Plate 23: Two Panels with four horizontal snapshots of the magnetic field as the loop emerges through the open boundary: Vertical magnetic field component (left) and magnetic energy density (right). (See Sect. 3.2 on page 216.)

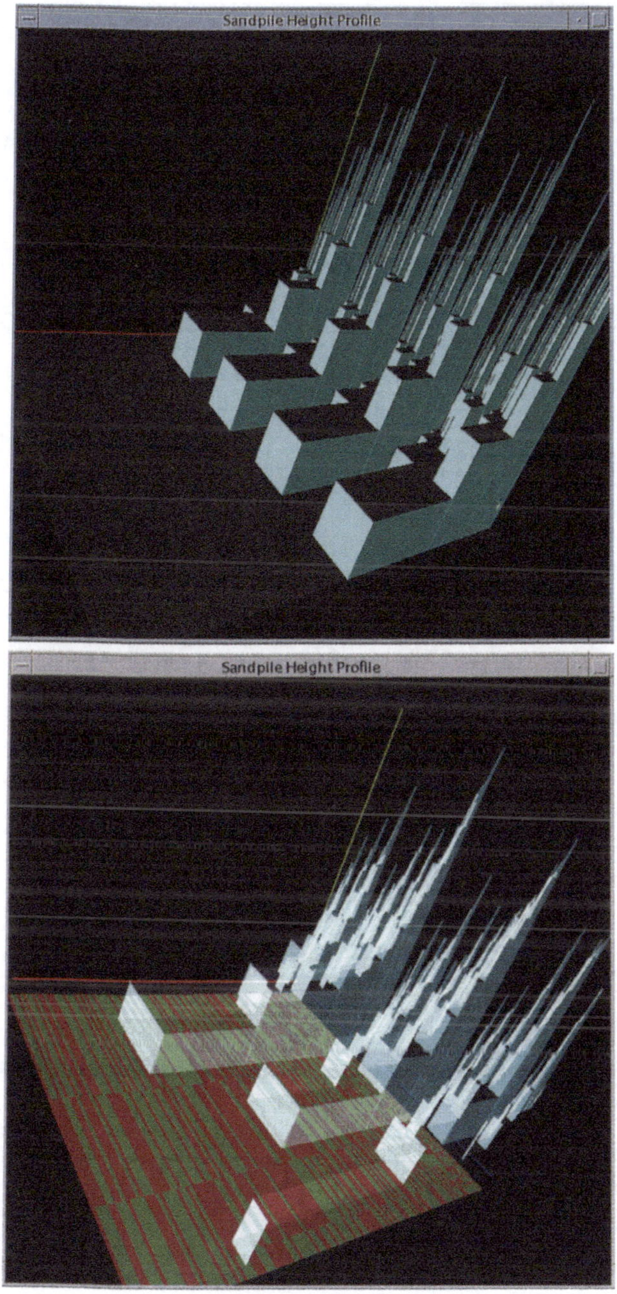

Color Plate 24: Self-repeating pattern of the analytic solution (left). Transparent surface superimposed over color coded dynamic parameter sheet (right). (See Sect. 2 on page 236.)

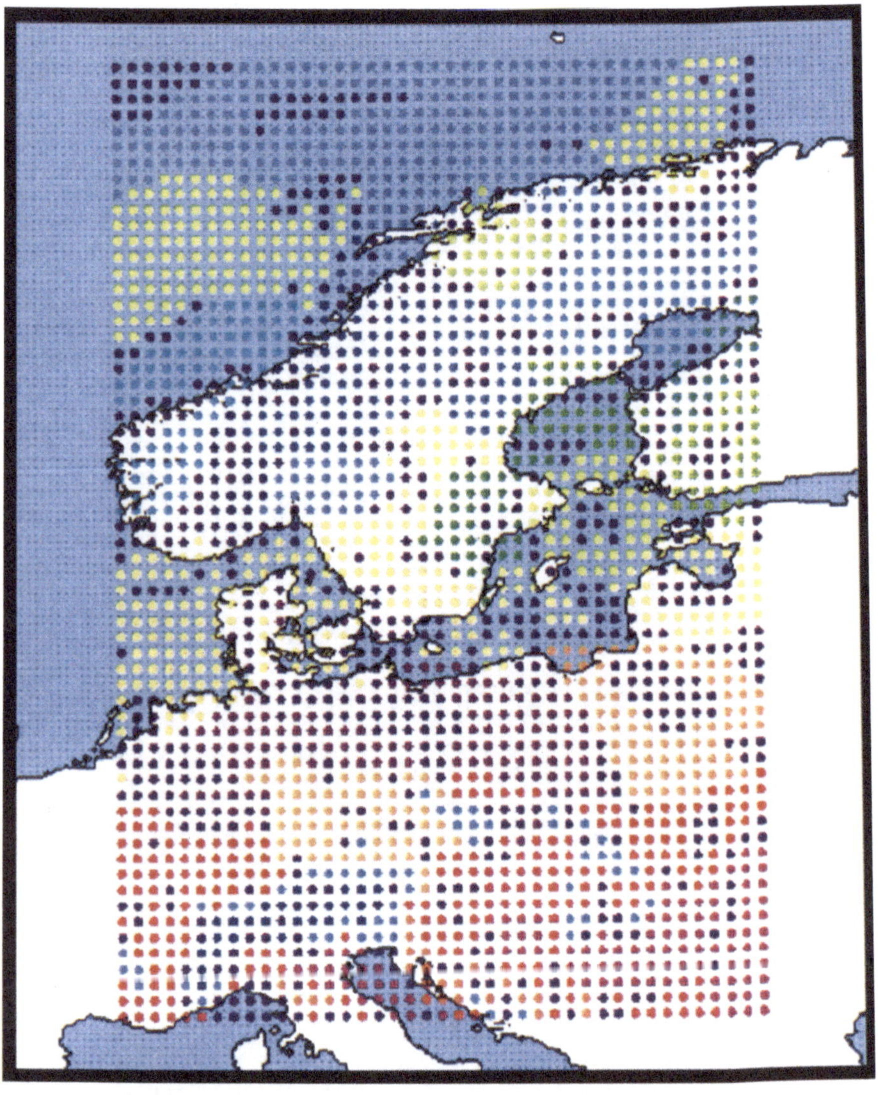

Color Plate 25: Graphical output of the visualization of the results of a cluster analysis. (See Sect. 4.3 on page 248.)

Author Index

Editorial Policy

§1. Volumes in the following four categories will be published in LNCSE:

i) Research monographs
ii) Lecture and seminar notes
iii) Conference proceedings
iv) Textbooks

Those considering a book which might be suitable for the series are strongly advised to contact the publisher or the series editors at an early stage.

§2. Categories i) and ii). These categories will be emphasized by Lecture Notes in Computational Science and Engineering. **Submissions by interdisciplinary teams of authors are encouraged.** The goal is to report new developments – quickly, informally, and in a way that will make them accessible to non-specialists. In the evaluation of submissions timeliness of the work is an important criterion. Texts should be well-rounded, well-written and reasonably self-contained. In most cases the work will contain results of others as well as those of the author(s). In each case the author(s) should provide sufficient motivation, examples, and applications. In this respect, Ph.D. theses will usually be deemed unsuitable for the Lecture Notes series. Proposals for volumes in these categories should be submitted either to one of the series editors or to Springer-Verlag, Heidelberg, and will be refereed. A provisional judgment on the acceptability of a project can be based on partial information about the work: a detailed outline describing the contents of each chapter, the estimated length, a bibliography, and one or two sample chapters – or a first draft. A final decision whether to accept will rest on an evaluation of the completed work which should include

- at least 100 pages of text;
- a table of contents;
- an informative introduction perhaps with some historical remarks which should be accessible to readers unfamiliar with the topic treated;
- a subject index.

§3. Category iii). Conference proceedings will be considered for publication provided that they are both of exceptional interest and devoted to a single topic. One (or more) expert participants will act as the scientific editor(s) of the volume. They select the papers which are suitable for inclusion and have them individually refereed as for a journal. Papers not closely related to the central topic are to be excluded. Organizers should contact Lecture Notes in Computational Science and Engineering at the planning stage.

In exceptional cases some other multi-author-volumes may be considered in this category.

§4. Category iv) Textbooks on topics in the field of computational science and engineering will be considered. They should be written for courses in CSE education. Both graduate and undergraduate level are appropriate. Multidisciplinary topics are especially welcome.

§5. Format. Only works in English are considered. They should be submitted in camera-ready form according to Springer-Verlag's specifications. Electronic material can be included if appropriate. Please contact the publisher. Technical instructions and/or TEX macros are available via http://www.springer.de/author/tex/help-tex.html; the name of the macro package is "LNCSE – LaTEX2e class for Lecture Notes in Computational Science and Engineering". The macros can also be sent on request.

General Remarks

Lecture Notes are printed by photo-offset from the master-copy delivered in camera-ready form by the authors. For this purpose Springer-Verlag provides technical instructions for the preparation of manuscripts. See also *Editorial Policy*.

Careful preparation of manuscripts will help keep production time short and ensure a satisfactory appearance of the finished book. The actual production of a Lecture Notes volume normally takes approximately 12 weeks.

The following terms and conditions hold:

Categories i), ii), and iii):
Authors receive 50 free copies of their book. No royalty is paid. Commitment to publish is made by letter of intent rather than by signing a formal contract. Springer-Verlag secures the copyright for each volume.

For conference proceedings, editors receive a total of 50 free copies of their volume for distribution to the contributing authors.

Category iv):
Regarding free copies and royalties, the standard terms for Springer mathematics monographs and textbooks hold. Please write to Peters@springer.de for details. The standard contracts are used for publishing agreements.

All categories:
Authors are entitled to purchase further copies of their book and other Springer mathematics books for their personal use, at a discount of 33,3 % directly from Springer-Verlag.

Addresses:

Professor M. Griebel
Institut für Angewandte Mathematik
der Universität Bonn
Wegelerstr. 6
D-53115 Bonn, Germany
e-mail: griebel@iam.uni-bonn.de

Professor D. E. Keyes
Computer Science Department
Old Dominion University
Norfolk, VA 23529–0162, USA
e-mail: keyes@cs.odu.edu

Professor R. M. Nieminen
Laboratory of Physics
Helsinki University of Technology
02150 Espoo, Finland
e-mail: rni@fyslab.hut.fi

Professor D. Roose
Department of Computer Science
Katholieke Universiteit Leuven
Celestijnenlaan 200A
3001 Leuven-Heverlee, Belgium
e-mail: dirk.roose@cs.kuleuven.ac.be

Professor T. Schlick
Department of Chemistry and
Courant Institute of Mathematical
Sciences
New York University
and Howard Hughes Medical Institute
251 Mercer Street, Rm 509
New York, NY 10012-1548, USA
e-mail: schlick@nyu.edu

Springer-Verlag, Mathematics Editorial
Tiergartenstrasse 17
D-69121 Heidelberg, Germany
Tel.: *49 (6221) 487-185
e-mail: peters@springer.de
http://www.springer.de/math/
peters.html

Lecture Notes in Computational Science and Engineering

Vol. 1 D. Funaro, *Spectral Elements for Transport-Dominated Equations.* 1997. X, 211 pp. Softcover. ISBN 3-540-62649-2

Vol. 2 H. P. Langtangen, *Computational Partial Differential Equations.* Numerical Methods and Diffpack Programming. 1999. XXIII, 682 pp. Hardcover. ISBN 3-540-65274-4

Vol. 3 W. Hackbusch, G. Wittum (eds.), *Multigrid Methods V.* Proceedings of the Fifth European Multigrid Conference held in Stuttgart, Germany, October 1-4, 1996. 1998. VIII, 334 pp. Softcover. ISBN 3-540-63133-X

Vol. 4 P. Deuflhard, J. Hermans, B. Leimkuhler, A. E. Mark, S. Reich, R. D. Skeel (eds.), *Computational Molecular Dynamics: Challenges, Methods, Ideas.* Proceedings of the 2nd International Symposium on Algorithms for Macromolecular Modelling, Berlin, May 21-24, 1997. 1998. XI, 489 pp. Softcover. ISBN 3-540-63242-5

Vol. 5 D. Kröner, M. Ohlberger, C. Rohde (eds.), *An Introduction to Recent Developments in Theory and Numerics for Conservation Laws.* Proceedings of the International School on Theory and Numerics for Conservation Laws, Freiburg / Littenweiler, October 20-24, 1997. 1998. VII, 285 pp. Softcover. ISBN 3-540-65081-4

Vol. 6 S. Turek, *Efficient Solvers for Incompressible Flow Problems.* An Algorithmic and Computational Approach. 1999. XVII, 352 pp, with CD-ROM. Hardcover. ISBN 3-540-65433-X

Vol. 7 R. von Schwerin, *Multi Body System SIMulation.* Numerical Methods, Algorithms, and Software. 1999. XX, 338 pp. Softcover. ISBN 3-540-65662-6

Vol. 8 H.-J. Bungartz, F. Durst, C. Zenger (eds.), *High Performance Scientific and Engineering Computing.* Proceedings of the International FORTWIHR Conference on HPSEC, Munich, March 16-18, 1998. 1999. X, 471 pp. Softcover. 3-540-65730-4

Vol. 9 T. J. Barth, H. Deconinck (eds.), *High-Order Methods for Computational Physics.* 1999. VII, 582 pp. Hardcover. 3-540-65893-9

Vol. 10 H. P. Langtangen, A. M. Bruaset, E. Quak (eds.), *Advances in Software Tools for Scientific Computing.* 2000. X, 357 pp. Softcover. 3-540-66557-9

Vol. 11 B. Cockburn, G. E. Karniadakis, C.-W. Shu (eds.), *Discontinuous Galerkin Methods.* Theory, Computation and Applications. 2000. XI, 470 pp. Hardcover. 3-540-66787-3

Vol. 12 U. van Rienen, *Numerical Methods in Computational Electrodynamics.* Linear Systems in Practical Applications. 2000. XIII, 375 pp. Softcover. 3-540-67629-5

Vol. 13 B. Engquist, L. Johnsson, M. Hammill, F. Short, (eds.), *Simulation and Visualization on the Grid.* Parallelldatorcentrum Seventh Annual Conference, Stockholm, December 1999, Proceedings. 2000. XIII, 301 pp. Softcover. 3-540-67264-8

Vol. 14 E. Dick, K. Riemslagh, J. Vierendeels, (eds.), *Multigrid Methods VI.* Proceedings of the Sixth European Multigrid Conference Held in Gent, Belgium, September 27-30, 1999. 2000. IX, 293 pp. Softcover. 3-540-67157-9

For further information on these books please have a look at our mathematics catalogue at the following URL: `http://www.springer.de/math/index.html`

GPSR Compliance

The European Union's (EU) General Product Safety Regulation (GPSR) is a set of rules that requires consumer products to be safe and our obligations to ensure this.

If you have any concerns about our products, you can contact us on ProductSafety@springernature.com

In case Publisher is established outside the EU, the EU authorized representative is:

Springer Nature Customer Service Center GmbH
Europaplatz 3
69115 Heidelberg, Germany

Batch number: 10761599

Printed by Printforce, the Netherlands